T0299214

Offshore Structure Hydrodynamics

Newly updated and translated into English for the first time, this standalone handbook perfectly combines background and theory with real-world experiments. All key topics are covered, including environmental conditions, wave theories, hydrostatics, and wave and current loads, with emphasis on nonlinear wave–body interaction. Focus is given to model testing, an important component in the design of offshore structures. Recent results on the hydrodynamics of perforated structures, moonpool and gap resonance, and third-order interaction effects have been added to this updated version.

Based on practical experience from multiple industry collaborations, combined with lectures that have been honed and improved over more than 30 years, the pedagogical, real-world approach in this book makes it an ideal companion for graduate students and researchers as well as ocean engineers.

Bernard Molin is Emeritus Professor at Ecole Centrale Marseille. He has published over 150 research papers, and in 2018 he received a Lifetime Achievement Award from the Ocean, Offshore and Arctic Engineering Division of ASME.

Cambridge Ocean Technology Series

Offshore Structure Hydrodynamics

BERNARD MOLIN
Ecole Centrale Marseille

CAMBRIDGE
UNIVERSITY PRESS

University Printing House, Cambridge CB2 8BS, United Kingdom

One Liberty Plaza, 20th Floor, New York, NY 10006, USA

477 Williamstown Road, Port Melbourne, VIC 3207, Australia

314–321, 3rd Floor, Plot 3, Splendor Forum, Jasola District Centre, New Delhi – 110025, India

103 Penang Road, #05–06/07, Visioncrest Commercial, Singapore 238467

Cambridge University Press is part of the University of Cambridge.

It furthers the University's mission by disseminating knowledge in the pursuit of education, learning, and research at the highest international levels of excellence.

www.cambridge.org
Information on this title: www.cambridge.org/9781009198042
DOI: 10.1017/9781009198059

© Bernard Molin 2023

First published 2023

A catalogue record for this publication is available from the British Library.

Library of Congress Cataloging-in-Publication Data
Names: Molin, Bernard, author.
Title: Offshore structure hydrodynamics / Bernard Molin.
Other titles: Hydrodynamique des structures offshore. English
Description: Cambridge ; New York, NY : Cambridge University Press, 2023. |
Series: Cambridge ocean technology series ; 10 | Translation of:
Hydrodynamique des structures offshore. | Includes bibliographical
references and index.
Identifiers: LCCN 2022026349 | ISBN 9781009198042 (hardback) |
ISBN 9781009198059 (ebook)
Subjects: LCSH: Offshore structures – Hydrodynamics. | Offshore oil well
drilling.
Classification: LCC TC1665 .M65 2023 | DDC 627/.98–dc23/eng/20221012
LC record available at https://lccn.loc.gov/2022026349

ISBN 978-1-009-19804-2 Hardback

Contents

Preface

I started my career as a research engineer at IFP (French Petroleum Institute, nowadays IFPEN). During the nearly 20 years that I spent there, I had the good fortune to be involved in numerous research projects, together with industrial partners; many of these projects were carried out within Clarom, a research organization led by IFP and Ifremer. After I quit IFP to join the academic world, I was asked by Clarom to write a handbook on the hydrodynamics of offshore structures, completing the series of Clarom "Guides Pratiques." This handbook was to be usable by engineers new to the field while retracing the achievements of all the Clarom projects. Quite a challenge! It seems I may have succeeded since the book has sold fairly well.

I wrote the book in French – starting from lecture notes of courses I was giving at ECN (Ecole Centrale de Nantes). Subsequently, it was often suggested that I should attempt an English version. Thanks to Bureau Veritas, this has now been done.

In the nearly 20 years that have gone by between the French and the English versions, I took part in many other research projects, some of them again with Clarom, and findings from these have contributed to enriching the book. I also gained experience from assisting Principia in industrial projects. Thanks to the GISHYDRO organization, with my colleagues at Ecole Centrale Marseille (ECM), I was lucky to have the opportunity to devise and perform numerous experimental campaigns at the BGO-FIRST tank at La Seyne-sur-Mer. Many of the experimental results used here as illustrations come from these model tests.

In the Acknowledgments section in the French edition, there is a long list of colleagues and friends. I want to thank all of them again, in particular, Olivier (Lili) Kimmoun, Fabien Remy, Yves-Marie Scolan, my colleagues at ECM. And add some new names: Christian Berhault, Thomas Coudray, Bruno Lécuyer, Alain Ledoux, Marie-Christine Rouault (colleagues from Principia); Thomas Sauder from SINTEF Ocean; and all the staff at Océanide, the company running BGO-FIRST, in particular, Jean-Pierre Aulanier, Alexandre Cinello, Thierry Rippol, and Bruno Thibault. Finally, I am grateful to my former colleague at ECM, Michael Paul, for checking and suggesting improvements to my English.

Bernard Molin

Acknowledgments

This book would not exist without Bureau Veritas (BV). BV gave me the incentive and the support to undertake the task of updating, and translating into English, my book *Hydrodynamique des Structures Offshore*, written more than 20 years ago.

I am particularly grateful to Pierre Besse, former head of BV's R&D department, who initiated this project, and to his successor, Quentin Derbanne, who pursued it to its end.

My warmest thanks go to Xiaobo Chen and Šime Malenica, who contributed to the writing of parts of the book. Together with Guillaume de Hauteclocque, they undertook a thorough re-reading of all the chapters and made many suggestions for improvements. BV also provided many illustrations.

Bernard Molin

Symbols

A	amplitude
a	radius
$\vec{A}$	angular motion
b	width
B	damping, center of buoyancy, width
b_d	normalized wave drift damping
B_d	wave drift damping
b_n	centered moment
c	wetted width
$\vec{C}$	moment
C_a	added mass coefficient
C_b	damping coefficient
C_D	drag coefficient
C_f	friction coefficient
C_G	group velocity
C_M	inertia coefficient: $C_M = 1 + C_a$
C_P	phase velocity
d	length, draft
D	diameter
E	complex or real envelope, energy
E	mean value
f	frequency
$\tilde{f}$	nondimensional frequency
$f*$	nondimensional frequency
$\vec{f}$	complex load
$\vec{f_d}$	normalized drift force
$\vec{F}$	load
$\vec{F_d}$	drift force
F_n	Froude number
g	gravitational acceleration
G	center of gravity
h	water depth
H	wave height, Kochin function

H_m	Hankel function
H_S	significant wave height
i	imaginary number $i^2 = -1$
I	inertia
J_m	Bessel function of the first kind
k	wave number
K	stiffness
K_C	Keulegan–Carpenter number
K_m	modified Bessel function
K_S	stability parameter
l, L	length, wave length
m, M	mass
m_a, M_a	added mass
m_n	spectral moment
M_α, M_β	metacenter
$\vec{n}$	normal vector
p	probability density function, pressure
P	cumulative probability, running point on the hull
Q	flow rate
r	velocity ratio
R	autocorrelation, radial coordinate
Re, re	Reynolds numbers
Re_o, re_o	oscillatory Reynolds numbers
s	curvilinear abscissa
S	spectral density, surface
S_B	hull
S_F	free surface
S_{WP}	waterplane area
S_t	Strouhal number
T	period
T_Z, T_2	mean up-crossing period: $T_2 = 2\pi \sqrt{m_0/m_2}$
T_P	peak period
U	velocity
U_C	current velocity
U_R	Ursell number, reduced velocity
$\vec{U}$	solid velocity
u, v, w	components of the fluid velocity
V	velocity, volume
$\vec{V}$	fluid velocity
$v_\star$	friction velocity
Y_m	Bessel function of the second kind
z_0	roughness height
α	rotation around x (roll, heel)

β	rotation around y (pitch, trim), wave direction, Stokes parameter, deadrise angle
γ	rotation around z (yaw), Euler constant
Γ	waterline
δ	boundary layer thickness, jet thickness
Δ	Laplace operator, stretching
ϵ	narrowness parameter
ε	small parameter identified with the steepness
ζ	nondimensional height $H/\sqrt{m_0}$, damping ratio, vertical motion at waterline
η	free surface elevation
θ	polar angle, phase angle
μ	discharge coefficient
ν	kinematic viscosity
ρ	density
σ	standard deviation
τ	Brard number, porosity
φ	complex potential
Φ	real potential
Ψ	stream function, auxiliary complex potential
ψ	heading
ω, Ω	angular frequency
ω_e	encounter frequency, equilibrium frequency
∇	gradient operator
∇_0	horizontal gradient
$\forall$	volumic displacement

Abbreviations

BIE	boundary integral equation
BVP	boundary value problem
CALM	catenary anchor leg mooring
CFD	computational fluid dynamics
FFT	fast Fourier transform
FPSO	floating production storage offloading
FSRU	floating storage regasification unit
GBS	gravity base structure
HDB	hydrodynamic Database
ISSC	International Ships and Offshore Structures Congress
ITTC	International Towing Tank Conference
JIP	joint industry project
JONSWAP	Joint North Sea Wave Project
LDV	laser doppler velocimetry
LNG	liquefied natural gas
MSL	mean sea level
OCIMF	Oil Companies International Marine Forum
OCV	offshore construction vessel
PIV	Particle image velocimetry
PTO	power takeoff
QTF	quadratic transfer function
RAO	response amplitude operator
SCR	steel catenary riser
TLP	tension leg platform
VIM	vortex-induced motion
VIV	vortex-induced vibration
VLCC	very large crude carrier
WEC	wave energy converter
WIO	wake-induced oscillation

1 Introduction

This book deals with structures, fixed or floating, in the open sea, unprotected or ill-protected from the environment, at fixed locations: Forward speed effects (i.e., ships under sailing conditions) are not considered. One of the main applications considered is the design of the supports used for oil exploration and exploitation (Figure 1.1). Many of them are represented in the accompanying figures. As the figures show these structures come in a wide variety of geometries and sizes. Reference will often be made to the following supports:

- *Jacket* (Figure 1.2): a fixed platform, nailed to the sea floor, consisting in an assembly of tubular members. Jackets are used for oil production in water depths up to 200 m (roughly) and also as supports for wind turbines.
- *Jack-up* or self-elevating unit: consisting in a buoyant hull and three or four movable legs, capable of raising the hull above the sea surface. Jack-ups are used for drilling at shallow depths. They are also widely used for work at sea, for instance, for installation of the hubs and blades of wind turbines.
- Gravity base structure (GBS): in concrete, resting on the seabed, also used for oil and gas production (Figure 1.3) or liquefied natural gas (LNG) storage. Small size GBSs are nowadays used as supports for offshore wind turbines, for instance, in the English Channel, off the coast of Normandy (Fécamp wind farm).
- *Floating production storage offloading* (FPSO): floating support used for oil production, usually shipshaped and moored via a *turret* around which they can freely rotate (Figure 1.4) or, in mild areas, anchored with a spread mooring (Figure 1.5).
- *Floating liquefied natural gas* (FLNG): similar to FPSO but used for gas production, with the liquefaction plant onboard. The Prelude FLNG (Figure 6.41), off the western coast of Australia, is the largest man-made floating structure, with a displacement of 600,000 tons.
- semi-submersible platform (*semi*): usually comprising an assembly of four *columns* piercing the free surface and two horizontal *pontoons*. Semis are used for drilling, occasionally for production or as *floatels*, and as supports for wind turbines (Figure 1.6).
- *Tension leg platform* (TLP): similar to the semis but in excess of buoyancy, anchored vertically by *tethers* (Figure 1.7). The vertical motions are thus

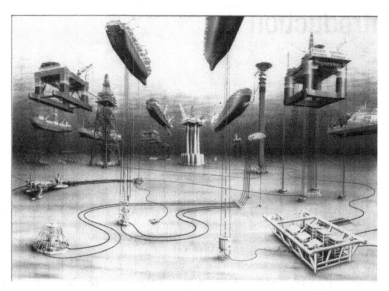

Figure 1.1 Offshore production systems (from Oilfield Publications Limited).

Figure 1.2 Bullwinkle jacket under tow (picture by Lanmon Aerial Photography, courtesy of Shell Exploration & Production Company).

suppressed, allowing location of the wellheads on the deck. TLPs are also used to support wind turbines (see Figure 1.8);

• Spar: truncated vertical cylinder of large draft, also used to support wind turbines.

Offshore oil is not the sole field of application considered here. Recent years have seen the rapid development of wind energy at sea, with the turbines first installed on fixed foundations and now on floating supports. The recovery of wave energy has been a long-standing research topic; even though the technology has not yet reached

Figure 1.3 The Troll GBS under tow (courtesy Equinor).

the economically viable stage, active research is still going on. There are many other operations taking place in the open sea, such as deep-sea mining, laying telephone cables, or oceanographic exploration. Fish farming is another marine activity gradually moving into deeper and less-sheltered areas (Figure 1.9).

To design structures intended to operate at a given location over many years, a precise knowledge of the prevailing sea conditions is required, not only of the most extreme waves, wind, and current conditions but also of the daily sea states that cause fatigue. In areas such as the North Sea, where extreme wave heights can reach 30 m, waves are the dominant loading factor: They exert cyclic loads that the structures and foundations must withstand. Floating and elastic structures respond dynamically to the wave loads. All these wave-induced stresses and responses need to be evaluated at the design stage. Currents also exert loads on the floaters and on their connections to the sea floor with a possible dynamic response due to alternate vortex shedding.

Figure 1.4 The Norne FPSO in the North Sea (courtesy Equinor).

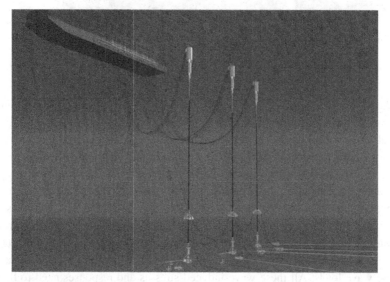

Figure 1.5 The Girassol FPSO off the coast of Angola and its riser towers (courtesy Doris Engineering).

Figure 1.6 The floating wind turbine Windfloat 1 during construction in Portugal. Note the heave plates whose function is to increase the vertical added mass, shifting the heave and pitch natural periods out of the wave period range. Additional benefit comes from decreased wave excitation loads and increased viscous damping (courtesy Principle Power, Inc.).

Model testing, covered in Chapter 9, is a route to assess the wave and current responses of offshore structures. Due to the requirements of offshore oil developments, theoretical and numerical analyses have made huge progress over the past 50 years. The main purpose of this book is to provide a state-of-the-art survey of present knowledge and of the numerical tools available to offshore engineers.

Given the wide variety of the considered structures, some kind of classification is required. This is proposed in the following section, with criteria based first on their geometry and then on their types of wave responses.

1.1   Classification of Offshore Structures

1.1.1   Large or Small Bodies

Among the constitutive elements of offshore structures, there is a fairly recurrent geometrical form: the circular cylinder. Circular cylinders come in a wide range of

Figure 1.7 The Heidrun Tension Leg Platform in the North Sea (courtesy Equinor).

diameters: a few centimeters in the case of umbilicals and mooring lines, around a meter for the risers and jacket bars, 10 to 30 m for the columns of semi-submersible platforms and TLPs.

These cylindrical elements are subjected to the flow induced by waves and currents.

A circular cylinder is poorly streamlined: When subjected to current alone massive separation occurs (see Figure 1.10). Whether the boundary layers are laminar or turbulent, and where they separate, depends on the Reynolds number $Re = U_C D / \nu$, where U_C is the current velocity, D the diameter, and ν the kinematic viscosity, $\nu \simeq 10^{-6} \, \text{m}^2 \, \text{s}^{-1}$ for sea water. Typically $U_C \sim 1$ m/s; therefore, the Reynolds numbers that we are concerned with range from 10^4 for cables and umbilicals 10 mm in diameter up to 10^7 for 10 m columns of semi-submersibles. It will be seen in Chapter 4 that it is for Reynolds numbers around 10^5 that the flow is the most intricate, with transition from laminar to turbulent taking place near the separation point.

In spite of its geometric simplicity, numerical modeling of a steady uniform flow around a circular cylinder at high Reynolds numbers (larger than 10^5) is still a challenge.

Figure 1.8 The wind turbine floater developed by SBM Offshore, to be installed at the Provence Grand Large site, off the Gulf of Fos near Marseille (courtesy SBM Offshore).

Wave-induced flow differs fundamentally from current, in that it reverses periodically as the waves pass by. If one considers a vertical pile subjected to a regular wave, and if we isolate a slice of this pile, the situation presents a strong similarity to the two-dimensional problem of a cylinder in uniform oscillatory flow of velocity $A\omega \cos \omega t$. Experiments show that a fundamental parameter is the ratio A/D of the amplitude of flow motion A to the diameter D.

When the amplitude A is large compared to the diameter D, the flow has a strong similarity to the steady current case: At each half cycle, a wake is emitted and carried far away. The difference is that, when the flow reverses, the wake travels back to the cylinder, meaning incoming vorticity and turbulence. Conversely, when the ratio A/D is small, the fluid particles at the cylinder wall do not travel a long enough distance for the boundary layer to separate (see Figure 1.11). The thickness of the boundary layer,

Figure 1.9 Dry transport of the Havfarm fish farm "Jostein Albert" (photo: Nordlaks/Deadline Media).

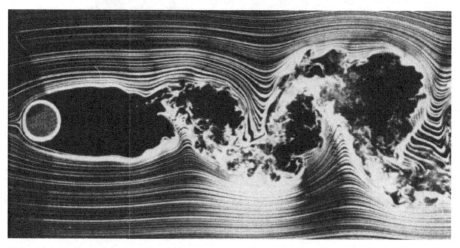

Figure 1.10 Uniform flow over a circular cylinder. Re = 10,000 (from van Dyke, 1982. Photograph by Thomas Corke and Hassan Nagib).

in laminar flow, is of the order $\sqrt{\nu/\omega}$, that is, for wave periods $T = 2\pi/\omega$ in-between 5 s and 20 s, 1 to 2 mm, quite negligible compared to the diameter in most cases. The outer flow is then adequately modeled by potential flow theory, that is, assuming perfect fluid and irrotationality, the streamlines being the same whether the flow is from left to right or right to left (Figure 1.12).

In place of the ratio A/D, the Keulegan–Carpenter number K_C defined as $K_C = A\omega T/D = 2\pi A/D$ is usually taken as the discriminating parameter. The boundary between attached flow and separated flow depends on the K_C value, but also on the Reynolds number or, equivalently, on the Stokes parameter $\beta = \text{Re}/K_C = D^2/(\nu T)$.

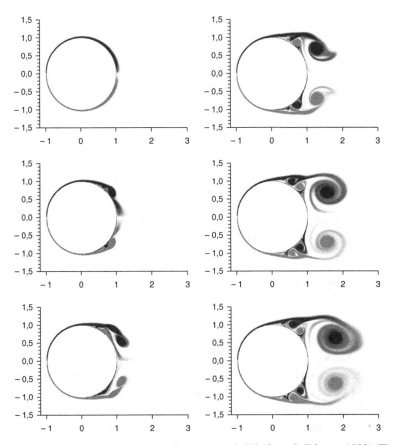

Figure 1.11 Navier–Stokes calculations at Re = 9,500 (from S. Etienne, 1999). The fluid is at rest and the flow starts impulsively at $t = 0$. The successive plots show the vorticity at times $U_C t/D$ going from 0.5 to 3. The boundary layer separates when the incoming flow has traveled between 0.5 and 1 diameter.

When the Stokes parameter is larger than about 10^5 (meaning a diameter larger than 1 m), separation does not occur until K_C exceeds 4 or 5. Referring to a semicolumn with a 20 m diameter, $K_C = 5$ means a wave amplitude around 15 m, close to the design wave in the North Sea!

It appears therefore legitimate to tackle the wave interaction with massive structures within the frame of a theory that assumes perfect fluid and irrotational flow: the potential flow theory. The characteristic dimensions of these "large bodies" are comparable to the wave lengths. As a result, when they interact with the structure, the incoming waves are significantly altered: they are "diffracted." When the structure responds to the waves, due to its motion another wave system is emitted, or "radiated." By means of some simplifying assumptions, potential flow theory offers the possibility to solve these diffraction and radiation problems, and to derive wave loads and responses.

In the absence of the body, it is also potential flow theory, which is used to describe the kinematics of the incoming waves.

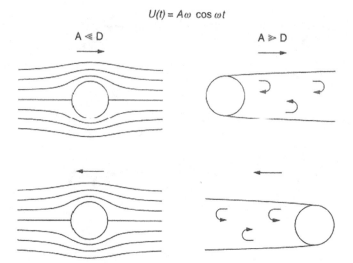

Figure 1.12 Oscillating flow around a circular cylinder. Limiting cases $A \ll D$ (left) and $A \gg D$ (right).

It is worth noting that, when a steady current coexists with an oscillatory flow, and the Keulegan–Carpenter number is low, the flow does not separate as long as the current velocity U_C is lower than the amplitude $A\omega$ of the oscillatory velocity. Potential flow theory is therefore applicable to the joint wave and current interaction with a large body, as long as the current velocity is smaller than the orbital velocities.

In contrast to large bodies, "small bodies" have characteristic dimensions smaller than or comparable to the amplitude of the fluid motion. They are therefore small compared to the wave lengths, with the result that diffraction effects are limited and that, locally, in the vicinity of a jacket bar slice, for example, the incident flow can be considered as uniform. The drawback is that the flow separates: Potential flow theory can no longer be applied; other numerical tools are required that solve the Navier–Stokes equations. The state-of-the-art is not to do Computational Fluid Dynamics (CFD) computations, but to relate, somewhat empirically, the local loads to the acceleration and velocity of the local incident flow, via the famous Morison formula. The inertia and drag coefficients that occur in this formulation are derived from representative tests.

The paradox is that the Morison equation, despite its imperfections, enables one to take into account a much more "refined" incident flow than the diffraction-radiation theory, limited to Stokes wave theories at first and second orders: The calculation of the hydrodynamic loading on jackets is usually done using third- or fifth-order Stokes theory or the stream function model.

It should not be concluded from the above that a given structure belongs to one of two categories. It can be both a large body and a small body, its constituent elements belonging to the two categories (the truss spars by example); it can be a small body under certain wave conditions and a large body in others (semi-submersibles).

1.1.2 Types of Loadings and Responses

Energetic wave periods cover a range going roughly from 3 to 20 s. At these periods (more especially in the range of 8–16 s), the waves exert high loads on floating (or deformable) structures, inducing responses at the same periods, and with amplitudes more or less linearly related to the wave amplitude. Catastrophic responses can result in cases of resonance.

A technique often used to limit this type of response is to shift the natural periods out of the range of the wave periods. This is one reason to employ soft moorings: The natural periods in surge, sway, and yaw of floating structures are typically higher than a minute. Likewise, semisubmersibles platforms are designed in such a way that their natural periods in heave, roll, and pitch be above the local wave periods. Conversely, tension leg platforms have very stiff vertical moorings, and natural periods in heave, roll, and pitch of usually less than 3 or 4 s.

It would be naive to believe that such a strategy eliminates all risks of resonance. Practically, one always observes some response at the natural periods, no matter how far they are from the wave periods. Nonlinear mechanisms are the underlying cause of these responses, which can be of very high amplitudes when the associated damping ratios are low.

The most well-known of these nonlinear mechanisms results from extending to "second order" wave loading. In irregular waves, when the free surface elevation is written as

$$\eta_I(x,y,t) = \sum_i A_i \, \cos(k_i \, x \, \cos \beta + k_i \, y \, \sin \beta - \omega_i \, t + \theta_i) \qquad (1.1)$$

(with $A_i^2 = 2 \, S(\omega_i) \, \Delta\omega$, $S(\omega)$ being the wave spectrum),
the linear, or first-order, loads are obtained as:

$$F^{(1)}(t) = \Re \left\{ \sum_i A_i \, f^{(1)}(\omega_i, \beta) \, e^{-i\omega_i \, t + i\theta_i} \right\} \qquad (1.2)$$

Equivalently, in the frequency domain:

$$S_{F^{(1)}}(\omega) = S(\omega) \, \| f^{(1)}(\omega, \beta) \|^2 \qquad (1.3)$$

$f^{(1)}(\omega, \beta)$ being the complex transfer function, or RAO (Response Amplitude Operator). The free surface elevation and the linear loading cover the same range of frequencies.

Proceeding to second order, supplementary loads are obtained that take place at the sums and differences of the carrier frequencies:

- difference frequency component:

$$F_-^{(2)}(t) = \Re \left\{ \sum_i \sum_j A_i \, A_j \, f_-^{(2)}(\omega_i, \omega_j, \beta) \, e^{i \, [-(\omega_i - \omega_j) \, t + \theta_i - \theta_j]} \right\} \qquad (1.4)$$

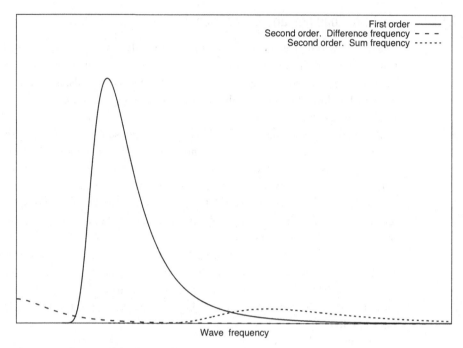

First order ——————
Second order. Difference frequency – – –
Second order. Sum frequency - - - - - -

Wave frequency

Figure 1.13 Spectra of the first- and second-order loads.

- sum frequency component:

$$F_+^{(2)}(t) = \mathfrak{R}\left\{\sum_i \sum_j A_i\ A_j\ f_+^{(2)}(\omega_i,\omega_j,\beta)\ \mathrm{e}^{\mathrm{i}\,[-(\omega_i+\omega_j)\,t+\theta_i+\theta_j]}\right\} \qquad (1.5)$$

The complex quantities $f_-^{(2)}$ and $f_+^{(2)}$ are known as **Quadratic Transfer Functions** (QTFs). In the frequency domain the spectra of the second-order loads take the forms:

$$S_{F_-^{(2)}}(\Omega) = 8\int_0^\infty S(\omega)\ S(\omega+\Omega)\ \|f_-^{(2)}(\omega,\omega+\Omega,\beta)\|^2\ \mathrm{d}\omega \qquad (1.6)$$

$$S_{F_+^{(2)}}(\Omega) = 8\int_0^{\Omega/2} S(\omega)\ S(\Omega-\omega)\ \|f_+^{(2)}(\omega,\Omega-\omega,\beta)\|^2\ \mathrm{d}\omega \qquad (1.7)$$

A qualitative illustration is provided by Figure 1.13, which shows typical shapes of $S_{F^{(1)}}$, $S_{F_-^{(2)}}$, and $S_{F_+^{(2)}}$ vs the frequency ω. It can be seen that $S_{F_-^{(2)}}$ covers the complete low-frequency domain, from $\omega = 0$ up to the wave frequencies. The second-order difference frequency loads actually are a first approximation of the low-frequency part of the nonlinear wave loads. In a similar way, $S_{F_+^{(2)}}$ covers a large part of the high-frequency range, beyond the wave frequencies.

As their names suggest, these second-order loads are much lower in magnitude than the first-order loads. But they can lead to very large responses, in resonant conditions. This is the case of the horizontal movements of moored structures that typically have natural periods longer than one minute. At such periods linear theory provides no excitation; but the second-order of approximation yields low-frequency loads,

Figure 1.14 The Eiko Maru storage tanker in the Arabian Gulf (from Molin & Bureau, 1980).

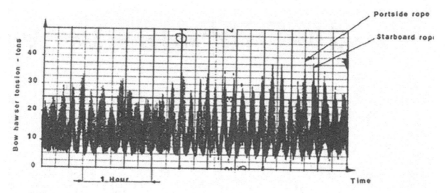

Figure 1.15 Eiko Maru tanker. Tension records in the bow hawsers (from Molin & Bureau, 1980).

so-called slowly-varying drift forces, that fluctuate in time roughly following the wave envelope signal.

Despite the fact that the loads are much weaker, the resulting slow-drift motions are much greater in amplitude than the first-order responses. Illustrations are provided in the following figures:

Figure 1.14 shows the storage tanker Eiko Maru in the Arabian Gulf. Figure 1.15 is a record of the bow hawsers tensions obtained by TotalEnergies in the late seventies. The dominant period in the record is close to 400 s, which is easily found to coincide with the natural period in surge, accounting for the combined stiffnesses of the hawsers and buoy anchoring (Molin & Bureau, 1980).

Figure 1.16 shows experimental records from tests, at Cehipar, on a large rectangular barge model (5 m long). Having their resonant frequencies in the wave frequency range, the time traces of the heave and roll responses look very similar to the wave elevation, whereas the sway motion is dominated by its slow-drift component, at its natural frequency.

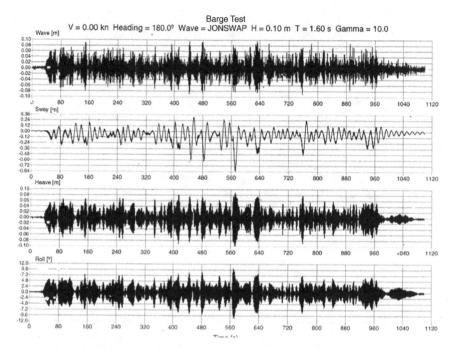

Figure 1.16 Model tests on a rectangular barge model in irregular beam waves. Time traces of the free surface elevation (top), sway, heave, and roll (bottom) motions.

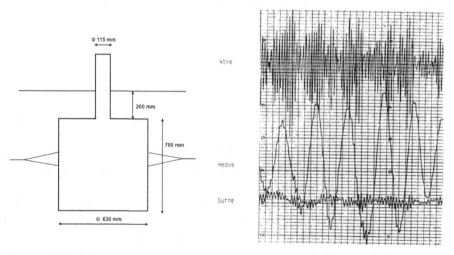

Figure 1.17 Model tests on a bottle-shaped model. Time traces of the free surface elevation (top), heave, and surge (bottom) motions.

Slow-drift motion may also occur for the vertical degrees of freedom, heave, roll, and pitch, when their natural frequencies are below the wave frequency range. Figure 1.17 shows experimental records of the heave and surge responses of a bottle-shaped

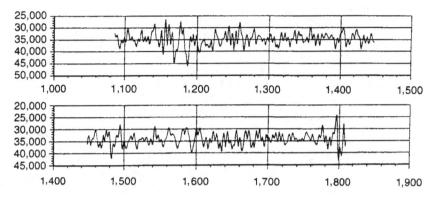

Figure 1.18 Model tests on a tension leg platform (PLTB 1,000). Time series of the tension in a tether.

model, where the low-frequency component of the heave response by far exceeds the wave frequency response.

The sum frequency second-order loads are considered responsible for the **springing**[1] behavior of Tension Leg Platforms, in sea states having peak periods around twice the heave, roll, and pitch natural periods. An example is shown in Figure 1.18 where tension fluctuations can be seen at a period of 5 s, when the peak period of the wave spectrum is 11 s. These experimental results are from an early TLP design, rather optimistic. Most existing TLPS have their natural periods somewhat lower, in the range 2–4 s. With these stiffer moorings, vibratory responses, known as **ringing**,[2] may still be observed in sea states having peak periods 4, 5, or 6 times larger than the natural periods, due to higher than second-order nonlinear wave loads. The ringing behavior was first discovered during the model tests undertaken for the Heidrun TLP and it was subsequently established that it is also a concern for deep water GBSs.

An example of ringing response is provided in Figure 1.20, which shows the effect of a steep wave group passing by an idealized GBS, a vertical cylinder, shown in Figure 1.19. The cylinder is connected to its foundation via an elastic steel plate, yielding a natural period of 0.44 s for the bending mode. The wave group has a mean period around 2 s; as it interacts with the cylinder, the third wave in the group, steeper and higher than the previous ones, triggers a vibratory response at the bending frequency.

With TLPs, springing is a concern to the fatigue life of the tethers, while ringing must be accounted for in design conditions. Obviously, to predict ringing numerically a second-order theory is insufficient when the ratio between resonant frequency and wave frequency is so wide. A third-order theory would exhibit loads taking place at frequencies $\omega_i + \omega_j + \omega_k$, probably still below the target frequency range.

It should be noted that other nonlinear phenomena, parametric instabilities, for instance, can induce resonant responses at frequencies different from the wave

---

[1] The coining springing was originally introduced to describe an elastic response of ship hulls.
[2] Another type of vibratory response of ship hulls.

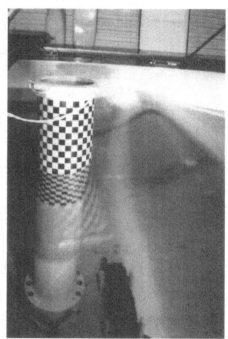

Figure 1.19 Model tests on a vibrating cylinder.

frequencies. A well-known case for ships is parametric roll that may occur when the wave encounter frequency is twice the roll natural frequency.

The determination, or at least the estimate, of its natural frequencies, is an important prerequisite in the hydrodynamic analysis of a system. It is also useful to have a proper estimate of the associated damping ratios.

1.2   Book Outline

The contents of this book are organized as follows:

Chapter 2 deals with environmental conditions. The main environmental factors are the waves (as measured, for instance, with a waverider buoy), the wind, and the current. The sea state concept is introduced, together with short-term and long-term statistics. The basic sea state parameters, that is, the significant wave height, mean wave periods, wave spectra, mean wind speeds, etc., are presented.

Chapter 3 presents the wave theories used in offshore engineering. Relatively deep water is assumed, so shallow water wave theories, such as cnoidal or solitary waves, are not covered. The classical Stokes perturbation method is introduced and first-, second-, third-, and fifth-order wave models are described, together with the stream function method.

Chapter 4 is devoted to the wave and current loads on small, or slender, bodies. The Morison equation is introduced and its limitations, in complex wave flows, and in waves and current, are highlighted.

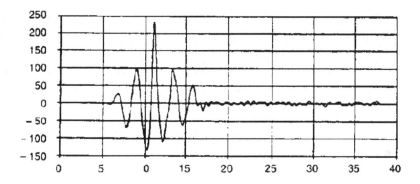

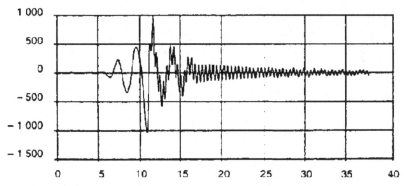

Figure 1.20 Model tests on a vibrating cylinder. Time series of the free surface elevation (top, in mm) and of the bending moment at the foundation (bottom, in m kN).

Chapter 5 addresses hydroelastic instabilities, such as Vortex-Induced Vibrations (VIVs), galloping, flutter, and Wake-Induced Oscillations (WIOs).

Chapters 6–8 are devoted to large bodies, and potential flow theory is applied throughout.

Chapter 6 deals with linear theory. The diffraction-radiation problem is introduced and different techniques of resolution are described. A wide range of applications are proposed, such as wave energy recovery, coupling between sloshing in tanks and sea-keeping, moonpool and gap resonances, etc. Illustrative comparisons with experimental results are presented.

Chapter 7 is devoted to second-order effects: drift forces in regular waves, sum and difference frequency loads in irregular waves, and high- and low-frequency wave responses. The different sources of damping involved in slow-drift motion are identified and tentatively quantified.

Chapter 8 covers some other types of nonlinear loads and responses: ringing loads, third-order runups, parametric instabilities, hydrodynamic impact, and hydrodynamics of perforated structures.

Finally, Chapter 9 covers model testing.

Four appendices follow. The first summarizes the foundations of potential flow theory, the second deals with hydrostatics, the third with damped mass spring oscillators, and the last with the boundary integral equation method.

The reader should have a good background in mathematics and, preferably, in fluid mechanics. For those not acquainted with potential flow theory, the first appendix should be sufficient to provide the basic knowledge.

1.3   References

ETIENNE S. 1999. Contribution à la modélisation de l'écoulement de fluide visqueux autour de faisceaux de cylindres circulaires, thèse de doctorat de l'université Aix-Marseille II (in French).

MOLIN B., BUREAU G. 1980. A simulation model for the dynamic behavior of tankers moored to single point moorings, in *Proc. Int. Symp. Ocean Eng. Ship Handling*, SSPA.

VAN DYKE M.D. 1982. *An Album of Fluid Motion*, The Parabolic Press.

2 Environmental Conditions

Offshore systems cannot be properly designed without accurate knowledge of the environmental conditions prevailing at the locations where they are to be installed. The main environmental parameters are the waves, the wind, and the current. In some places other factors may have to be taken into account such as earthquakes, internal waves, and ice (icebergs, floes, or frost).

2.1   Waves

2.1.1   The Concept of Sea State

While they operate, offshore structures must withstand millions of waves, in weather conditions that are changing constantly.

One may distinguish two time scales, one associated with the succession of the waves, that is a few seconds, and the other related to the changing sea state conditions, of the order of a few hours.

Somewhat arbitrarily it is postulated that the duration of a sea state is 3 hours, that is to say during this time lapse the wave conditions remain stationary: The mean values of the wave heights or wave periods or direction do not change. The life of a structure at sea then consists of a succession of sea states, at the rate of $365 \times 8 = 2920$ per year. Each sea state consists of a succession of waves, of varying heights and periods. These waves must be described in a sufficiently accurate way in order to be able to calculate the associated hydrodynamic loads.

One particular problem is to define an extreme sea state, or an extreme wave, that the system will be designed to withstand. Overall knowledge of the sea states may also be necessary to perform fatigue analysis, especially in areas like the North Sea, where storms are frequent.

The first step in a new offshore development project is to acquire precise knowledge of the prevailing sea states. There are methods that offer a basis for predicting sea conditions from meteorological records, but the most reliable way is to perform in situ measurements, for instance, with an accelerometric buoy. The usual procedure is to record the buoy's vertical acceleration (which is then integrated twice) during a 20-minute time span every 3 hours, at a frequency of 1 or 2 Hz.

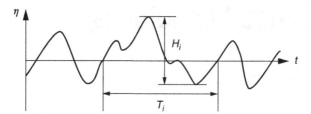

Figure 2.1 Time record of the free surface elevation. Definition of crests-to-troughs H_i and associated periods T_i.

There are two possible ways to analyze the time series: statistical (wave-by-wave) analysis or spectral analysis.

2.1.2   Wave-by-Wave Analysis

Consider a time series of the free surface elevation $\eta(t)$ such as that sketched in Figure 2.1. The process η is assumed to be centered, that is its time-averaged value is zero. The record may exhibit negative maxima and positive minima, so it is rather difficult to define what is a wave without any ambiguity.

By convention, a wave is defined as a portion of the time series bounded by two crossings of the zero value with a positive slope (so-called up-crossing).[1] The wave period is the time interval between the two up-crossings, and its height is the vertical distance from the lowest minimum up to the highest maximum.

In this way one obtains, over the duration of the record, N_w waves with heights H_i ($i = 1$, N_w) and periods T_i. Then one may derive averaged values $\overline{H}$ and $\overline{T}$ (usually written T_Z or T_{uc}), maximum values (H_{max}, T_{max}), and other statistical parameters.

It is customary to rearrange the wave heights by decreasing values, and then to retain only the first third (or tenth) for averaging. This leads to the definition of new parameters such as $H_{1/3}$ (averaged height of the highest third) and $T_{1/3}$ (associated mean wave period).

Another important parameter that will be encountered subsequently is the narrowness parameter defined as

$$\epsilon^2 = 1 - \left(\frac{N_w}{N_{max}} \right)^2 \tag{2.1}$$

N_w being the number of waves and N_{max} the number of maxima; $\epsilon = 0$ means that there are no secondary maxima (the signal is close to a modulated sinusoid); $\epsilon = 1$ means that there are infinitely more secondary maxima than waves.

---

[1] Some people prefer down-crossing.

2.1.3 Spectral Analysis

To the process $\eta(t)$, still assumed to be centered, one associates the autocorrelation function $R(\tau)$ defined as:

$$R(\tau) = \mathrm{E}\{\eta(t)\,\eta(t+\tau)\} \tag{2.2}$$

where, strictly, E means ensemble averaging. The process $\eta(t)$ being stationary, or at least assumed to be, $R(\tau)$ only depends on τ. In practice, $R(\tau)$ is estimated by calculating the time-averaged value of $\eta(t)\,\eta(t+\tau)$ (ergodicity assumption).

Taking the Fourier transform of $R(\tau)$ one obtains the spectral density $G(\omega)$:

$$G(\omega) = \frac{1}{2\pi}\int_{-\infty}^{\infty} R(\tau)\,\mathrm{e}^{-i\omega\tau}\,\mathrm{d}\tau \tag{2.3}$$

And, conversely, $R(\tau)$ may be derived from $G(\omega)$ through inverse Fourier transform:

$$R(\tau) = \int_{-\infty}^{\infty} G(\omega)\,\mathrm{e}^{i\omega\tau}\,\mathrm{d}\omega \tag{2.4}$$

Since $R(\tau)$ is an even ($R(\tau) = R(-\tau)$) and real function, $G(\omega)$ is even and real as well. In order to avoid negative frequencies, it is usual to use, in place of $G(\omega)$, the one-sided spectral density $S(\omega)$ defined as:

$$S(\omega) = \quad 2\,G(\omega) \qquad\qquad \text{for } \omega \geq 0 \tag{2.5}$$
$$S(\omega) = \quad 0 \qquad\qquad\qquad \text{for } \omega < 0 \tag{2.6}$$

The relationships between autocorrelation function and spectral density then become:

$$S(\omega) \quad = \quad \frac{2}{\pi}\int_{0}^{\infty} R(\tau)\,\cos\omega\tau\,\mathrm{d}\tau \tag{2.7}$$

$$R(\tau) \quad = \quad \int_{0}^{\infty} S(\omega)\,\cos\omega\tau\,\mathrm{d}\omega \tag{2.8}$$

In particular, the variance of the process η is obtained as the area below the spectral density curve:

$$\mathrm{E}(\eta^2) = R(0) = \int_{0}^{\infty} S(\omega)\,\mathrm{d}\omega \tag{2.9}$$

It will be established further that, in a regular wave of amplitude A, the energy per unit area is $1/2\,\rho g\,A^2$. In irregular seas, the spectral density $S(\omega)$ ($\times \rho g$) represents the frequency distribution of the energy in the wave system. This can easily be seen when considering the free surface elevation as the sum of a large number of regular waves (this is a usual way to synthesize an irregular sea state):

$$\eta(t) = \sum_{i} A_i\,\cos(\omega_i t + \theta_i) \tag{2.10}$$

Then it comes

$$\eta(t)\,\eta(t+\tau) = \sum_i \sum_j A_i\,A_j\,\cos(\omega_i t + \theta_i)\,\cos(\omega_j(t+\tau)+\theta_j) \tag{2.11}$$

$$\eta(t)\,\eta(t+\tau) = \frac{1}{2}\sum_i \sum_j A_i\,A_j\cos[(\omega_i - \omega_j)t - \omega_j\tau + \theta_i - \theta_j]$$

$$+ \frac{1}{2}\sum_i \sum_j A_i\,A_j\cos[(\omega_i + \omega_j)t + \omega_j\tau + \theta_i + \theta_j] \tag{2.12}$$

which has, as a time-averaged value:

$$R(\tau) = \mathrm{E}\{\eta(t)\,\eta(t+\tau)\} = \frac{1}{2}\sum_i A_i^2\,\cos\omega_i\tau \tag{2.13}$$

Through identification with equation (2.8) one deduces

$$A_i^2 = 2S(\omega_i)\,\mathrm{d}\omega_i \tag{2.14}$$

The moments of the spectral density $S(\omega)$ are defined as:

$$m_n = \int_0^\infty \omega^n\,S(\omega)\,\mathrm{d}\omega \tag{2.15}$$

from which an averaged period T_m can be defined

$$T_m = 2\pi\,\frac{m_0}{m_1} = T_1 \tag{2.16}$$

This mean period is different from the mean up-crossing period T_Z (or T_{uc}) from the wave-by-wave analysis. It is shown further that T_Z is related to the moments of orders 0 and 2 through

$$T_Z = 2\pi\,\sqrt{\frac{m_0}{m_2}} = T_2. \tag{2.17}$$

Another reference period is $T_P = 2\pi/\omega_P$, where ω_P is the frequency of the peak of the spectral density.

It must be borne in mind that these three periods (T_1, T_2, T_P) take different values.

2.1.4   Short-Term Statistics

By "short-term" what is meant here is "during the considered sea state," assumed to be stationary, usually 3 hours long. This is in contrast to "long-term" statistics that are considered in the following section, long-term meaning over the full range of sea states encountered during the operational life of the structure.

The wave-by-wave analysis considered above provides a way to build up, empirically and for a given sea state, the statistical distribution of wave heights and up-crossing periods (or any other related quantity).

There is another method, purely statistical, that relies on the assumption that the wave elevation is a Gaussian process. This feature is closely related to the possibility of decomposing the free surface elevation as the sum of a large number of **independent** sinusoids, as in (2.10). According to the Central Limit Theorem, the resulting process is Gaussian (or normal).

A random variable X is Gaussian if its probability density function (pdf) $p(x)$ writes:

$$p(x) = \frac{1}{\sqrt{2\pi}\sigma}\, e^{-x^2/2\sigma^2} \tag{2.18}$$

where σ is the standard deviation: $\sigma^2 = E(X^2) = m_0$

(the considered process is still assumed to be centered, that is, its mean value is nil). We recall that the pdf is defined by:

$$\text{Probability }\{x < X \le x + dx\} = p(x)\, dx \tag{2.19}$$

We will also make use of the cumulative probability $P(x)$ defined through:

$$P(x) = \text{Probability }\{X < x\} = \int_{-\infty}^{x} p(\xi)\, d\xi \tag{2.20}$$

For a Gaussian process:

$$P(x) = \frac{1}{\sqrt{2\pi}\,\sigma} \int_{-\infty}^{x} e^{-\xi^2/2\sigma^2}\, d\xi = \frac{1}{2}\left[1 + \operatorname{erf}\left(\frac{x}{\sqrt{2}\sigma}\right)\right] \tag{2.21}$$

where "erf" is the "error function":

$$\operatorname{erf}(x) = \frac{2}{\sqrt{\pi}} \int_{0}^{x} e^{-u^2}\, du \tag{2.22}$$

Figure 2.2 shows the pdf p and the cumulative probability P for a Gaussian process.

The knowledge of the probability density function of the free surface elevation is not enough to draw statistical information on zero up-crossings or wave heights. The derivative processes (with respect to time) $\dot{\eta}(t)$ and $\ddot{\eta}(t)$ (both centered and Gaussian) need to be introduced as well and the joint pdf $p(\eta, \dot{\eta}, \ddot{\eta})$ has to be derived.

Given N centered Gaussian processes $X_1, \ldots, X_N$, it can be established that the joint pdf $p(X_1, \ldots, X_N)$ is obtained as:

$$p(X_1, \ldots, X_N) = \frac{1}{(2\pi)^{N/2}}\, \frac{1}{\sqrt{\det \mathbf{C}}}\, e^{-\frac{1}{2}(X_1,\ldots,X_N)\,\mathbf{C}^{-1}\,{}^t(X_1,\ldots,X_N)} \tag{2.23}$$

where $\mathbf{C}$ is the covariance matrix $\mathbf{C}_{ij} = E(X_i X_j)$.

When $(X_1, X_2, X_3 = \eta, \dot{\eta}, \ddot{\eta})$, one obtains

$$\mathbf{C} = \begin{pmatrix} E(\eta^2) & E(\eta\dot{\eta}) & E(\eta\ddot{\eta}) \\ E(\eta\dot{\eta}) & E(\dot{\eta}^2) & E(\dot{\eta}\ddot{\eta}) \\ E(\eta\ddot{\eta}) & E(\dot{\eta}\ddot{\eta}) & E(\ddot{\eta}^2) \end{pmatrix} = \begin{pmatrix} m_0 & 0 & -m_2 \\ 0 & m_2 & 0 \\ -m_2 & 0 & m_4 \end{pmatrix} \tag{2.24}$$

m_0, m_2, and m_4 being the moments of the order 0, 2, and 4 of the spectral density $S(\omega)$.

To obtain the different terms of the matrix, it suffices to start from the representation (2.10) of the free surface elevation and to derive the time-averaged values, as was done for the autocorrelation.

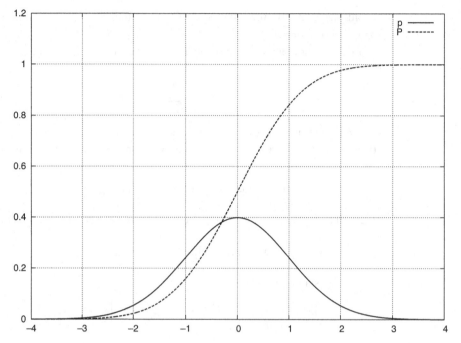

Figure 2.2 Probability density function p and cumulative probability P, as functions of η/σ.

Considering first η and $\dot{\eta}$ alone, one obtains that they are independent processes, with joint pdf $p(\eta,\dot{\eta})$ given by:

$$p(\eta,\dot{\eta}) = p(\eta)\,p(\dot{\eta}) = \frac{1}{2\pi}\,\frac{1}{\sqrt{m_0 m_2}}\,e^{-\eta^2/2m_0}\,e^{-\dot{\eta}^2/2m_2} \qquad (2.25)$$

Rice Formula for Level Crossing

Rice formula gives the average frequency at which the process $\eta(t)$ crosses a level η_R with positive slope:

$$v^+(\eta_R) = \int_0^\infty \dot{\eta}\,p(\eta_R,\dot{\eta})\,d\dot{\eta} \qquad (2.26)$$

(To obtain this expression, one simply states that a crossing will occur being t and $t + dt$ if the slope $\dot{\eta}$ is positive and if η is in-between $\eta_R - \dot{\eta}\,dt$ and η_R; then one integrates over all acceptable values of η and $\dot{\eta}$, in the limit $dt \rightarrow 0$).

One obtains, for a given level η_R:

$$v^+(\eta_R) = \frac{1}{2\pi}\,\sqrt{\frac{m_2}{m_0}}\,e^{-\eta_R^2/2m_0} \qquad (2.27)$$

When η_R is taken to be equal to zero, one obtains the up-crossing period:

$$T_Z(0) = \frac{1}{v^+(0)} = 2\pi\,\sqrt{\frac{m_0}{m_2}} = T_2 \qquad (2.28)$$

For a level different from zero, the up-crossing period is deduced from $T_Z(0)$ through

$$T_Z(\eta_R) = e^{\eta_R^2/2m_0} T_Z(0) \tag{2.29}$$

The level η_{RN} that is crossed (on average) every N waves is given by:

$$\eta_{RN} = \sqrt{2 m_0 \ln N} \tag{2.30}$$

Distribution of Maxima

Rice formula, applied to $\dot{\eta}$ and $\ddot{\eta}$, gives the mean frequency of maxima (a maximum being the zero crossing of $\dot{\eta}$ with negative slope):

$$\mu = \frac{1}{2\pi} \sqrt{\frac{m_4}{m_2}} \tag{2.31}$$

from which the narrowness parameter ϵ introduced previously can be given a theoretical value:

$$\epsilon^2 = 1 - \left(\frac{N_w}{N_{\max}}\right)^2 = 1 - \left(\frac{\nu^+}{\mu}\right)^2 = 1 - \frac{m_2^2}{m_0 m_4} \tag{2.32}$$

Finally, the probability density function of the maxima (or minima) is obtained through the following expression, where the reduced variable y is defined as $y = \eta_{\max}/\sqrt{m_0}$:

$$p(y) = \frac{1}{\sqrt{2\pi}} \left\{ \epsilon\, e^{-y^2/2\epsilon^2} + \sqrt{1-\epsilon^2}\, y\, e^{-y^2/2} \int_{-\infty}^{y\sqrt{1-\epsilon^2}/\epsilon} e^{-u^2/2}\, du \right\} \tag{2.33}$$

Figure 2.3 shows the pdfs given by (2.33), for different values of the narrowness parameter ϵ, from 0 up to 1. When $\epsilon \simeq 0$, meaning that there is only one maximum between two successive zero up-crossings, the probability density function is the **Rayleigh law**:

$$\epsilon = 0 \quad \Rightarrow \quad p(y) = y\, e^{-y^2/2} \qquad (y \geq 0)$$
$$p(y) = 0 \qquad (y < 0) \tag{2.34}$$

When $\epsilon \simeq 1$, meaning that there is a very large number of maxima in-between two successive zero up-crossings, the pdf follows the same Gaussian law as the process η itself:

$$\epsilon = 1 \quad \Rightarrow \quad p(y) = \frac{1}{\sqrt{2\pi}} e^{-y^2/2} \tag{2.35}$$

(as many negative as positive maxima).

It can be observed in Figure 2.3 that the pdfs obtained for ϵ less than 0.4 are very close to the Rayleigh law. Many sea states are consistent with this narrowness criterion. Another argument for assuming narrowness is that the crest-to-trough value is then simply twice the maximum value. As a consequence, the wave heights H also follow the Rayleigh law, now given by:

$$p(H) = \frac{H}{4 m_0} e^{-H^2/8 m_0} \tag{2.36}$$

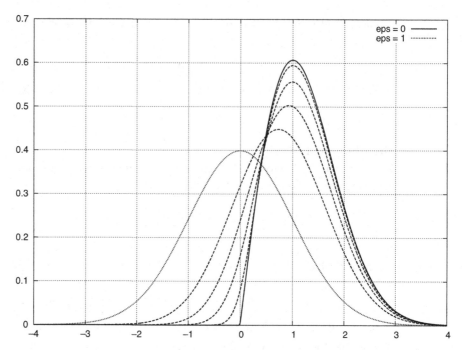

Figure 2.3 Probability density function of the maxima $\eta_{\max}/\sqrt{m_0}$ for ϵ from 0 to 1 with steps equal to 0.2.

The associated cumulative probability is

$$P(H) = 1 - e^{-H^2/8 m_0} \tag{2.37}$$

It is important to bear in mind that this wave height distribution assumes a narrow-banded and Gaussian process.

From the pdf a mean wave height $\overline{H}$ can be obtained:

$$\overline{H} = \int_0^\infty H \, p(H) \, \mathrm{d}H = \sqrt{2\pi} \, \sqrt{m_0} \tag{2.38}$$

One may now give theoretical values to $H_{1/3}$ and $H_{1/10}$ introduced in Section 2.1.2. First one introduces the wave height $\widetilde{H}_{1/N}$ which has a risk $1/N$ of being exceeded and therefore verifies:

$$P\left(\widetilde{H}_{1/N}\right) = 1 - \frac{1}{N} \tag{2.39}$$

From which

$$\widetilde{H}_{1/N} = \sqrt{8 m_0 \ln N} \tag{2.40}$$

The mean value of the wave heights greater than $\widetilde{H}_{1/N}$ is given by

$$H_{1/N} = N \int_{\widetilde{H}_{1/N}}^\infty H \, p(H) \, \mathrm{d}H \tag{2.41}$$

As a result one gets:

$$H_{1/10} = 5.092 \sqrt{m_0} \tag{2.42}$$

$$H_{1/3} = 4.004 \sqrt{m_0} \tag{2.43}$$

When N increases, $H_{1/N}$ is asymptotically given by

$$H_{1/N} \simeq 2 \left[\sqrt{2 \ln N} + \frac{1}{\sqrt{2 \ln N}} \right] \sqrt{m_0} \tag{2.44}$$

$H_{1/3}$ and H_S:

It may be wondered why $H_{1/3}$ is most often used as a reference for the mean wave height of a sea state in place of $\overline{H}$. The main reason, apparently, is that it has been found that $H_{1/3}$ offers a better agreement with visual estimates, for instance, from ship masters. Another reason is that $H_{1/3}$ is simply related to the standard deviation $\sqrt{m_0}$ of the free surface elevation:

$$H_{1/3} \simeq 4 \sqrt{m_0} \tag{2.45}$$

This, as already mentioned, is under the assumption of a narrow-banded Gaussian process. To make clear this restriction, **by definition** 4 $\sqrt{m_0}$ is the **significant** wave height H_S (also often noted H_{m0}).

It is usual to parametrize the sea states with the significant wave height H_S and the mean up-crossing period T_2 which can equally well be obtained through wave-by-wave or spectral analyses.

Maximum Wave Height

A 3-hour sea state means about 1,000 waves, for a mean up-crossing period around 10 s. What is the expected value of the highest wave in the record? Formula (2.44) above with N equal to the total number of waves cannot be applied because the highest wave in the record may be smaller than $\widetilde{H}_{1/N}$.

If one assumes that the N waves that constitute the time series are **independent** random variables with the same cumulative probability $P(H)$, then the cumulative probability P_N of the maximum height H_N is given by

$$P_N(H_N) = P^N(H_N) \tag{2.46}$$

When the wave heights follow the Rayleigh law it is

$$P_N(H_N) = \left(1 - e^{-H_N^2/8m_0} \right)^N \tag{2.47}$$

which behaves asymptotically as

$$P_N(H_N) = \exp \left\{ -N \, e^{-H_N^2/8m_0} \right\} \tag{2.48}$$

Taking the derivative with respect to H_N we get the probability density function of the maximum wave height:

$$p_N(H_N) = \frac{N \, H_N}{4 \, m_0} \, e^{-H_N^2/8m_0} \, \exp \left\{ -N \, e^{-H_N^2/8m_0} \right\} \tag{2.49}$$

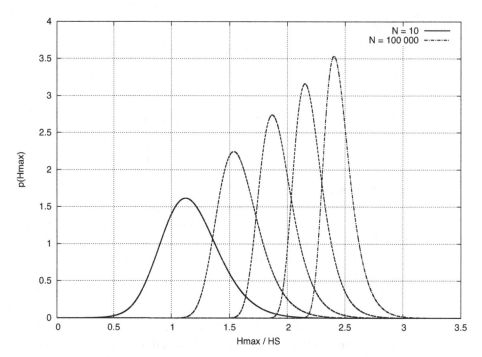

Figure 2.4 Probability density function of the height of the highest wave in a sea state, for different numbers of waves.

Figure 2.4 shows the pdf p_N, for $N = 10$, 100, 1000, 10 000 and 100 000.

The expected value of the highest wave is obtained through calculating $\int_0^\infty H_N \ p_N(H_N) \ dH_N$. Asymptotically, when N increases to infinity, it can be approximated as

$$E(H_N) = 2\left[\sqrt{2 \ln N} + \frac{\gamma}{\sqrt{2 \ln N}}\right] \sqrt{m_0} \tag{2.50}$$

where γ is the Euler constant: $\gamma = 0.5772$.

The standard deviation of H_N is asymptotically given by

$$\sigma(H_N) = \frac{\pi}{\sqrt{3} \ln N} \sqrt{m_0} = \frac{\pi}{4 \sqrt{3} \ln N} H_S \tag{2.51}$$

For $N = 1000$ it is 17% of the significant wave height.

Let us consider a 3-hour sea state with a 10-second mean period, that is 1080 waves. Then

$$\overline{H_{\text{max}}} = E(H_{1080}) = 7.78 \ \sqrt{m_0} = 1.94 \ H_{1/3} \tag{2.52}$$

When we increase the duration to 6 hours we get

$$\overline{H_{\text{max}}} = 8.13 \ \sqrt{m_0} = 2.03 \ H_{1/3} \tag{2.53}$$

a value that is little different from(2.52). One then understands that it does not make that much difference to take 3 hours or 6 hours as the reference duration of a sea state. In both cases, the maximum wave height is about twice the significant wave height.

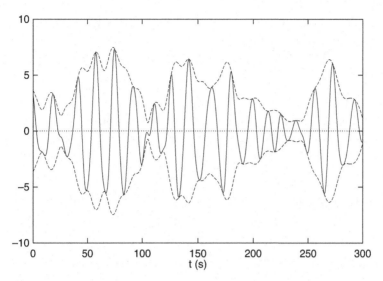

Figure 2.5 Free surface elevation (full line) and associated pseudo-envelope (dash line).

Pseudo-Envelope and Wave Grouping

We consider again the free surface elevation as

$$\eta(t) = \sum_i A_i \cos(\omega_i t + \theta_i) \tag{2.54}$$

and we assume the wave spectrum to be narrow-banded, around a mean frequency ω_0. The elevation can be rewritten as

$$\eta(t) = \sum_i A_i \cos[(\omega_i - \omega_0 + \omega_0)t + \theta_i] \tag{2.55}$$

$$= \sum_i A_i \cos[(\omega_i - \omega_0)t + \theta_i] \cos \omega_0 t$$

$$- \sum_i A_i \sin[(\omega_i - \omega_0)t + \theta_i] \sin \omega_0 t \tag{2.56}$$

$$= Y_c(t) \cos \omega_0 t + Y_s(t) \sin \omega_0 t \tag{2.57}$$

where $Y_c(t)$ and $Y_s(t)$ are slowly varying in time.

The pseudo-envelope $E(t)$ (see Figure 2.5) is defined by

$$E^2(t) = Y_c^2(t) + Y_s^2(t) \tag{2.58}$$

It can be noted that $E(t)$ can be defined independently of the carrier frequency ω_0 since

$$E^2(t) = \sum_i \sum_j A_i A_j \cos[(\omega_i - \omega_j)t + \theta_i - \theta_j] = \eta^2(t) + \zeta^2(t) \tag{2.59}$$

where $\zeta(t) = \sum_i A_i \sin(\omega_i t + \theta_i)$ is the Hilbert transform of $\eta(t)$:

$$\zeta(t) = \frac{1}{\pi} \int_{-\infty}^{\infty} \frac{\eta(s)}{t - s} \, \mathrm{d}s \tag{2.60}$$

It will be seen in Chapter 7 that the signal $E^2(t)$ bears much resemblance to the second-order slowly varying loads induced by the waves.

It can easily be obtained that the spectral density of $E^2(t)$ is

$$S_{E^2}(\Omega) = 8 \int_0^\infty S(\omega)\, S(\omega + \Omega)\, d\omega \qquad (2.61)$$

$E^2(t)$ is the sum of the squares of two independent random Gaussian processes with identical probability densities. Therefore its probability density function is given by

$$p(E^2) = \frac{1}{2m_0}\, e^{-E^2/2m_0} \qquad (2.62)$$

where from the pdf of $E(t)$ follows the Rayleigh law as already established:

$$p(E) = \frac{E}{m_0}\, e^{-E^2/2m_0} \qquad (2.63)$$

It can be derived that the joint pdf of $E(t)$ and $\dot{E}(t)$ is given by

$$p(E, \dot{E}) = \frac{1}{m_0\,\sqrt{2\pi b_2}}\, E\, e^{-E^2/2m_0}\, e^{-\dot{E}^2/2b_2} \qquad (2.64)$$

where b_2 is the second centered moment

$$b_2 = \int_0^\infty (\omega - \overline{\omega})^2\, S(\omega)\, d\omega = m_2 - \frac{m_1^2}{m_0} \qquad (2.65)$$

Therefore $b_2/m_2 = 1 - m_1^2/m_0 m_2$ is another parameter representative of the narrowness of the spectrum which, compared to ϵ, offers the advantage of not involving the 4th moment m_4.

Applying Rice formula, one may deduce the level crossing frequency of $E(t)$:

$$v^+(E) = \int_0^\infty p(E, \dot{E})\, \dot{E}\, d\dot{E} = \sqrt{\frac{b_2}{2\pi}}\, \frac{E}{m_0}\, e^{-E^2/2m_0} \qquad (2.66)$$

This frequency reaches its maximum at a level equal to $\sqrt{m_0}$:

$$v^+(\sqrt{m_0}) = \frac{1}{\sqrt{2\pi e}}\, \sqrt{\frac{b_2}{m_0}} \qquad (2.67)$$

meaning an average number of waves equal to

$$N_v = \sqrt{\frac{e}{2\pi}}\, \frac{1}{\sqrt{1 - \dfrac{m_1^2}{m_0 m_2}}} \qquad (2.68)$$

This number is significative of the mean length of wave groups. It shows that wave groups are all the longer as the parameter $\sqrt{1 - m_1^2/m_0 m_2} = \sqrt{1 - T_2^2/T_1^2}$ is smaller. For a JONSWAP spectrum, this parameter varies from 0.28 when γ is equal to 1 down to 0.24 for $\gamma = 6$, therefore within a rather narrow range. In the case of residual swells it can reach much lower values, typically between 0.05 and 0.15 according to Longuet-Higgins (1984).

Unfortunately it is not possible to obtain, at a given crossing level, the distribution of the length of the groups around this mean value. It is often assumed, in the literature, that this distribution follows the Poisson law. In fact, the Poisson law is only verified for asymptotically small crossing levels, so that successive crossings (up then down) are distant and uncorrelated (Boudet & Molin, 1995).

Another approach to wave grouping is to consider that successive waves constitute a Markov process. The key information is then the correlation coefficient between two successive waves. This correlation coefficient has been studied by many authors. For example, Arhan & Ezraty (1978) obtain theoretical and measured values (in the North Sea) of the order of 0.3. For more information and a comparison between the two methods, reference is made to Longuet-Higgins (1984).

Effects of Nonlinearities

All these results have been obtained under the assumption that the free surface elevation can be written as

$$\eta(t) = \sum_i A_i \, \cos(\omega_i \, t + \theta_i) \tag{2.69}$$

where the phases $\omega_i \, t + \theta_i$ are **independent** random variables. That the signal is Gaussian is a direct consequence of the independence assumption.

It will be seen later that the evolution of water waves is governed by nonlinear equations, in particular, at the free surface. It is only for waves of very low steepness that these equations can be linearized, justifying that the different components in (2.69) do not interact.

When the steepness increases, free surface nonlinearities play an increasingly important role. In Chapter 3 the Stokes perturbation procedure will be introduced whereby the free surface elevation η is decomposed as

$$\eta = \varepsilon \, \eta^{(1)} + \varepsilon^2 \, \eta^{(2)} + \varepsilon^3 \, \eta^{(3)} + \dots \tag{2.70}$$

where the small parameter ε is identified with the wave steepness. The results obtained so far assume that the leading term $\varepsilon \, \eta^{(1)}$ is dominant and that the following ones may be ignored.

The second-order correction $\varepsilon^2 \, \eta^{(2)}$ consists in components taking place at the sum and difference frequencies $\omega_i \pm \omega_j$ of the initial signal, with phases $\theta_i \pm \theta_j$. The hypothesis of independence of its constitutive components is no longer verified, and the pdf of the free surface elevation deviates from the Gaussian law, with higher crests and flatter troughs. These effects are amplified in shallow water. When accounting for these second-order nonlinearities good agreement has been found between simulated and measured distributions of crest elevations (Forristall, 2000).

Third-order nonlinearities result in modifications of the wave lengths (keeping the frequencies the same) and exchanges of energy between the constitutive components. They lead to such phenomena as the Benjamin–Feir instability of regular wave trains, and, according to many researchers, to the occurrence of rogue waves.

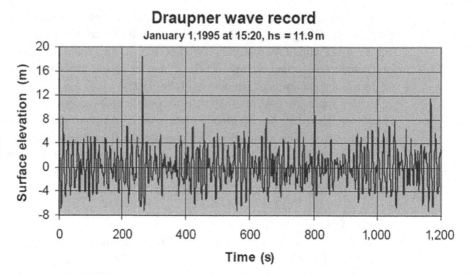

Figure 2.6 The New Year wave event recorded at Draupner (from Haver, 2004).

Higher-order nonlinearities, coming into play when the wave steepness exceeds some threshold value, quickly lead to three-dimensional instabilities and/or wave breaking.

Abnormal Waves

Phenomena such as wave breaking may suggest that the Rayleigh law is conservative and that it is unlikely that, in a sea state, waves higher than twice the significant wave height may be encountered.

In fact, there have been observations of waves higher than twice the significant wave height! These waves are known as "rogue waves" or "freak waves."[2] A famous case is the "New Year wave," recorded from the Draupner platform in the North Sea, on January 1, 1995. The time series of the measured free surface elevation (obtained with a downward-pointing laser-based wave sensor) is shown in Figure 2.6. A 25.6 m wave appears "out of nowhere," 2.2 times the significant wave height, which is 11.9 m. The wave is strongly asymmetric with a crest more than 18 m above MSL (mean sea level).

There has been much debate on whether such heights can be met just through linear processes, or whether nonlinear effects are at hand. From Figure 2.4, the probability of obtaining, among 1000 waves, a wave with a height more than 2.1 times H_S is not strictly zero, although it is very low. Many researchers acknowledge that third-order nonlinearities play an important, if not dominant, role. However, there can be other phenomena leading to local increases in the wave heights, such as refraction by current or bathymetry. A review is provided by Kharif and Pelinovsky (2003). In the case of the New Year Draupner wave, an effect of crossing seas has been advocated (McAllister *et al.*, 2019).

Many accidents at sea have been attributed to rogue waves.

---

[2] An acknowledged definition is that rogue waves are more than 2.1 times H_S in height.

2.1.5   Some Usual Spectra

Theoretical considerations and empirical adjustments have led to analytical expressions of wave spectra. Some are based on wind velocity and duration, and/or fetch length, etc. Other expressions directly involve sea parameters such as the significant wave height and the mean (or peak) wave period. The latter are more frequently used in marine and offshore engineering.

Pierson-Moskowitz Spectrum

This spectral shape results from observations in the Northern Atlantic Ocean and assumes a fully developed sea state, that is to say, the fetch is infinite and the wind has been blowing for a sufficiently long time for a stationary sea state to have been established. When parametrized with the significant wave height H_S and mean up-crossing period T_2, this spectrum takes the form:

$$S_{PM}(\omega) = \frac{1}{4\pi}\, H_S^2\, \left(\frac{2\pi}{T_2}\right)^4\, \omega^{-5}\, e^{-\frac{1}{\pi}\left(\frac{2\pi}{T_2}\right)^4\, \omega^{-4}} \tag{2.71}$$

The moments can be expressed through the Γ function:

$$\Gamma(x) = \int_0^\infty u^{x-1}\, e^{-u}\, du \tag{2.72}$$

In particular, it can readily be verified that the following relationships are fulfilled:

$$H_S \;=\; 4\,\sqrt{m_0} \tag{2.73}$$

$$T_2 \;=\; 2\pi\,\sqrt{\frac{m_0}{m_2}} \tag{2.74}$$

while the peak period T_P is 1.408 times the mean up-crossing period T_2.

When expressed with respect to the peak frequency ω_P the Pierson-Moskowitz spectrum takes the form

$$S_{PM}(\omega) \simeq \frac{5}{16}\, H_S^2\, \omega_P^4\, \omega^{-5}\, e^{-\frac{5}{4}\left(\frac{\omega}{\omega_P}\right)^{-4}} \tag{2.75}$$

The Pierson-Moskowitz spectrum is also known as ITTC or ISSC spectrum.

JONSWAP Spectrum

JONSWAP is an acronym for *JOint North Sea WAve Project*, a measurement campaign carried out in the early 1970s that resulted in a form more general than the Pierson-Moskowitz spectrum (see Hasselmann *et al.*, 1973). The JONSWAP spectrum has the form

$$S_J(\omega) = \alpha\, S_{PM}(\omega)\, \gamma^a \tag{2.76}$$

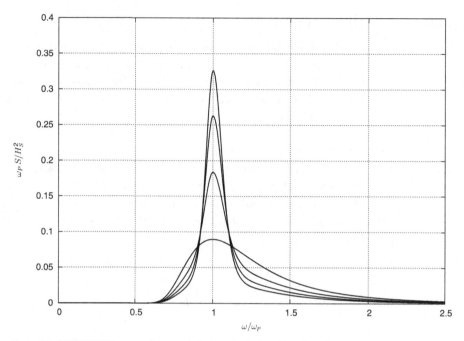

Figure 2.7 JONSWAP spectra for $\gamma = 1, 3, 6$, and 10. Same peak periods.

where a is given by

$$a = e^{-\dfrac{(\omega - \omega_P)^2}{2\sigma^2\omega_P^2}} \tag{2.77}$$

with $\sigma = 0.07$ for $\omega < \omega_P$ and $\sigma = 0.09$ for $\omega > \omega_P$, ω_P being the peak frequency.

When the parameter γ is greater than one, the peak of the JONSWAP spectrum is narrower and higher than with the Pierson-Moskowitz spectrum. The measurement campaign in the North Sea resulted in an average value of γ equal to 3.3.

In equation (2.76) the coefficient α is to be adjusted so that the identity

$$H_S^2 = 16 \int_0^\infty S(\omega)\, d\omega \tag{2.78}$$

remains verified. An approximate value, valid for $0.5 \leq \gamma \leq 6$, is $\alpha = 1 - 0.287 \ln \gamma$.

Figures 2.7 and 2.8 show JONSWAP spectra for different values of the enhancement factor γ. In Figure 2.7, the peak period T_P is kept the same, in Figure 2.8 it is the mean up-crossing period T_2.

Comments

At high frequency, the Pierson-Moskowitz and JONSWAP spectra offer the same asymptotic law in ω^{-5}. This negative fifth power is based on the assumption that, for a fully developed sea state, the high-frequency part of the spectrum does not depend

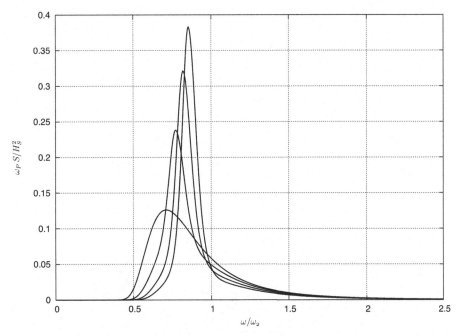

Figure 2.8 JONSWAP spectra for $\gamma = 1, 3, 6$, and 10. Same mean up-crossing periods.

on wind velocity but only on gravity. From dimensional analysis, it is deduced that the tail of the spectrum is of the form

$$S(\omega) \sim \alpha \, g^2 \, \omega^{-5} \tag{2.79}$$

where α is known as the Phillips constant.

It is now believed that the wind friction velocity $v^\star$ still plays some role and that the asymptotic form of the spectra is closer to ω^{-4}.

Strictly speaking, the ω^{-5} tail means that the 4th moment m_4 of the Pierson-Moskowitz and JONSWAP spectra is infinite, and so the narrowness parameter ϵ is equal to 1, apparently contradicting the idea that the wave heights follow the Rayleigh distribution. In practice, when used to synthesize wave elevation, the spectra will be truncated to some upper limit ω_T, and the high-frequency part of the spectra only means that free surface profiles have a somewhat bumpy shape, with little effect on the crest-to-trough values.

Directional Distribution

Under the form $S(\omega)$ wave spectra give information on the free surface elevation at one point. To reconstruct the free surface elevation in time and space, additional information must be given on the angular distribution. Directional spectra are usually introduced as $S(\omega, \beta)$:

$$S(\omega, \beta) = S(\omega) \, G(\omega, \beta) \tag{2.80}$$

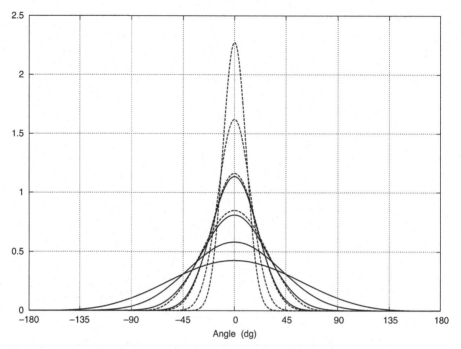

x-axis: Angle (dg), ranging from −180 to 180

Figure 2.9 Spreading functions $G_1(\beta)$ (full lines) and $G_2(\beta)$ (dash lines) for s and s' equal to 2, 4, 8, and 16.

where the function $G(\omega, \beta)$ must verify:

$$\int_0^{2\pi} G(\omega, \beta) \, d\beta = 1 \qquad (2.81)$$

Practically, in the absence of further information, one usually assumes that the spreading function $G(\omega, \beta)$ only depends on β. The following functions are often used:

$$G_1(\beta) = C_1(s) \, \cos^{2s} \frac{\beta - \overline{\beta}}{2} \qquad -\pi \leq \beta - \overline{\beta} \leq \pi \qquad (2.82)$$

$\overline{\beta}$ being the main propagation direction, or:

$$G_2(\beta) = C_2(s') \, \cos^{2s'}(\beta - \overline{\beta}) \qquad |\beta - \overline{\beta}| < \frac{\pi}{2} \qquad (2.83)$$

$$G_2(\beta) = 0 \qquad |\beta - \overline{\beta}| > \frac{\pi}{2} \qquad (2.84)$$

$C_1(s)$ and $C_2(s')$ being constants ensuring:

$$\int_0^{2\pi} G_1(\beta) \, d\beta = \int_0^{2\pi} G_2(\beta) \, d\beta = 1 \qquad (2.85)$$

Figure 2.9 shows the spreading functions G_1 and G_2 when s and s' take the same values 2, 4, 8, and 16. It can be seen that they are somewhat equivalent when s and s' are related through $s = 4 \, s'$.

$S(\omega)$ or $S(f)$

Here we have chosen to express the spectrum with respect to the angular frequency ω, following the most widespread usage. The frequency f is sometimes used. When switching from angular frequency to frequency (or vice versa), one must keep in mind that the spectra are energy **densities**, so the following relationship must be fulfilled

$$S_f(f)\, df = S(\omega)\, d\omega \tag{2.86}$$

As a result,

$$S_f(f) = 2\pi\, S(2\pi f) \tag{2.87}$$

More rarely, the spectrum may be expressed as $S_T(T)$ with T the wave period. Then the rule is the same:

$$- S_T(T)\, dT = S(\omega)\, d\omega \tag{2.88}$$

and

$$S_T(T) = \frac{2\pi\, S(2\pi/T)}{T^2} \tag{2.89}$$

2.1.6   Long-Term Statistics

Wave measurements over a long period enable us to provide a basis for building up so-called scatter diagrams, double entry tables with the significant wave height H_S on one side and the peak period T_P or the mean up-crossing period T_2 on the other. In each box $H_S \times T_P$ the number of occurrences is shown. Each occurrence corresponds to one sea state, 3 hours long. One year therefore means 2,920 observations. In order to increase the number of observations, hindcasting methods, based on archived meteorological data, are customarily used.

An example of a scatter diagram, corresponding to a site in the North Sea, is given in Table 2.1.

The problem that arises is, from these observations limited in time, how to deduce a design sea state, defined, for example, as the significant height that has a probability of 10% to be exceeded within 20 years.

The general principle of the methods employed is to build up, on the basis of the scatter diagram, an empirical cumulative distribution $P(H_S)$ (or $P(H_S, T_P)$). This empirical law is then fitted with a theoretical asymptotic law. Design values are derived by extrapolation beyond the measured values.

Among the different statistical laws representative of the tails of cumulative distributions, the most commonly used in offshore engineering is the Weibull law, expressed by

$$P(H) = 1 - \mathrm{e}^{-H^\alpha/\rho} \tag{2.90}$$

and which therefore requires the calibration of two parameters α and ρ.

This is achieved by plotting, on so-called Weibull paper, $\ln\,[-\ln(1 - P(H))] = \alpha \ln H - \ln \rho$ vs $\ln H$, and by fitting the obtained points to a straight line. The slope of the line gives α and its intercept with the vertical axis gives ρ.

Table 2.1 Example of North Sea scatter diagram.

$H_S \backslash T_P$	1–3	3–5	5–7	7–9	9–11	11–13	13–15	15–17	17–19	19–21 (s)	TOTAL
0–1	27	1,112	3,092	2,468	1,084	357	102	27	7	2	8,278
1–2	3	939	7,350	11,515	7,753	3,309	1,099	318	85	30	32,401
2–3	0	83	2,387	8,081	8,591	4,684	1,726	505	128	38	26,223
3–4	0	3	373	3,050	5,498	3,977	1,638	476	111	28	15,154
4–5	0	0	36	855	2,902	2,987	1,429	416	88	18	8,731
5–6	0	0	2	165	1,189	1,899	1,117	337	66	11	4,786
6–7	0	0	0	20	358	981	763	250	45	6	2,423
7–8	0	0	0	1	78	407	455	171	30	3	1,145
8–9	0	0	0	0	12	133	234	109	20	1	509
9–10	0	0	0	0	1	34	102	63	12	1	213
10–11	0	0	0	0	0	7	37	33	8	1	86
11–12	0	0	0	0	0	1	12	15	4	0	32
12–13	0	0	0	0	0	0	3	6	3	0	12
13–14	0	0	0	0	0	0	1	2	1	0	4
14–15 (m)	0	0	0	0	0	0	0	1	1	0	2
TOTAL	30	2,137	13,240	26,155	27,466	18,776	8,718	2,729	609	139	99,999

Return Period

By definition, the return period of a value H is such that this H value is exceeded once on average. For example, on the basis of $8 \times 365 = 2920$ sea states per year, a return period of 100 years corresponds to a value of the cumulative distribution equal to

$$1 - \frac{1}{100 \times 2920} = 0.99999658 \qquad (2.91)$$

The associated sea state is coined as centenal.

Similarly, a 10-year return period means a cumulative distribution function equal to 0.9999658, a return period of one year gives 0.999658.

This obviously does not mean that one is sure to meet a sea state equal to or more severe than the centenal sea state over a time span of a century (the probability of not exceeding being $1/e$, i.e., 37%). On the other hand, the average value and the most likely value of the most severe sea state encountered over a duration of a century are larger than the centenal sea state.

Figure 2.10 shows, on Weibull paper, the empirical cumulative distribution of the significant wave height, derived from the scatter diagram in Table 2.1. It is striking that the points actually align in a straight line for H_S values larger than 3 or 4 m! When one extends the rectilinear part, a significant height of about 16 m is reached at a return period of 100 years.

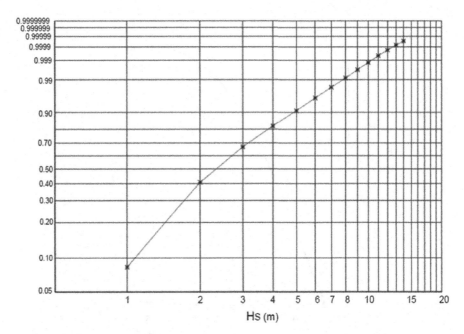

Figure 2.10 Empirical cumulative distribution of the significant wave heights, based on Table 2.1.

Associated Mean or Peak Wave Period

Once the significant centenal wave height is determined, there arises the problem of associating it with an average up-crossing or peak period. One may, from the scatter diagram, build a conditional distribution function $P(T_Z|H_S)$ or $P(T_P|H_S)$ and extrapolate.

Environmental contours, such as shown in Figure 2.11, can also be derived. Computations are then performed following the environmental contour (or part of the contour) associated with the return period. This method is known as IFORM (Inverse First Order Reliability Method).

Another approach, known as Response Based Design (RBD), consists in running systematic calculations for all the recorded sea states on the site, and deriving statistics based on the obtained responses. This is feasible when the local environmental conditions have been sufficiently documented and when the numerical models used are sufficiently efficient (for a comparison between IFORM, RBD, and more traditional design methods, see Fontaine *et al.*, 2013).

2.2   Wind

Offshore structures often have large superstructures, of considerable heights, and they are sensitive to wind. The wind exerts mean loads that mooring systems must withstand, and also dynamic loads, in a wide frequency range.

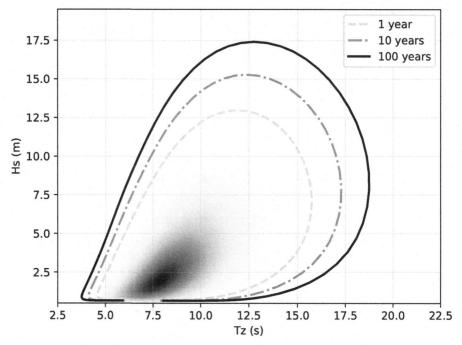

Figure 2.11 Example of environmental contours (H_S, T_Z) (courtesy Bureau Veritas).

The structure of the wind above the sea is very similar to that of a turbulent flow over a rough plate, the thickness of the boundary layer being several tens (or hundreds) of meters. Some specific features complicate the matter:

- the roughness (the waves) is in motion;
- the air density varies with height, temperature, and humidity;
- Coriolis forces result in a variation of the wind direction with height (Ekman spiral). However, this effect is relatively negligible over the first tens of meters;
- depending on the vertical temperature gradient, the atmosphere may be stable or unstable. The difference in temperature between air and water contributes to this (in)stability.

As with waves, the analysis is based on the assumption of stationary conditions over the duration of the sea state, usually 3 hours. Different averaged velocities, denoted $V_m(z, T)$, are defined, with z the height above Mean Sea Level (MSL) and T the duration over which the averaged value is derived. The reference is usually $V_m(10, 600)$, the wind velocity being measured 10 m above MSL and averaged over 600 s (10 mn). Note that averaging over a number N of seconds actually means gradually displacing a time window of N seconds over the 3 hours time series, averaging for each position of the time window, and retaining the maximum of all values obtained.

2.2.1   Variation of the Mean Wind Velocity with Height and Averaging Time

In stable (or neutral) atmospheric conditions, it is usually considered that the mean wind profile follows the logarithmic law of turbulent boundary layers:

$$V_m(z) = \frac{v^\star}{K} \ln \frac{z}{z_0} \tag{2.92}$$

where $v^\star$ is the **friction velocity**, K the **von Karman constant** ($K \simeq 0.4$), and z_0 a **roughness height** that depends on the sea parameters. Typical z_0 values are over a very wide range, from less than a millimeter to several centimeters!

Forristall (1988) has proposed the following law relating the roughness height to the friction velocity:

$$z_0 = 0.0144 \frac{v^{\star 2}}{g} \tag{2.93}$$

while the friction velocity is related to the reference mean wind speed (usually $V_m(10,600)$) through

$$v^{\star 2} = C_d V_m^2 \tag{2.94}$$

with C_d the friction coefficient.

In practice, one uses logarithmic or power laws that relate $V_m(z,t)$ to $V_m(10,t)$. For instance,

$$\frac{V_m(z,t)}{V_m(10,t)} = \frac{\ln z/z_0}{\ln 10/z_0} \tag{2.95}$$

or

$$\frac{V_m(z,t)}{V_m(10,t)} = \left(\frac{z}{10}\right)^\alpha \tag{2.96}$$

where typically α lies in-between 0.1 and 0.2.

Different empirical formulas have been proposed to relate $V_m(z,T)$ to $V_m(z,600)$. For instance,

$$\frac{V_m(z,t)}{V_m(z,600)} = 1.58 - 0.09 \ln t \tag{2.97}$$

or

$$\frac{V_m(z,t)}{V_m(z,600)} = 0.62 + \frac{1.36}{t^{1/5}} \tag{2.98}$$

DNVGL (2017) has proposed the following general law relating $V_m(z,T)$ to $V_m(10,600)$:

$$V_m(z,T) = V_m(10,600) \left[1 + 0.137 \ln \frac{z}{10} - 0.047 \ln \frac{T}{600}\right] \tag{2.99}$$

2.2.2   Wind Spectra

Measured wind velocities, at a given height, and over a duration T, can be separated into a mean value and a fluctuating part:

$$V(t) = V_m(T) + V'(t) \qquad (2.100)$$

while the transverse and vertical components also fluctuate around a zero mean value.

The autocorrelation function $R(\tau)$ can be derived from

$$R(\tau) = E\{V'(t)\, V'(t + \tau)\} \qquad (2.101)$$

The **turbulence intensity** I_v is defined as the ratio σ_V/V_m with σ_V the standard deviation of the in-line velocity:

$$\sigma_V^2 = R(0) = E\{V'^2(t)\} \qquad (2.102)$$

Typical values of the turbulence intensity are in the range 5%–15%.

Through Fourier transform of $R(\tau)$ one obtains the spectral density of the fluctuating part of the velocity. It is customary to express the spectral density via the frequency f (at variance with waves), and to put it in nondimensional form:

$$\frac{f\, S(f)}{V_m^2} \quad \left(\text{or}\ \frac{f\, S(f)}{\sigma_V^2} \right) = F(\widetilde{f}) \qquad (2.103)$$

with

$$\widetilde{f} = \frac{f\, L_V}{V_m} \qquad (2.104)$$

L_V being a reference length.

As the frequency goes to infinity, Kolmogorov's similarity hypothesis is usually applied, meaning that the asymptotic form of $S(f)$ should be some constant times $f^{-5/3}$. All standard forms of wind spectra follow this asymptotic high-frequency law. However, they widely differ on the low-frequency part, particularly at frequencies in the range of the slow-drift natural frequencies of moored structures. This is due to the fact that historically these spectra were calibrated with regards to the high-frequency response of slender buildings, prone to elastic vibrations under wind loads.

Figure 2.12, taken from Ochi & Shin (1988), shows some reference spectra, from literature, compared to measurements. They are shown as $S(f^*) = f\, S(f)/v^{\star 2}$ plotted vs $f^* = f\, z/V_m(z)$ with z the reference height.

The analytical form of the Davenport spectrum is usually given as

$$\frac{f\, S(f)}{\sigma_V^2} = \frac{2}{3}\, \frac{\widetilde{f}^2}{(1 + \widetilde{f}^2)^{4/3}} \qquad (2.105)$$

with the wind velocity referred to 10 m above MSL.

The Harris spectrum has a similar form

$$\frac{f\, S(f)}{\sigma_V^2} = \frac{4\, \widetilde{f}}{\left(1 + 70.8\, \widetilde{f}^2\right)^{5/6}} \qquad (2.106)$$

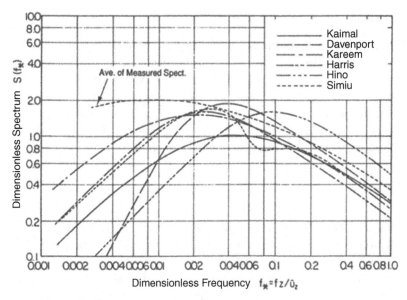

Figure 2.12 Wind spectra (from Ochi & Shin, 1988).

The Davenport and Harris spectra are known to underpredict the wind fluctuations in the low-frequency range.

The Kaimal spectrum is given by

$$\frac{f\,S(f)}{\sigma_V^2} = \frac{6.868\,\tilde{f}}{\left(1 + 10.32\,\tilde{f}\right)^{5/3}} \tag{2.107}$$

There are many other formulations (Hino, Kareem, Simiu, Leigh, etc.). Based on their measurements, Ochi & Shin (1988) have proposed the following empirical form

$$\frac{f\,S(f,z)}{V_m^2(10)} = \begin{cases} 583\,C_d\,f^* & \text{for}\quad 0 \le f^* \le 0.003 \\[2ex] C_d\,\dfrac{420\,f^{*0.7}}{\left(1 + f^{*0.35}\right)^{11.5}} & \text{for}\quad 0.003 \le f^* \le 0.1 \\[2ex] C_d\,\dfrac{838\,f^*}{\left(1 + f^{*0.35}\right)^{11.5}} & \text{for}\quad 0.1 \le f^* \end{cases} \tag{2.108}$$

where f^* is defined as

$$f^* = \frac{f\,z}{V_m(z)} \tag{2.109}$$

and where the friction coefficient is obtained from the 10 m wind velocity as

$$C_d \simeq 10^{-3} + 6 \cdot 10^{-5}\,V_m(10) \tag{2.110}$$

(with V_m given in m/s).

2.2.3   Squalls

In equatorial and tropical areas such as the Gulf of Guinea, the wave conditions are usually mild. However, sudden storms frequently occur, of short duration, where the wind velocity increases rapidly and dies out. A typical record is shown in Figure 2.13 where the wind velocity goes from 10 m/s up to nearly 30 m/s in less than 5 minutes. The abrupt rise in velocity is accompanied by a change in direction of more than 90 degrees. These squalls are of concern for the mooring systems of FPSOs, which may respond dynamically and undergo large excursions.

2.3   Current

The current at a given offshore site results from the superposition of oceanic currents, tidal currents, and wind-generated surface currents. The periodicity of the oceanic currents is of the order of a month, while tides have diurnal or semi-diurnal cycles. Even though not quite fulfilled in the case of semi-diurnal tidal currents, the stationarity assumption over a time span of 3 hours is customarily made.

Strong surface currents may be observed nearshore, next to the mouths of rivers, for instance, the Congo river in the Gulf of Guinea.

Current velocities vary widely depending on location. In the North Sea the centenal current velocity is less than 1 m/s; in the English Channel tidal currents reach 3 or 4 m/s in some places.

Current profiles over the depth are usually strongly sheared. The current may also vary in direction over the water column (the Ekman spiral). The current profile has a major effect on the elastic response of risers, under vortex shedding.

2.4   Internal Waves

The density of sea water is not strictly constant. It varies with salinity and temperature. In particular, over the water column, there can be a significant jump in density between the warmer surface waters and the colder deep waters. The boundary between the two layers can be quite thin, to the point where it can be considered as an interface between two fluids of slightly different densities. This phenomenon is particularly marked in the tropical and equatorial zones, where the differences in temperatures between the upper and lower layers can reach twenty degrees. Their interface, the thermocline, is typically at depths of the order of 50 m.

Stratification can also be related to a difference in salinity, for example, near an estuary, or in Norwegian fjords (or in the vicinity of melting pack ice.)

Just as waves appear at the air–water interface, **internal waves** can appear at the interface between the two layers. In deep ocean, their phase velocity is approximately given by

$$C_P = \sqrt{g \, h \, \frac{\Delta \rho}{\rho}} \qquad (2.111)$$

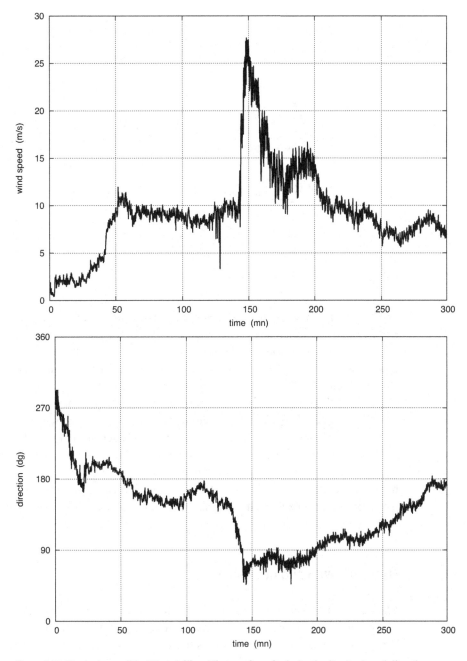

Figure 2.13 Typical squall in West Africa. Time series of wind velocity (top) and direction (bottom) (courtesy TotalEnergies).

with h the upper layer thickness and $\Delta\rho/\rho$ the relative density difference, typically in the range 10^{-4} to 10^{-3}. Internal waves propagate slowly but the horizontal components of the associated orbital velocities can reach notable values. To an observer, they

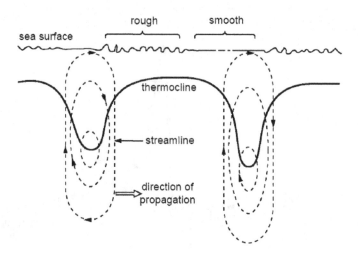

Figure 2.14 Schematic view of an internal wave.

appear as a time-varying current, strongly sheared, with a change in direction through the interface (see Figure 2.14).

The associated free surface elevations are quite negligible. It is not these that make it possible to visually detect the presence of an internal wave but, indirectly, the associated current: It interacts with the short surface waves by modifying, locally, their heights and their wave lengths, accentuating or neutralizing their breaking. Internal waves thus appear very clearly on some satellite images (see Figure 2.15).

For internal waves to be generated, the body of water must already be in motion, under the effect of tidal or oceanic currents, and it must encounter a change in bathymetry, for example, the continental slope, or an underwater ridge. When the required conditions (stratification, strong currents, variable bathymetry) are satisfied, the currents associated with the internal waves produced can reach appreciable values, of the order of 1 m/s or 2 m/s.

2.5   Marine Growth and Corrosion

Marine growth quickly occurs on any object submerged or occasionally submerged in the sea (Figure 2.16). It may consist in relatively soft material (algae, seaweeds, barnacles) or quasi-solid appurtenances (mussels, oysters, corals). Marine growth adds to the size and to the weight of marine structures.

In design it is accounted for as an extra thickness given to all members, typically 50 to 100 mm. The density is usually taken as about 1300 kg/m^3.

Corrosion is an important problem for steel structures designed to stay at sea for many years. Cathodic protection can be ensured thanks to sacrificial anodes. As can be seen from Figure 2.17, anodes increase the frontal areas and they need to be included in the calculation of the hydrodynamic loads.

Figure 2.15 Internal waves seen from space [Envisat ASAR image of Strait of Gibraltar (January 7, 2007). Copyright ESA 2007].

Figure 2.16 Marine growth (courtesy TotalEnergies).

Figure 2.17 Cathodic protection of a steel jacket (courtesy TotalEnergies).

2.6   References

ARHAN M., EZRATY R. 1978. Statistical relations between successive wave heights, *Oceno-logica Acta*, **1**, 151–158.

BOUDET L., MOLIN B. 1995. Analyse numérique et expérimentale du groupage des vagues, in *Actes des 5èmes Journées de l'Hydrodynamique*, Rouen, 255–269 (in French; http://website.ec-nantes.fr/actesjh/).

DNVGL 2017. Environmental conditions and environmental loads. Recommended practice RP-C205.

FONTAINE E., ORSERO P., LEDOUX A., NERZIC R., PREVOSTO M., QUINIOU V. 2013. Reliability analysis and Response Based Design of a moored FPSO in West Africa, *Structural Safety*, **41**, 82–96.

FORRISTALL G.Z. 1988. Wind spectra and gust factors over water, *Proc. 20th Offshore Techn. Conf.*, paper No 5735.

FORRISTALL G.Z. 2000. Wave crest distributions: observations and second-order theory, *J. Physical Oceanography*, 1931–1943.

HASSELMANN K. *et al.* 1973. Measurements of wind-wave growth and swell decay during the Joint North Sea Wave Project (JONSWAP), *Deutsche Hydr. Zeit*, Reihe **A, Nr.12**, 7–95.

HAVER S. 2004. A possible freak wave event measured at the Draupner jacket January 1 1995, in *Proc. Rogue Waves 2004*, Brest (www.ifremer.fr/web-com/stw2004/rw/).

KHARIF C. & PELINOVSKY E. 2003. Physical mechanisms of the rogue wave phenomenon, *European Journal of Mechanics - B/Fluids*, **22**, 603–634.

LONGUET-HIGGINS M.S. 1984. Statistical properties of wave groups in a random sea state, *Phil. Trans. R. Soc. Lond. A*, **312**, 219–250.

MCALLISTER M.L., DRAYCOTT S., ADCOCK T.A.A., TAYLOR P.Y., VAN DEN BRE-MER T.S. 2019. Laboratory recreation of the Draupner wave and the role of breaking in crossing seas, *J. Fluid Mech.*, **860**, 767–786.

OCHI M.K., SHIN Y.S. 1988. Wind turbulent spectra for design consideration of offshore structures, *Proc. 20th Offshore Technology Conf.*, paper No 5736.

3 Wave Theories

3.1 Introduction

There are almost innumerable wave theories. Here we confine ourselves to the wave theories that are applied for the design of marine structures, in relatively deep water, away from shore, over flat seabed. Phenomena such as shoaling, refraction, and diffraction over varying bathymetries are not considered.

Wave models follow several classifications, one of them being deep water or shallow water wave theory. In the simple case of a regular wave, there are three characteristic lengths: the mean water depth h, the crest to trough height $H \simeq 2A$, with A the amplitude, and the wave length L (see Figure 3.1). A key parameter is the Ursell number usually defined as

$$U_R = \frac{H}{h}\left(\frac{L}{h}\right)^2 \simeq 8\pi^2 \frac{kA}{(kh)^3} \tag{3.1}$$

with k the wave number $k = 2\pi/L$. The Ursell number is a combination of the "nonlinearity parameter" H/h and of the "dispersion parameter" h/L (or kh). It makes it possible to discriminate between two main families of wave theories: when U_R is high, shallow water theories must be applied such as the Korteweg-de Vries or Boussinesq equations. These models are based on the assumption that the dispersion parameter is low. They apply to such waves as tidal waves or tsunamis, and to wind waves in shallow areas. When the Ursell number is not too high (less than 26 according to Holthuijsen, 2007), Stokes theories, based on the assumption that the steepness kA is small, may be applied. With a wave length L of

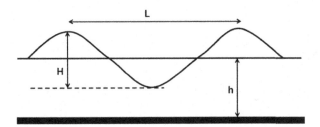

Figure 3.1 Characteristic lengths.

150 m and a height H of 10 m, $U_R \leq 26$ means a water depth larger than 20 m.

Here we are concerned with fixed or floating structures in not too shallow depths and we follow the Stokes model.

Water waves are generated by the wind, which creates a shear at the free surface. Hence they have some vorticity. Nevertheless we shall assume irrotationality as this assumption allows us to use the powerful frame of potential flow theory. This applies to waves propagating away from their generation areas, or to waves produced in a wave tank.

3.2 General Equations

We make use of a rectangular coordinate system $Oxyz$ with $z = 0$ the mean free surface and $z = -h$ the seabed (Figure 3.2). The free surface elevation (assumed to be single valued) is given by

$$z = \eta(x, y, t) \tag{3.2}$$

We assume the fluid to be incompressible and the flow to be irrotational, and we use potential flow theory. The fluid velocity $\vec{V}(x, y, z, t)$ is related to the velocity potential $\Phi(x, y, z, t)$ via $\vec{V} = \nabla\Phi$ (see Appendix A).

The velocity potential Φ verifies the following equations:

1. The Laplace equation (expressing incompressibility), in the whole fluid domain:

$$\Delta\Phi = 0 \tag{3.3}$$

2. The no-flow condition at the sea floor, assumed to be impermeable:

$$\frac{\partial\Phi}{\partial z} = \Phi_z = 0 \qquad\qquad z = -h \tag{3.4}$$

3. The free surface equations:

   – The "kinematic" condition:

$$\eta_t + \Phi_x\,\eta_x + \Phi_y\,\eta_y = \Phi_z \qquad\qquad z = \eta(x, y, t) \tag{3.5}$$

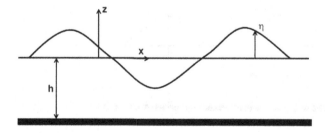

Figure 3.2 Geometry.

– The "dynamic" condition:

$$\Phi_t + \frac{1}{2}(\nabla\Phi)^2 + g\, z = 0 \qquad\qquad z = \eta(x,y,t) \qquad (3.6)$$

The kinematic condition expresses that the free surface is a material surface: a particle at the free surface remains at the free surface. The dynamic condition expresses that the pressure at the free surface is equal to the atmospheric pressure, assumed to be constant. Surface tension, which plays a role only for waves of very small wave lengths (less than a few centimeters), is ignored.

The main difficulty is that the upper boundary of the fluid domain (the free surface) is poorly known, and that the related equations are nonlinear. Part of the remedy is to "work" in a constant fluid domain $-h \le z \le 0$ and to assume that the velocity potential can be extended via Taylor development from $z = 0$ up to $z = \eta$:

$$\Phi(x,y,z,t) = \Phi(x,y,0,t) + z\,\Phi_z(x,y,0,t) + \ldots \qquad 0 \le z \le \eta(x,y,t) \quad (3.7)$$

Correlatively Φ is written as a power series of a small parameter ε, which will turn out to be the wave steepness:

$$\Phi = \Phi^{(1)} + \Phi^{(2)} + \ldots \qquad (3.8)$$

where:

$$\Phi^{(1)} = \varepsilon\, \phi^{(1)} \qquad\qquad \Phi^{(2)} = \varepsilon^2\, \phi^{(2)} \qquad \ldots \qquad (3.9)$$

In a similar way, the free surface elevation is written

$$\eta = \eta^{(1)} + \eta^{(2)} + \ldots \qquad (3.10)$$

where $\eta^{(1)}$ is of order ε, $\eta^{(2)}$ of order ε^2, etc.

The procedure then is to insert the developments (3.7), (3.8), (3.10) into the free surface equations (3.5) and (3.6), and to sort out the terms of order ε, of order ε^2, etc. Then we obtain:

At first order:

$$\Phi_{tt}^{(1)} + g\,\Phi_z^{(1)} = \qquad 0 \qquad\qquad z = 0 \qquad (3.11)$$

$$\eta^{(1)} = \quad -\frac{1}{g}\,\Phi_t^{(1)} \qquad\qquad z = 0 \qquad (3.12)$$

At second order:

$$\Phi_{tt}^{(2)} + g\,\Phi_z^{(2)} = -\eta^{(1)}\,(\Phi_{ttz}^{(1)} + g\,\Phi_{zz}^{(1)}) - 2\,\nabla\Phi^{(1)}\cdot\nabla\Phi_t^{(1)} \qquad z = 0 \quad (3.13)$$

$$\eta^{(2)} = -\frac{1}{g}\,\eta^{(1)}\,\Phi_{zt}^{(1)} - \frac{1}{2g}\,\left(\nabla\Phi^{(1)}\right)^2 - \frac{1}{g}\,\Phi_t^{(2)} \qquad z = 0 \quad (3.14)$$

It is possible to continue to order 3 and beyond, but the expressions quickly become very intricate.

3.3   First-Order Wave Theory

3.3.1   Regular Waves (Airy)

We make an assumption of periodicity, in time (at a period T) and in space (at a wave length L). We also assume the wave system to be a plane wave, propagating along the x axis, that is there is no y dependence.

We introduce the complex "reduced potential" $\varphi(x,z)$

$$\Phi^{(1)}(x,z,t) = \Re \left\{ \varphi(x,z)\, e^{-i\omega t} \right\} \tag{3.15}$$

where $\omega = 2\pi/T$ is the angular frequency. The equations satisfied by φ are:

$$
\begin{aligned}
\Delta\varphi = \varphi_{xx} + \varphi_{zz} &= 0 & &\text{fluid domain} \\
g\,\varphi_z - \omega^2\,\varphi &= 0 & &z = 0 \\
\varphi_z &= 0 & &z = -h \\
&& &\text{periodicity in } x
\end{aligned}
\tag{3.16}
$$

The fluid domain is rectangular: $x_0 < x \le x_0 + L \quad -h \le z \le 0$. Hence one may seek for solutions in the form $\varphi(x,z) = F(x)\,G(z)$.

The Laplace equation becomes:

$$F''\,G + F\,G'' = 0 \tag{3.17}$$

or, dividing by $F\,G$:

$$\frac{F''}{F} + \frac{G''}{G} = 0 \tag{3.18}$$

F''/F is a function of x alone, G''/G a function of z alone; hence they must be two opposite constants:

$$\frac{F''}{F} = \pm k^2 \qquad\qquad \frac{G''}{G} = \mp k^2 \tag{3.19}$$

The assumption of periodicity in x leads to the choice $F''/F = -k^2$, which has the solutions

$$F(x) = e^{\pm i k x} \tag{3.20}$$

Associated solutions for G are exponential functions:

$$G(z) = e^{\pm k z} \tag{3.21}$$

regrouped as

$$G(z) = \cosh k(z + h) \tag{3.22}$$

in order to fulfill the no-flow condition at the sea floor.

Finally, the potential φ has the form

$$\varphi = C\,\cosh k\,(z + h)\, e^{\pm i k x} \tag{3.23}$$

with C a constant.

It remains to fulfill the free surface condition, which gives:

$$C \, g \, k \, \sinh k \, h - C \, \omega^2 \, \cosh k \, h = 0 \qquad (3.24)$$

that is

$$\omega^2 = g \, k \, \tanh k \, h \qquad (3.25)$$

This condition, the **dispersion equation**, links the wave length L and the period T:

$$L = \frac{g \, T^2}{2\pi} \, \tanh \frac{2\pi \, h}{L} \qquad (3.26)$$

In the general case where T is given, this equation needs to be solved through iterations. When the water depth h is large enough compared to the wave length L (in practice, when $h > L/2$), the hyperbolic tangent is very close to 1 and the wave length is given by

$$L = \frac{g \, T^2}{2\pi} \simeq 1.56 \, T^2 \qquad (3.27)$$

(with T in seconds and L in meters).

The free surface elevation is obtained through

$$\eta^{(1)}(x,t) = \Re \left\{ \frac{i\omega}{g} \, \varphi \right\}_{z=0} = \Re \left\{ \frac{i\omega}{g} \, C \, \cosh k \, h \, e^{\pm i k \, x} \, e^{-i\omega \, t} \right\} \qquad (3.28)$$

When we take

$$C = -i \, \frac{A \, g}{\omega \, \cosh k \, h} \qquad (3.29)$$

then

$$\eta^{(1)}(x,t) = A \, \cos(\pm k \, x - \omega \, t) \qquad (3.30)$$

with A the wave amplitude, and the associated velocity potential is

$$\Phi^{(1)}(x,z,t) = \frac{A \, g}{\omega} \, \frac{\cosh k \, (z + h)}{\cosh k \, h} \, \sin(\pm k \, x - \omega \, t) \qquad (3.31)$$

The free surface profile is a sine curve. When a plus sign is used in (3.30) and (3.31), the wave propagates from left to right, with a minus sign it propagates from right to left.

More generally $A \, \cos(k \, x \, \cos \beta + k \, y \, \sin \beta - \omega t)$ is a wave that propagates at an angle β with respect to the Ox axis.

The superimposition of two waves of identical amplitudes and traveling in opposite direction is also a mathematical solution to the initial problem:

$$A \, \cos(k \, x - \omega t) + A \, \cos(-k \, x - \omega t) = 2 \, A \, \cos k \, x \, \cos \omega \, t \qquad (3.32)$$

This is no longer a progressive wave but a standing wave, such as results, for instance, from full reflection by a vertical wall.

Phase Velocity and Group Velocity

By definition, the **phase velocity** C_P is the velocity of the crests and troughs of the wave profile. Their positions are given by $k\,x - \omega\,t = n\,\pi$. Hence

$$C_P = \frac{\omega}{k} = \sqrt{\frac{g}{k}} \, \tanh kh \qquad (3.33)$$

When $kh \geq 3$ (deep water):

$$C_P = \sqrt{\frac{g}{k}} = \frac{g}{\omega} = \frac{g\,T}{2\pi} \qquad (3.34)$$

When $kh \ll 1$ (shallow water depth), then $\tanh kh \simeq kh$ and

$$C_P = \sqrt{g\,h} \qquad (3.35)$$

The phase velocity does not depend on the period any longer, but only on the water depth.

The concept of **group velocity** can be introduced by considering the superimposition of two regular waves of same amplitude and close frequencies:

$$
\begin{aligned}
\eta^{(1)} &= A\,\cos(k_1\,x - \omega_1\,t) + A\,\cos(k_2\,x - \omega_2\,t) \\
&= 2\,A\,\cos\left[\frac{k_1 - k_2}{2}\,x - \frac{\omega_1 - \omega_2}{2}\,t\right]\cos\left[\frac{k_1 + k_2}{2}\,x - \frac{\omega_1 + \omega_2}{2}\,t\right]
\end{aligned}
$$

that is a wave of frequency $(\omega_1 + \omega_2)/2$ and wave number $(k_1 + k_2)/2$ modulated by an envelope of frequency $(\omega_1 - \omega_2)/2$ and wave number $(k_1 - k_2)/2$. This is a schematic representation of a wave group, which propagates at the velocity

$$C_G = \frac{\omega_1 - \omega_2}{k_1 - k_2} \qquad (3.36)$$

Going to the limit $\omega_1 - \omega_2 \to 0$, the group velocity is defined by

$$C_G = \frac{\partial \omega}{\partial k} \qquad (3.37)$$

Differentiating the dispersion equation $\omega^2 = g\,k\,\tanh kh$, we obtain

$$2\,\omega\,\mathrm{d}\omega = g\left(\tanh kh + \frac{kh}{\cosh^2 kh}\right)\mathrm{d}k \qquad (3.38)$$

giving

$$\frac{C_G}{C_P} = \frac{\partial\omega/\partial k}{\omega/k} = \frac{1}{2} + \frac{k\,h}{\sinh 2kh} \qquad (3.39)$$

In deep water, then $k\,h/\sinh 2kh \simeq 0$, the group velocity is half the phase velocity. This explains why, when watching a wave crest within a group, one sees it progressively move to the front of the group, while decreasing in height, and disappear, to the benefit of the following waves. The group velocity is also the velocity at which the wave energy travels.

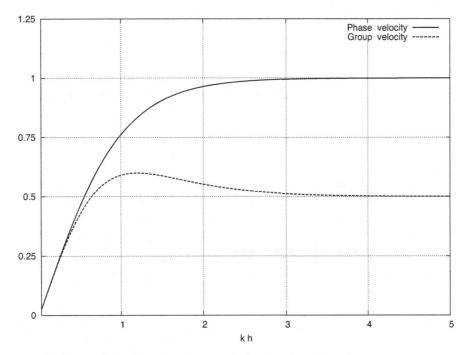

Figure 3.3 Phase velocity $C_P\,\omega/g$ and group velocity $C_G\,\omega/g$ vs kh.

In shallow depth ($\sinh 2\,kh \simeq 2\,kh$), the group and phase velocities are equal, all wave components traveling at the same velocity $\sqrt{g\,h}$. The medium is no longer dispersive (in a linear theory). Differences in velocities will result from differences in crest elevations.

Figure 3.3 shows the nondimensional phase velocity $C_P\,\omega/g \equiv \tanh kh$ and group velocity $C_G\,\omega/g$, vs kh. It can be checked visually that, when kh becomes larger than π, they reach their infinite depth asymptotes.

Orbital Velocities and Trajectories

Taking the gradient of the velocity potential

$$\Phi^{(1)}(x,z,t) = \frac{A\,g}{\omega}\,\frac{\cosh k(z+h)}{\cosh kh}\,\sin(k\,x - \omega\,t) \tag{3.40}$$

the following horizontal and vertical components of the flow velocity are obtained:

$$u = \Phi_x^{(1)} = \frac{A\,g\,k}{\omega}\,\frac{\cosh k(z+h)}{\cosh kh}\,\cos(k\,x - \omega\,t)$$

$$= A\,\omega\,\frac{\cosh k(z+h)}{\sinh kh}\,\cos(k\,x - \omega\,t) \tag{3.41}$$

$$w = \Phi_z^{(1)} = \frac{A\,g\,k}{\omega}\,\frac{\sinh k(z+h)}{\cosh kh}\,\sin(k\,x - \omega\,t)$$

$$= A\,\omega\,\frac{\sinh k(z+h)}{\sinh kh}\,\sin(k\,x - \omega\,t) \tag{3.42}$$

In deep water $(kh > 3)$ they reduce to

$$u = A\,\omega\,e^{kz}\,\cos(k\,x - \omega t) \tag{3.43}$$

$$w = A\,\omega\,e^{kz}\,\sin(k\,x - \omega t) \tag{3.44}$$

Trajectories are obtained by integrating in time the coupled equations:

$$\frac{dx}{dt} = u(x, z, t) \tag{3.45}$$

$$\frac{dz}{dt} = w(x, z, t) \tag{3.46}$$

Within the scope of the first-order theory followed here, where the fluid motion is assumed to be "infinitesimal," the fluid velocity (u, w) can be taken at the reference point (x_0, z_0). One then has to integrate in time:

$$\frac{dx}{dt} = u(x_0, z_0, t) = A\omega\,\frac{\cosh k(z_0 + h)}{\sinh kh}\,\cos(k\,x_0 - \omega t) \tag{3.47}$$

$$\frac{dz}{dt} = w(x_0, z_0, t) = A\omega\,\frac{\sinh k(z_0 + h)}{\sinh kh}\,\sin(k\,x_0 - \omega t) \tag{3.48}$$

giving:

$$x = x_0 - A\,\frac{\cosh k(z_0 + h)}{\sinh kh}\,\sin(k\,x_0 - \omega t) \tag{3.49}$$

$$z = z_0 + A\,\frac{\sinh k(z_0 + h)}{\sinh kh}\,\cos(k\,x_0 - \omega t) \tag{3.50}$$

The trajectories are ellipses, becoming flatter and flatter by the sea floor. In deep water $(kh > \pi)$, the trajectories are circular. Streamlines and trajectories are illustrated in Figure 3.4, for different depth cases.

Mass Transport

What has been derived above is the first order of approximation of the trajectories. In an inconsistent way, but that can be justified for time-averaged quantities, a second order of approximation can be derived by writing:

$$\frac{dx}{dt}(x, z, t) = u(x_0 + x - x_0, z_0 + z - z_0, t) \tag{3.51}$$

$$\simeq u(x_0, z_0, t) + (x - x_0)\,\frac{\partial u}{\partial x}(x_0, z_0, t) + (z - z_0)\,\frac{\partial u}{\partial z}(x_0, z_0, t) \tag{3.52}$$

Inserting there the $x - x_0$ and $z - z_0$ expressions previously derived, we get

$$\frac{dx^{(2)}}{dt} = A^2 k\omega\,\frac{\cosh^2 k(z_0 + h)}{\sinh^2 kh}\,\sin^2(k\,x_0 - \omega t)$$

$$+ A^2 k\omega\,\frac{\sinh^2 k(z_0 + h)}{\sinh^2 kh}\,\cos^2(k\,x_0 - \omega t) \tag{3.53}$$

which has a nonzero time-averaged value:

$$\overline{\frac{dx^{(2)}}{dt}} = \frac{1}{2}\,A^2 k\omega\,\frac{\cosh 2k(z_0 + h)}{\sinh^2 kh} \tag{3.54}$$

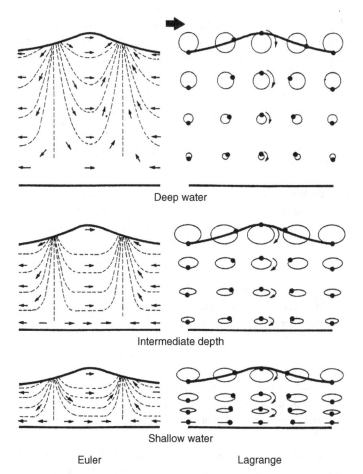

Deep water

Intermediate depth

Shallow water

Euler   Lagrange

Figure 3.4 Streamlines (left) and trajectories (right) in deep water (top), intermediate depth (middle) and shallow depth (bottom) (from Susbielles & Bratu, 1981, *Vagues et ouvrages pétroliers en mer*, Éditions Technip).

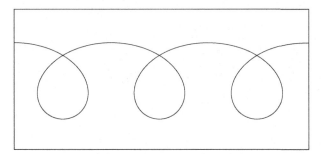

Figure 3.5 Trajectory obtained through integrating equations (3.45) and (3.46).

The trajectories are not strictly closed: the water particles drift in the direction of wave propagation. Figure 3.5 shows an example of a trajectory obtained by exactly integrating in time equations (3.45) and (3.46). The particle is taken at the free surface, the water depth is infinite, the steepness $k A$ is $0.1 \times \pi$, that is $H/L = 10\%$, close to breaking condition. It can be observed that the drift velocity is appreciable.

At the free surface, from equation (3.54), the drift velocity (known as Stokes drift) is $\overline{U} = k A^2 \omega = A^2 \omega^3/g$ (in deep water). In a long-crested irregular wave system, it is therefore $\overline{U} = 2 m_3/g$ where m_3 is the third moment of the wave spectrum. The Stokes drift is an important component to the drift of objects lost at sea.

The flow rate is obtained by integrating $dx^{(2)}/dt$ over the depth:

$$Q = \int_{-h}^{0} \frac{1}{2} A^2 k \omega \, \frac{\cosh 2k(z_0 + h)}{\sinh^2 kh} \, dz_0 = \frac{1}{2} A^2 \omega \coth kh \tag{3.55}$$

The same result can be obtained (much faster!) by taking an Eulerian viewpoint, that is integrating $u(x,z,t)$ from $-h$ up to $\eta^{(1)}(x,t)$, and taking the time averaged value, which, to second order of approximation, reduces to:

$$Q = \overline{u(x,0,t) \, \eta(x,t)} = \frac{1}{2} A^2 \omega \coth kh \tag{3.56}$$

When waves are generated in a tank of finite length, mass transport is necessarily compensated by a return current. The only possible profile, according to potential theory, is uniform over the depth, with velocity

$$u_R = -\frac{1}{2} \frac{A^2 \omega}{h} \coth kh \tag{3.57}$$

This velocity can be appreciable. Taking for instance an amplitude A of 20 cm, a period $T = 2\pi/\omega$ of 2 seconds, and a depth of one meter, the velocity of the return current is 7.5 cm/s. On a scale of 1:50, this would mean 53 cm/s, or about 1 knot, at full scale!

Figure 3.6 shows the return current obtained with a numerical wave tank that solves successively the first-order, second-order, and third-order problems, in space

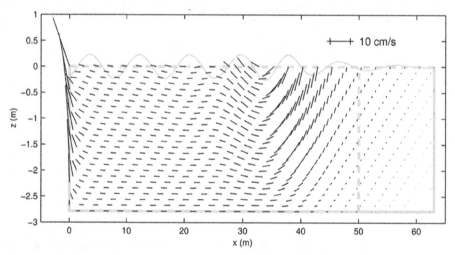

Figure 3.6 Generation of the return current at the wave front of a regular wave system (from Stassen, 1999).

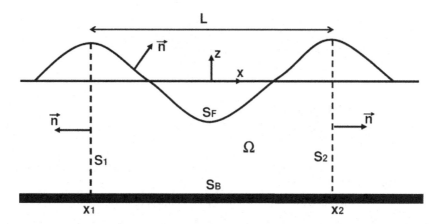

Figure 3.7 Geometry.

and time (Stassen, 1999). The figure shows the low-frequency component of the velocity field obtained at second order. A downward flow appears at the wave front, followed by a uniform return flow toward the wavemaker, finally sucked up by a sink at the intersection of the free surface with the wavemaker (this is the second-order idealization).

Energy and Energy Flux

Consider a fluid domain Ω bounded by the sea bottom (S_B), the free surface (S_F) and two vertical cuts (S_1, S_2), one wave length apart (Figure 3.7).

The fluid energy in Ω divides into potential energy E_P and kinetic energy E_C.

The potential energy writes:

$$E_P = \int\int_\Omega \rho g \, z \, dS = \int_{x_1}^{x_2} \frac{1}{2} \rho g \, (\eta^2 - h^2) \, dx \tag{3.58}$$

giving (eliminating the static contribution $-1/2 \, \rho g \, h^2$):

$$E_P = \frac{1}{2} \rho g \, A^2 \int_{x_1}^{x_2} \cos^2(kx - \omega t) \, dx = \frac{1}{4} \rho g \, A^2 \, L \tag{3.59}$$

The kinetic energy

$$E_C = \frac{1}{2} \rho \int\int_\Omega (\nabla\Phi)^2 \, dS \tag{3.60}$$

can be transformed into

$$E_C = \frac{1}{2} \rho \int_{S_1 \cup S_B \cup S_2 \cup S_F} \Phi \, \nabla\Phi \cdot \vec{n} \, dl \tag{3.61}$$

$\vec{n}$ being the outward normal vector.

The integral over the sea bottom is nil. The integrations over S_1 and S_2 cancel out due to periodicity. There only remains the integral over the free surface, that is approximated as

$$E_C = \frac{1}{2} \int_{x_1}^{x_2} \Phi \, \Phi_z \Big|_{z=0} \, dx \qquad (3.62)$$

Then the same value as for E_P is obtained:

$$E_C = \frac{1}{4} \rho g \, A^2 \, L \qquad (3.63)$$

and the total energy is

$$E = \frac{1}{2} \rho g \, A^2 \, L \qquad (3.64)$$

Therefore, in a regular wave system, the averaged energy density per unit area is $1/2 \rho g \, A^2$.

The time derivative of the energy in the fluid domain Ω is

$$\frac{dE}{dt} = \frac{d}{dt} \left\{ \rho \iint_\Omega \left[\frac{1}{2} (\nabla \Phi)^2 + g \, z \right] dS \right\} \qquad (3.65)$$

$$= \rho \iint_\Omega \frac{\partial}{\partial t} \left[\frac{1}{2} (\nabla \Phi)^2 + g \, z \right] dS + \int_{S_F} \left[\frac{1}{2} (\nabla \Phi)^2 + g \, z \right] \vec{U} \cdot \vec{n} \, dl \qquad (3.66)$$

where $\vec{U}$ is the free surface velocity.

The first term can be transformed in the following way:

$$\rho \iint_\Omega \frac{\partial}{\partial t} \left[\frac{1}{2} (\nabla \Phi)^2 + g \, z \right] dS = \rho \iint_\Omega \nabla \Phi \cdot \nabla \Phi_t \, dS \qquad (3.67)$$

$$= \rho \int_{S_1 \cup S_F \cup S_2 \cup S_B} \Phi_t \, \nabla \Phi \cdot \vec{n} \, dl \qquad (3.68)$$

so that, the integral over the sea floor being nil:

$$\frac{dE}{dt} = \rho \int_{S_1 \cup S_2} \Phi_t \, \nabla \Phi \cdot \vec{n} \, dl + \rho \int_{S_F} \left[\frac{1}{2} (\nabla \Phi)^2 + g \, z + \Phi_t \right] \vec{U} \cdot \vec{n} \, dl \qquad (3.69)$$

The term between brackets in the free surface integral is the pressure, assumed to be zero. As a result,

$$\frac{dE}{dt} = \rho \int_{S_1 \cup S_2} \Phi_t \, \nabla \Phi \cdot \vec{n} \, dl \qquad (3.70)$$

When S_1 and S_2 are one wave length apart, the integrals over S_1 and S_2 cancel out, so that $dE/dt \equiv 0$. It may sound as if the exercise is of no interest. In fact, it shows that the quantity of energy that enters the domain by unit time, or **energy flux**, is given by:

$$F_L(x) = -\rho \int_{-h}^{\eta} \Phi_t \, \Phi_x \, dz \simeq -\rho \int_{-h}^{0} \Phi_t \, \Phi_x \, dz \qquad (3.71)$$

Carrying out the integration, and taking the time-averaged value, one obtains

$$\overline{F}_L = \frac{1}{4} \rho \, A^2 \, g \, \frac{\omega}{k \, \sinh 2kh} (2 \, kh + \sinh 2kh) = \frac{1}{2} \rho g \, C_G \, A^2 \qquad (3.72)$$

meaning that the energy travels at the group velocity. $\overline{F}_L$ is the allowable power, per unit width, that can be recovered by a wave energy converter.

3.3.2   Irregular Waves

At first order of approximation, an irregular sea state can be modeled as the superposition of a large number of Airy components. The free surface elevation is written as

$$\eta^{(1)}(x,y,t) = \sum_i A_i \; \cos[k_i \; (x \cos \beta_i + y \sin \beta_i) - \omega_i t + \theta_i], \qquad (3.73)$$

with $A_i^2 = 2S(\omega_i, \beta_i) \, \Delta\omega_i \, \Delta\beta_i$ where $S(\omega, \beta)$ is the directional spectrum. The phases θ_i are taken as equally distributed random variables and the frequencies ω_i and wave numbers k_i are linked by the dispersion equation $\omega_i^2 = g \, k_i \, \tanh k_i h$.

It is often argued that the amplitudes A_i are themselves random variables, Rayleigh distributed around the mean square value $2S(\omega_i, \beta_i) \, \Delta\omega_i \, \Delta\beta_i$. What actually matters is that the number of wave components be large enough. Another issue is whether directional sea states should be represented via a single summation as in (3.73) (one wave component per direction β_i and per frequency ω_i) or via a double summation. It is generally acknowledged that the single summation is preferable. Other points of discussion are whether the frequencies ω_i and directions β_i should be equally spaced or randomized, etc.

Because of linearity, the velocity potential is obtained by summing up all Airy components:

$$\Phi^{(1)}(x,y,z,t) = \sum_i \frac{A_i \, g}{\omega_i} \frac{\cosh k_i (z + h)}{\cosh k_i h} \; \sin\left[k_i \; (x \cos \beta_i + y \sin \beta_i) - \omega_i t + \theta_i \right]$$

$$(3.74)$$

From this the fluid kinematics can be simply derived. For instance, the x component of the velocity at a point (x, y, z) is

$$u(x,y,z,t) = \Phi_x^{(1)}(x,y,z,t) = \sum_i A_i \, \omega_i \, \cos \beta_i \; \frac{\cosh k_i (z + h)}{\sinh k_i h}$$

$$\cos\left[k_i \; (x \cos \beta_i + y \sin \beta_i) - \omega_i t + \theta_i \right] \qquad (3.75)$$

When one prefers to remain within the frequency domain, from (3.75) the spectrum of the x velocity at a vertical coordinate z is given by

$$S_u(\omega, z) = \omega^2 \, \frac{\cosh^2 k(z + h)}{\sinh^2 kh} \int_0^{2\pi} S(\omega, \beta) \; \cos^2 \beta \; d\beta \qquad (3.76)$$

Likewise, the energy flux in a long-crested sea state is given by

$$F_L = \sum_i \frac{1}{2}\rho \, g \, C_G(\omega_i) \, A_i^2 = \rho \, g \int_0^\infty C_G(\omega) \, S(\omega) \, d\omega \qquad (3.77)$$

In deep water $C_G = g/(2\omega)$. Then the energy flux can be written as

$$F_L = \frac{1}{64} \rho \, g^2 \, H_S^2 \, T_{-1} \qquad (3.78)$$

where T_{-1} is a new mean period defined by

$$T_{-1} = 2\pi \, \frac{\int_0^\infty S(\omega) \, \omega^{-1} \, d\omega}{m_0} \qquad (3.79)$$

3.3.3   How Good Is the First-Order Approximation?

Strictly speaking the Airy wave model is valid for waves of infinitely small ("infinitesimal") amplitude. It is legitimate to ask the question of how fast it degrades when the wave steepness increases and from what degree of steepness higher-order wave models should be applied. (As written earlier in this Chapter we assume the water to be deep enough so that the Ursell number is small and we are not concerned with limitations due to the depth being too shallow.)

Regular Wave Case

For the sake of simplicity, we assume infinite depth. The velocity potential is given by

$$\Phi^{(1)}(x,z,t) = \frac{Ag}{\omega} e^{kz} \sin(kx - \omega t) \tag{3.80}$$

and the free surface elevation by

$$\eta^{(1)}(x,t) = A \cos(kx - \omega t) \tag{3.81}$$

The Laplace equation and the decay condition for $z \to -\infty$ are exactly satisfied. What must be checked is how well the free surface equations are verified, that is, considering first the dynamic condition, compare

$$\eta^{(1)} \quad \text{and} \quad -\frac{1}{g} \Phi_t^{(1)} - \frac{1}{2g} \left(\nabla\Phi^{(1)}\right)^2 \quad \text{at } z = \eta \tag{3.82}$$

$$A \cos\theta \stackrel{?}{=} A \cos\theta \, e^{k A \cos\theta} - \frac{1}{2} k A^2 \, e^{2 k A \cos\theta} \tag{3.83}$$

$$A \cos\theta \stackrel{?}{=} A \left[\cos\theta + \frac{1}{2} k A \cos 2\theta + O(k^2 A^2)\right] \tag{3.84}$$

where $\theta = k x - \omega t$

The relative difference of the two sides is of order $k A$. It can be checked, likewise, that the relative error committed in the kinematic condition is also of order $k A$. Figure 3.8 shows the errors committed in the dynamic and kinematic conditions for $k A = 0.2$.

Bichromatic Seas

Let us now consider a wave system consisting in the superposition of two Airy components. The free surface elevation is

$$\eta^{(1)}(x,t) = A_1 \cos(k_1 x - \omega_1 t) + A_2 \cos(k_2 x - \omega_2 t) = A_1 \cos\theta_1 + A_2 \cos\theta_2 \tag{3.85}$$

and the velocity potential

$$\Phi^{(1)} = \frac{A_1 g}{\omega_1} e^{k_1 z} \sin\theta_1 + \frac{A_2 g}{\omega_2} e^{k_2 z} \sin\theta_2 \tag{3.86}$$

The dynamic condition at the free surface now writes:

$$A_1 \cos\theta_1 + A_2 \cos\theta_2 \stackrel{?}{=} A_1 \exp[k_1(A_1 \cos\theta_1 + A_2 \cos\theta_2)] \cos\theta_1$$
$$+ A_2 \exp[k_2(A_1 \cos\theta_1 + A_2 \cos\theta_2)] \cos\theta_2 \tag{3.87}$$

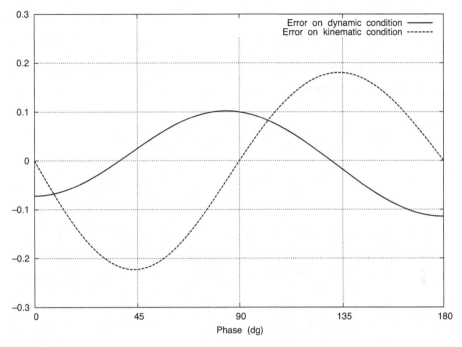

Figure 3.8 Airy wave model in deep water. Relative errors committed on the free surface conditions for $kA = 0.2$.

$$-\frac{1}{2}k_1 A_1^2 \exp[2k_1(A_1 \cos\theta_1 + A_2 \cos\theta_2)] - \frac{1}{2}k_2 A_2^2 \exp[2k_2(A_1 \cos\theta_1 + A_2 \cos\theta_2)]$$
(3.88)

$$-A_1 A_2 \frac{\omega_1 \omega_2}{g} \exp[(k_1 + k_2)(A_1 \cos\theta_1 + A_2 \cos\theta_2)]\ \cos(\theta_1 - \theta_2)$$
(3.89)

In the exponentials there are now terms involving $k_1 A_1$, $k_1 A_2$, $k_2 A_1$, and $k_2 A_2$. For the error committed on the dynamic condition to be acceptable, it is not sufficient that the "individual" steepnesses $k_1 A_1$ and $k_2 A_2$ be small, it is also necessary that the "cross" steepnesses $k_1 A_2$ and $k_2 A_1$ be small.

This supplementary requirement is difficult to satisfy when the frequencies of the two wave components are far apart. Consider for instance, the case of two regular waves with periods 15 and 5 seconds, and amplitudes 5 and 0.5 m. The steepnesses $k_1 A_1$ and $k_2 A_2$ are respectively equal to 0.09 and 0.08, but the cross-steepness $k_2 A_1$ is equal to 0.8! Obviously, the Airy wave model is not valid in this case.

3.3.4   Stretching Models

The inadequacy of the simple summation of Airy components clearly appears when one wishes to express the fluid kinematics in the crests of a multichromatic wave system. The horizontal velocities arising from the high-frequency components take unrealistic values and the resulting kinematics are highly exaggerated. As a result, the drag loads on structures such as jackets or compliant towers are overestimated.

Part of the problem, as noted just above, results from using the same reference level $z = 0$ (the undisturbed free surface) for all wave components when the short waves actually ride the long waves.

Empirical methods, known as "stretching models," have been proposed to correct the crest kinematics.

The crudest method consists in applying the direct summation:

$$u(x,z,t) = \sum_i A_i \, \omega_i \, \frac{\cosh k_i(z + h)}{\sinh k_i h} \, \cos[k_i \, x - \omega_i t + \theta_i] \qquad (3.90)$$

for $-h \le z \le 0$, and

$$u(x,z,t) = \sum_i A_i \, \omega_i \, \coth k_i h \, \cos[k_i \, x - \omega_i t + \theta_i] \qquad (3.91)$$

for $0 \le z \le \eta^{(1)}(x,t)$: the horizontal velocity calculated at the mean water level is taken to be constant up to the free surface. The same procedure is applied to the vertical velocity and to the accelerations.

As a result, the velocity profile has an unsightly discontinuous slope at $z = 0$. A variant consists in taking a velocity profile varying linearly from $z = 0$ up to $z = \eta^{(1)}$, keeping a constant slope:

$$u(x,z,t) \simeq u(x,0,t) + z \, \frac{\partial u}{\partial z}(x,0,t) \qquad (3.92)$$

This method is known as **linear extrapolation**.

The **Wheeler** (1970) model stretches the vertical coordinate (hence the coinage "stretching"), in such a way that, at $z = \eta^{(1)}$, the same kinematics as given by the direct summation in $z = 0$ be retrieved. To the true vertical coordinate z is associated a fictitious one z' given by

$$z' = h \, \frac{z - \eta^{(1)}}{h + \eta^{(1)}} \qquad \text{or} \qquad z' + h = h \, \frac{z + h}{h + \eta^{(1)}} \qquad (3.93)$$

where the kinematics are calculated.

As a result the horizontal velocity profile becomes

$$u(x,z,t) = \sum_i A_i \, \omega_i \, \frac{\cosh [k_i h \, (z + h)/(h + \eta^{(1)})]}{\sinh k_i h} \, \cos[k_i \, x - \omega_i t + \theta_i] \qquad (3.94)$$

for $-h \le z \le \eta^{(1)}(x,t)$.

A similar procedure, devised by **Chakrabarti**, consists in playing with the water depth:

$$u(x,z,t) = \sum_i A_i \, \omega_i \, \frac{\cosh k_i(z + h)}{\sinh k_i(h + \eta^{(1)})} \, \cos[k_i \, x - \omega_i t + \theta_i] \qquad (3.95)$$

One may note that, in all these models, the Laplace equation is violated: mass is no longer conserved! Then it becomes difficult to assess their adequacy otherwise than through direct comparison with experimental velocity profiles. There have been numerous experimental campaigns performed with this aim, generally concluding that the crest velocities are underestimated by the Wheeler and Chakrabarti models, and overestimated by the linear extrapolation model.

The **Delta-stretching**, proposed by Rodenbusch et Forristall (1986), yields profiles intermediate between Wheeler and linear extrapolation. The principle is the same as in the Wheeler model, that is, to the true coordinate z one associates a fictitious coordinate z' from which the kinematics are derived, but this is done only in the upper layer, from $-h_\Delta$ up to the free surface, where the coordinated z' is obtained through

$$z' = (z + h_\Delta) \frac{h_\Delta + \Delta\eta}{h_\Delta + \eta} - h_\Delta \tag{3.96}$$

z' varies from $-h_\Delta$ up to $\Delta\eta$ when z goes from $-h_\Delta$ to η.

When z' is larger than zero, the kinematics is obtained through linear extrapolation.

Taking $h_\Delta = h$ and $\Delta = 0$ the Wheeler model is recovered; with $h_\Delta = h$ and $\Delta = 1$ linear extrapolation is obtained. Rodenbusch and Forristall (1986) suggest a Δ value of 0.3.

Other stretching models have been proposed. In particular, the Wheeler model can be improved by taking advantage of the idea that short waves ride long waves, and by associating to each wave component ω_i a reference level η_i given by

$$\eta_i = \sum_{j=1}^{i-1} A_j \cos(k_j x - \omega_j t + \theta_j) \tag{3.97}$$

(the frequencies being ordered in increasing values), and a fictitious z_i' coordinate:

$$z_i' = h \frac{z - \eta_i}{h + \eta_i} \tag{3.98}$$

giving:

$$u(x,z,t) = \sum_i A_i \, \omega_i \, \frac{\cosh[k_i h \, (z + h)/(h + \eta_i)]}{\sinh k_i h} \cos[k_i \, x - \omega_i t + \theta_i] \tag{3.99}$$

Figure 3.9 shows different velocity profiles obtained numerically below a high crest in a sea state of 15 m significant wave height and 18 s peak period, in 100 m water depth. In this case linear extrapolation gives a crest velocity 40% higher than the Wheeler model. This means drag loads twice as high!

Figure 3.10 shows a comparison between the Wheeler and Chakrabarti models and experimental measurements, performed with Laser Doppler Velocimetry (LDV). The tests are done in deep water (1.8 m) at wave periods around 1.5 s ($kh \sim 3$). In such case, as is shown in the following sections, in regular waves the second- and third-order corrections to the first-order kinematics are nil: a good agreement is found between the experimental and theoretical velocities from linear theory, up to the actual free surface. In irregular waves, the experimental measurements lie in between linear theory and the stretching models considered.

An alternative to these stretching models is to use the kinematics from second-order theory, presented in the next section. Encouraging comparisons with experiments are presented by Stansberg *et al.* (2008).

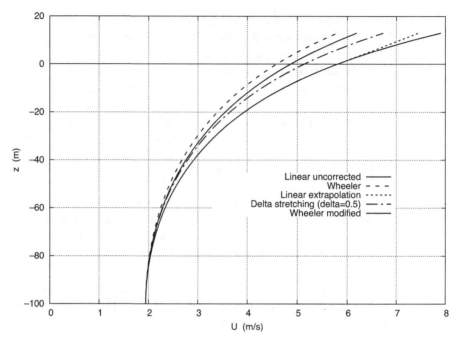

Figure 3.9 Comparison of stretching models. Horizontal velocity profile below a high crest.

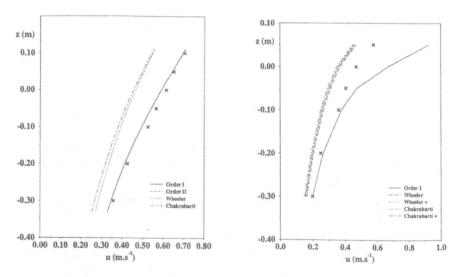

Figure 3.10 Comparison of stretching models with LDV measurements, in regular (left) and irregular (right) waves (from Coche, 1997).

3.4   Second-Order of Approximation

In Section 3.2 the following equations have been established for the second-order potential $\Phi^{(2)}$ and second-order free surface elevation $\eta^{(2)}$:

Laplace equation:

$$\Delta\Phi^{(2)} = 0 \qquad\qquad -h \le z \le 0 \tag{3.100}$$

No-flow equation at the sea floor:

$$\Phi_z^{(2)} = 0 \qquad\qquad z = -h \tag{3.101}$$

Free surface equation for $\Phi^{(2)}$ in $z = 0$:

$$\Phi_{tt}^{(2)} + g\,\Phi_z^{(2)} = -\eta^{(1)}\,(\Phi_{ttz}^{(1)} + g\,\Phi_{zz}^{(1)}) - 2\,\nabla\Phi^{(1)} \cdot \nabla\Phi_t^{(1)} \tag{3.102}$$

The second-order free surface elevation being then obtained as:

$$\eta^{(2)} = -\frac{1}{g}\,\eta^{(1)}\,\Phi_{zt}^{(1)} - \frac{1}{2g}\,\left(\nabla\Phi^{(1)}\right)^2 - \frac{1}{g}\,\Phi_t^{(2)} \qquad z = 0 \tag{3.103}$$

To obtain the second-order potential associated with a given first-order wave, one needs to input its velocity potential $\Phi^{(1)}$ and elevation $\eta^{(1)}$ into the right-hand side of equation (3.102) above, and solve the $\Phi^{(2)}$ problem.

3.4.1 Regular Waves (Stokes Order 2)

From the first-order free surface elevation and velocity potential

$$\eta^{(1)}(x,y,t) = A\,\cos(kx - \omega t) \tag{3.104}$$

$$\Phi^{(1)}(x,y,z,t) = \frac{A\,g}{\omega}\,\frac{\cosh k(z+h)}{\cosh kh}\,\sin(kx - \omega t) \tag{3.105}$$

we get the free surface condition to be verified by $\Phi^{(2)}$ in $z = 0$ as

$$\Phi_{tt}^{(2)} + g\,\Phi_z^{(2)} = -\frac{3}{2}\,\frac{A^2\,\omega^3}{\sinh^2 kh}\,\sin 2(kx - \omega t) \tag{3.106}$$

A particular solution verifying the Laplace equation, equation (3.106), and the no-flow condition in $z = -h$, is:

$$\Phi^{(2)}(x,y,t) = \frac{3}{8}\,\frac{A^2\,\omega}{\sinh^4 kh}\,\cosh 2k(z+h)\,\sin 2(kx - \omega t) - g\,\delta^{(2)}\,t \tag{3.107}$$

The second-order contribution to the free surface elevation, from (3.103), is then

$$\eta^{(2)}(x,t) = \frac{k\,A^2}{4}\,(3\,\coth^3 kh - \coth kh)\,\cos 2(kx - \omega t) - \frac{k\,A^2}{2\,\sinh 2kh} + \delta^{(2)} \tag{3.108}$$

When $\delta^{(2)} = kA^2/(2\,\sinh 2kh)$ the mean water level is zero. This would correspond to waves generated in a tank of constant volume.

When the water is deep ($kh > \pi$) then $\Phi^{(2)} \simeq 0$ and the second-order free surface correction is simply

$$\eta^{(2)} = \frac{k\,A^2}{2}\,\cos 2(kx - \omega t) \tag{3.109}$$

This second-order correction is of relative order kA.

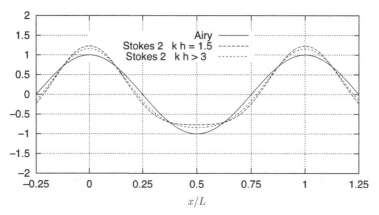

Figure 3.11 Free surface profiles of a regular wave at first and second orders of approximation. $kA = 0.1 \times \pi$, $kh = 1.5$ and $kh = 3$.

It is important to realize that the double frequency 2ω and the double wave number $2k$ are not linked by the dispersion equation:

$$(2\omega)^2 \neq g\,(2k)\,\tanh(2kh) \tag{3.110}$$

except, asymptotically, when the water depth goes to zero. The wave system described by $\Phi^{(2)}$ cannot exist on its own, independently of the first-order wave system. Such accompanying wave systems are known as **locked** or **bound** waves. It is only when the wave system interacts with a structure, or a change in bathymetry, that second-order **free** waves, with wave number k_2 linked to 2ω through the linear dispersion equation, can be released.

For the Stokes development to be convergent, the contribution of $\Phi^{(2)}$ must be small compared to $\Phi^{(1)}$. Taking the ratio of the amplitudes of $\Phi^{(2)}$ and $\Phi^{(1)}$, we obtain, as $kh \to 0$:

$$\frac{|\Phi^{(2)}|}{|\Phi^{(1)}|} \simeq \frac{3}{8}\frac{kA}{(kh)^3} = \frac{3}{64\,\pi^2}\,U_R \tag{3.111}$$

where U_R is the Ursell number. It is verified here that the convergence of the Stokes perturbation procedure supposes the steepness kA to be sufficiently low compared to the dispersion parameter kh.

Figure 3.11 shows the free surface profiles obtained at orders 1 and 2, for a steepness $2A/L$ equal to 10%, and for two kh values: $kh = 1.5$ and $kh = 3$. The main effect of the second-order correction is that the crests have become higher and steeper, while the troughs have become flatter and more shallow.

3.4.2   Bichromatic Seas

A bichromatic wave system consists in the superposition of two regular waves of different amplitudes, frequencies, and directions of propagation. At second order it is the base case to consider in order to be able to generalize to the case of an irregular sea state.

Let us consider a wave system consisting, at first-order of approximation, in two Airy waves superimposed. Without loss of generality, we can assume that one component, of amplitude A_1 and frequency ω_1, propagates along the Ox axis, while the second component, of amplitude A_2 and frequency ω_2, travels at an angle β. The first-order free surface elevation is

$$\eta^{(1)}(x,y,t) = A_1 \cos(k_1 x - \omega_1 t) + A_2 \cos(k_2 x \cos \beta + k_2 y \sin \beta - \omega_2 t) \quad (3.112)$$

and the first-order potential:

$$\Phi^{(1)}(x,y,z,t) = \frac{A_1 g}{\omega_1} \frac{\cosh k_1(z+h)}{\cosh k_1 h} \sin(k_1 x - \omega_1 t)$$
$$+ \frac{A_2 g}{\omega_2} \frac{\cosh k_2(z+h)}{\cosh k_2 h} \sin(k_2 x \cos \beta + k_2 y \sin \beta - \omega_2 t) \quad (3.113)$$

The right-hand side, be Q, of the free surface condition (3.13) verified by $\Phi^{(2)}$, being quadratic with respect to $\Phi^{(1)}$ and $\eta^{(1)}$, has the general form:

$$Q = \Re \left\{ q_{11} e^{-2i\omega_1 t} + q_{22} e^{-2i\omega_2 t} + q_{12+} e^{-i(\omega_1 + \omega_2)t} + q_{12-} e^{-i(\omega_1 - \omega_2)t} \right\} \quad (3.114)$$

The first two terms account for the interaction of each component with itself, it is the regular wave case studied in the previous section. We are concerned here with the cross-interaction that gives rise to two terms at the sum ($\omega_1 + \omega_2$) and difference ($\omega_1 - \omega_2$) frequencies. With all calculations done, we obtain for Q_{12+} and Q_{12-}:

$$Q_{12+} = q_+ \sin\left[(k_1 + k_2 \cos \beta)x + k_2 \sin \beta y - (\omega_1 + \omega_2)t\right] \quad (3.115)$$

where

$$q_+ = -\frac{1}{2} A_1 A_2 \left(\frac{\omega_1^3}{\sinh^2 k_1 h} + \frac{\omega_2^3}{\sinh^2 k_2 h} \right)$$
$$- A_1 A_2 \omega_1 \omega_2 (\omega_1 + \omega_2) \left(\frac{\cos \beta}{\tanh k_1 h \ \tanh k_2 h} - 1 \right) \quad (3.116)$$

$$Q_{12-} = q_- \sin\left[(k_1 - k_2 \cos \beta)x - k_2 \sin \beta y - (\omega_1 - \omega_2)t\right] \quad (3.117)$$

where

$$q_- = -\frac{1}{2} A_1 A_2 \left(\frac{\omega_1^3}{\sinh^2 k_1 h} - \frac{\omega_2^3}{\sinh^2 k_2 h} \right) - A_1 A_2 \omega_1 \omega_2 (\omega_1 - \omega_2)$$
$$\left(\frac{\cos \beta}{\tanh k_1 h \ \tanh k_2 h} + 1 \right) \quad (3.118)$$

In deep water ($k_1 h$ and $k_2 h$ larger than π) these expressions simplify into:

$$q_+ = -A_1 A_2 \omega_1 \omega_2 (\omega_1 + \omega_2) (\cos \beta - 1) \quad (3.119)$$
$$q_- = -A_1 A_2 \omega_1 \omega_2 (\omega_1 - \omega_2) (\cos \beta + 1) \quad (3.120)$$

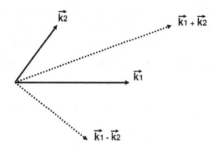

Figure 3.12 Wave number vectors $\vec{k}_1 + \vec{k}_2$ and $\vec{k}_1 - \vec{k}_2$ of the second-order wave system.

The associated potentials are:
Sum frequency:

$$\Phi_+^{(2)}(x,y,z,t) = \frac{q_+}{-(\omega_1 + \omega_2)^2 + gk_+ \ \tanh k_+ h} \ \frac{\cosh k_+(z+h)}{\cosh k_+ h}$$

$$\times \sin\left[(k_1 + k_2 \cos \beta)x + k_2 \sin \beta y - (\omega_1 + \omega_2)t\right] \quad (3.121)$$

where:

$$k_+ = |\vec{k}_1 + \vec{k}_2| = \sqrt{k_1^2 + k_2^2 + 2k_1 k_2 \cos \beta} \quad (3.122)$$

Difference frequency:

$$\Phi_-^{(2)}(x,y,z,t) = \frac{q_-}{-(\omega_1 - \omega_2)^2 + gk_- \ \tanh k_- h} \ \frac{\cosh k_-(z+h)}{\cosh k_- h}$$

$$\times \sin\left[(k_1 - k_2 \cos \beta)x - k_2 \sin \beta y - (\omega_1 - \omega_2)t\right] \quad (3.123)$$

where:

$$k_- = |\vec{k}_1 - \vec{k}_2| = \sqrt{k_1^2 + k_2^2 - 2k_1 k_2 \cos \beta} \quad (3.124)$$

These results merit some comment:

1. Whatever the frequencies and whatever the angle β, it can be checked that $\omega_1 \pm \omega_2$ and $|\vec{k}_1 \pm \vec{k}_2|$ cannot fulfill the dispersion equation (except in the limiting case when the water depth goes to zero):

$$(\omega_1 \pm \omega_2)^2 \neq g \ |\vec{k}_1 \pm \vec{k}_2| \ \tanh |\vec{k}_1 \pm \vec{k}_2| h \quad (3.125)$$

    The wave systems associated with $\Phi_+^{(2)}$ and $\Phi_-^{(2)}$ are **bound** (or **locked**) to the first-order waves.
2. These "waves" propagate in different directions, as shown in Figure 3.12.
3. Two particular cases occur when the two frequencies ω_1 and ω_2 are near:

For $\beta \sim \pi$, the sum wave number k_+ is very small. The decay of $\Phi_+^{(2)}$ with the vertical coordinate is very slow. In the case $\omega_1 = \omega_2$ and $\beta = \pi$, $\Phi_+^{(2)}$ does not depend any more on the space variables x, y, z. As a result, the second-order pressure associated with $\Phi_+^{(2)}$ is felt in the whole water column and is believed, by oceanographers, to be responsible for microseisms (Longuet-Higgins, 1950). In the case of wave–body interaction, an analogous effect is encountered in the second-order diffraction problem where the interaction between incoming and reflected wave systems leads to second-order pressures slowly decaying with the vertical coordinate.

For $\beta \sim 0$ an analogous situation occurs for the second-order potential at the difference frequency, which also decays slowly with the vertical coordinate. The importance of this component is very sensitive to the water depth: in the case $\beta = 0$, the associated horizontal acceleration behaves as $h^{-5/2}$ when $h \rightarrow 0$ (Molin & Fauveau, 1984), with the consequence that the second-order potential becomes an important contribution to the low-frequency loading on moored structures (see Chapter 7).

An important conclusion is that, from second-order, wave effects can be felt at depths greater than half a wave length.

3.4.3 Irregular Seas

All that one has to do is sum up the previous results for all couples (ω_i, ω_j) of frequencies present in the wave signal. In the general case of a short-crested sea in finite depth, the developments become very tedious but they are straightforward. For the sake of simplicity, in the following we confine ourselves to the case of a **long-crested** sea in a water depth deep enough that the first-order velocity potential can be taken as

$$\Phi^{(1)}(x,z,t) = \sum_i \frac{A_i g}{\omega_i} e^{k_i z} \sin(k_i x - \omega_i t + \theta_i) \tag{3.126}$$

From the analysis given above, it results that the sum frequency second-order potential is nil. The difference frequency second-order potential is obtained as

$$\Phi_-^{(2)}(x,z,t) = \sum_i \sum_j -\frac{A_i A_j \omega_i \omega_j (\omega_i - \omega_j)}{g(k_i - k_j) \tanh(k_i - k_j)h - (\omega_i - \omega_j)^2}$$
$$\frac{\cosh(k_i - k_j)(z + h)}{\cosh(k_i - k_j)h} \sin\left[(k_i - k_j)x - (\omega_i - \omega_j)t + \theta_i - \theta_j\right] \tag{3.127}$$

(each couple (ω_i, ω_j) occurring twice the coefficient q_- has been halved).

The second-order contribution to the free surface elevation is then:

$$\eta^{(2)}(x,t) = \sum_i \sum_j A_i A_j \frac{k_i + k_j}{4} \cos\left[(k_i + k_j)x - (\omega_i + \omega_j)t + \theta_i + \theta_j\right]$$
$$+ \sum_i \sum_j A_i A_j \frac{(\omega_i - \omega_j)^2}{g}$$
$$\left(\frac{1}{4} - \frac{\omega_i \omega_j}{g(k_i - k_j) \tanh(k_i - k_j)h - (\omega_i - \omega_j)^2}\right)$$
$$\times \cos\left[(k_i - k_j)x - (\omega_i - \omega_j)t + \theta_i - \theta_j\right] \tag{3.128}$$

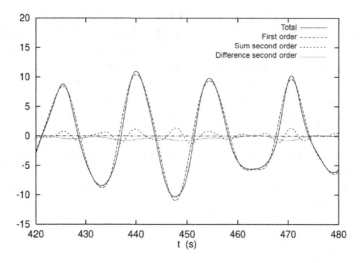

Figure 3.13 First- and second-order components of the free surface elevation in a group of high waves.

In the unrealistic case where the water depth would be really infinite, that is such that

$$| \tanh(k_i - k_j)h | = 1 \quad \forall k_i, k_j, \quad i \neq j \tag{3.129}$$

$\eta^{(2)}$ can be further simplified into:

$$
\begin{aligned}
\eta^{(2)}(x,t) \ = \ & \sum_i \sum_j A_i A_j \, \frac{k_i + k_j}{4} \, \cos\left[(k_i + k_j)x - (\omega_i + \omega_j)t + \theta_i + \theta_j\right] \\
& - \sum_i \sum_j A_i A_j \, \frac{|k_i - k_j|}{4} \, \cos\left[(k_i - k_j)x - (\omega_i - \omega_j)t + \theta_i - \theta_j\right]
\end{aligned}
$$

$$\tag{3.130}$$

An example of the decomposition of the free surface elevation into its first- and second-order components is provided in Figure 3.13.

The difference frequency component of the second-order elevation in (3.130) has much analogy with the wave envelope (actually the pseudo-envelope, see Chapter 2) squared, which writes

$$E^2(t) = \sum_i \sum_j A_i A_j \, \cos\left[(k_i - k_j)x - (\omega_i - \omega_j)t + \theta_i - \theta_j\right] \tag{3.131}$$

Because of the minus sign in (3.130), this component creates a depression, or **set-down**, of the free surface under the groups of high waves. The associated velocity is opposite to the direction of wave propagation, in order to compensate for the local excess of mass transport. Conversely, in the calm areas in between groups of high waves, the mean elevation is positive and the second-order low-frequency velocity is in the direction of wave propagation. As for the sum frequency component in (3.130), its effect, as in the regular wave case, is to render the crests steeper and the troughs flatter.

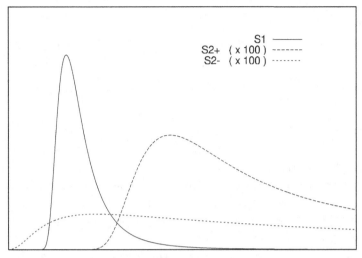

Figure 3.14 First- and second-order wave spectra of the free surface elevation in an irregular sea state.

The depression of the mean free surface under groups of high waves is beneficial to airgap issues. However, this feature does not hold anymore in the case of short-crested or cross seas, where a setup, in place of a set-down, may be observed. This is the case of the New Year wave at Draupner (see Figure 2.6), where a setup appears coinciding with the rogue wave event. Hindcast studies suggest that two wave systems, nearly orthogonal, coexisted at the time (e.g., see McAllister *et al.*, 2019).

From (3.130) spectra of the sum and difference frequency components of the second-order elevation can be derived, as follows (in the infinite depth long-crested case):

Sum frequency:

$$S_{\eta_+^{(2)}}(\Omega) = \frac{1}{2} \int_0^{\Omega/2} S(\omega)\, S(\Omega - \omega) \left(\frac{\omega^2 + (\Omega - \omega)^2}{g} \right)^2 d\omega \qquad (3.132)$$

Difference frequency:

$$S_{\eta_-^{(2)}}(\Omega) = \frac{1}{2} \int_0^\infty S(\omega)\, S(\omega + \Omega) \left(\frac{\omega^2 - (\omega + \Omega)^2}{g} \right)^2 d\omega \qquad (3.133)$$

An illustration is given in Figure 3.14.

3.5   Third-Order of Approximation

Pursuing to order ε^3, the double development introduced in Section 3.2, the following free surface condition is obtained for the third-order component of the velocity potential:

$$\Phi_{tt}^{(3)} + g\,\Phi_z^{(3)}\Big|_{z=0} = -\eta^{(2)}\,\frac{\partial}{\partial z}\left(\Phi_{tt}^{(1)} + g\,\Phi_z^{(1)}\right) - \eta^{(1)}\,\frac{\partial}{\partial z}\left(\Phi_{tt}^{(2)} + g\,\Phi_z^{(2)}\right)$$

$$-\frac{1}{2}\eta^{(1)2}\,\frac{\partial^2}{\partial z^2}\left(\Phi_{tt}^{(1)} + g\,\Phi_z^{(1)}\right) - \eta^{(1)}\,\frac{\partial}{\partial z}\left(\frac{\partial}{\partial t}\nabla\Phi^{(1)2}\right)$$

$$-\Phi_x^{(1)2}\,\Phi_{xx}^{(1)} - \Phi_y^{(1)2}\,\Phi_{yy}^{(1)} - \Phi_z^{(1)2}\,\Phi_{zz}^{(1)}$$

$$-2\Phi_x^{(1)}\,\Phi_y^{(1)}\,\Phi_{xy}^{(1)} - 2\Phi_x^{(1)}\,\Phi_z^{(1)}\,\Phi_{xz}^{(1)} - 2\Phi_y^{(1)}\,\Phi_z^{(1)}\,\Phi_{yz}^{(1)}$$

$$-2\frac{\partial}{\partial t}\left(\nabla\Phi^{(1)}\cdot\nabla\Phi^{(2)}\right)\tag{3.134}$$

The right-hand side is a cubic expression of the initial first-order velocity potential $\Phi^{(1)}$ and free surface elevation $\eta^{(1)}$ (given that $\Phi^{(2)}$ and $\eta^{(2)}$ are themselves quadratic expressions). This means that, if we start from a regular wave with frequency ω and wave number k, we will get forcing terms in $(3\omega, 3k)$ and, also, some terms back to (ω, k) (since $\cos^3\theta = \cos 3\theta/4 + 3\cos\theta/4$).

In multichromatic waves, starting with frequencies ω_i and wave number vectors $\vec{k_i}$, we will get all combinations $\omega_i \pm \omega_j \pm \omega_k$ associated with $\vec{k_i} \pm \vec{k_j} \pm \vec{k_k}$.

3.5.1 Regular Waves (Stokes Order 3)

For the sake of simplicity, we assume infinite depth. Then the second-order potential $\Phi^{(2)}$ is nil and many terms disappear from the right-hand side of equation (3.134) which reduces to:

$$\Phi_{tt}^{(3)} + g\,\Phi_z^{(3)}\Big|_{z=0} = -\Phi_x^{(1)2}\,\Phi_{xx}^{(1)} - \Phi_y^{(1)2}\,\Phi_{yy}^{(1)} - \Phi_z^{(1)2}\,\Phi_{zz}^{(1)}$$

$$-2\Phi_x^{(1)}\,\Phi_y^{(1)}\,\Phi_{xy}^{(1)} - 2\Phi_x^{(1)}\,\Phi_z^{(1)}\,\Phi_{xz}^{(1)} - 2\Phi_y^{(1)}\,\Phi_z^{(1)}\,\Phi_{yz}^{(1)}\tag{3.135}$$

Taking

$$\Phi^{(1)} = \frac{A\,g}{\omega}\,e^{kz}\,\sin(kx - \omega t)\tag{3.136}$$

we obtain

$$\Phi_{tt}^{(3)} + g\,\Phi_z^{(3)}\Big|_{z=0} = -A^3\,\omega^3\,k\,\sin(kx - \omega t)\tag{3.137}$$

(no triple frequency terms).

It turns out that starting from (in order to fulfill the Laplace equation and the decay for $z \to -\infty$) an expression such as

$$\Phi^{(3)}(x, z, t) = C\,e^{kz}\,\cos(kx - \omega t)\tag{3.138}$$

does not work, unless C be a linear function of time, conflicting with the assumption of periodicity.

This is a well-known situation when dealing with nonlinear oscillatory systems. The remedy is to relax the assumptions of constant amplitude and phase, and to start from an initial solution of the form

$$\eta(x, t) = A(x, t)\,\cos(kx - \omega t - \alpha(x, t))\tag{3.139}$$

where the amplitude A and the phase α are slowly varying in time and space.

Applying this procedure to the problem considered here, it turns out that it suffices to vary the phase, and that such simple expressions as $\alpha = \omega^{(2)}t$ or $\alpha = k^{(2)}x$ work.

If we keep the frequency unchanged (as would be the case when waves are generated in a tank with a wavemaker), taking

$$\eta(x,t) \;=\; A \, \cos\left[\left(k + k^{(2)}\right)x - \omega t\right] \tag{3.140}$$

$$\Phi(x,z,t) \;=\; \frac{A g}{\omega}\, e^{(k+k^{(2)})z}\, \sin\left[\left(k + k^{(2)}\right)x - \omega t\right] \tag{3.141}$$

where $\omega^2 = gk$, we get $k^{(2)} = -k^3 A^2$: the wave number decreases, meaning that the wave length increases. The correction is of second order in the wave steepness kA.

Conversely, if we keep the wave length unchanged, we get an increase of the wave frequency given by $\omega^{(2)} = k^2 A^2 \omega/2$.

To the third-order included the free surface elevation is obtained as

$$\eta(x,t) = \left(1 - \frac{3}{8} A^2 k^2\right) A \, \cos\theta + \frac{1}{2} A^2 k \, \cos 2\theta + \frac{3}{8} A^3 k^2 \, \cos 3\theta \tag{3.142}$$

where $\theta = \left(k + k^{(2)}\right)x - \omega t$

Compared to the linear solution, the crest-to-trough value is unchanged but the amplitude of the first-harmonic component has been modified.

In Figure 3.15 we show, at a given time and as a function of x/L where L is the wave length calculated at first-order, the free surface profiles according to first-, second-, and third-order theories, for a steepness $2A/L$ equal to 10%. It can be observed that the third-order increase of the wave length is appreciable.

In Figure 3.16 we show the errors committed on the dynamic-free surface condition according to the three models, at the same steepness. It can be checked that the error is effectively reduced as the order of approximation increases, even though we are dealing here with a high steepness case.

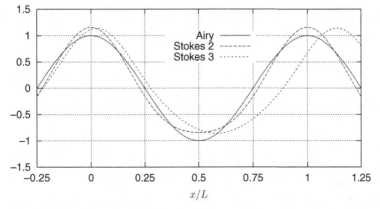

Figure 3.15 Regular wave in infinite depth. Free surface profiles at orders 1, 2, and 3 vs x/L. Steepness $2A/L = 0.1$.

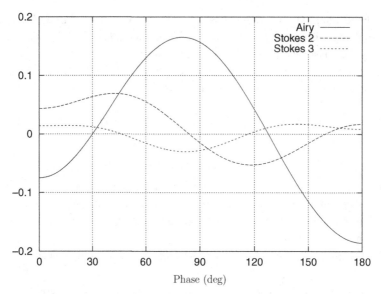

Figure 3.16 Regular wave in infinite depth. Relative error on the free surface dynamic condition at orders 1, 2, and 3. Steepness $2A/L = 0.1$.

3.5.2　Bichromatic Seas

Let us take, as in Section 3.4.2, one component (A_1, ω_1, k_1) propagating along the Ox axis, and the second one (A_2, ω_2, k_2) propagating at an angle β:

$$\eta^{(1)}(x,y,t) = A_1 \cos(k_1 x - \omega_1 t) + A_2 \cos(k_2 x \cos\beta + k_2 y \sin\beta - \omega_2 t) \quad (3.143)$$

For the sake of simplicity, again we assume infinite depth. From the results obtained above in the regular wave case, each wave component interacts with itself and sees its wave number modified according to

$$k_1^{(2)} = -k_1^3 A_1^2 \qquad\qquad k_2^{(2)} = -k_2^3 A_2^2 \quad (3.144)$$

However, other terms at frequencies ω_1 and ω_2 appear in the right-hand side of (3.134), obtained through

$$\omega_1 = \omega_1 \pm \omega_2 \mp \omega_2 \qquad\qquad \omega_2 = \omega_2 \pm \omega_1 \mp \omega_1 \quad (3.145)$$

and result in complementary second-order corrections to the wave numbers. The wave number of the first component is therefore modified according to

$$k_1^{(2)} = -k_1^3 A_1^2 + k_2^2 A_2^2 k_1 \, f(\omega_1, \omega_2, \beta) \quad (3.146)$$

where f is a complicated function, first given by Longuet–Higgins & Phillips (1962).

In the particular case where the two frequencies are equal ($\omega_2 = \omega_1 = \omega$, $k_2 = k_1 = k$) the wave number correction is given by

$$k_1^{(2)} = \frac{1}{2} k^3 A_1^2 \, f(0) + k^3 A_2^2 \, f(\beta) = -k^3 A_1^2 + k^3 A_2^2 \, f(\beta) \quad (3.147)$$

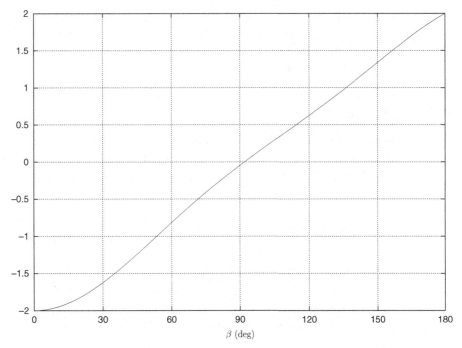

Figure 3.17 Interaction function $f(\beta)$ in infinite depth.

where the function $f(\beta)$ is given by

$$f(\beta) = -\left\{(1 - \cos\beta)\,\sqrt{2 + 2\cos\beta} + 2\cos\beta + \frac{1}{2}\sin^2\beta\right.$$

$$\left. + \frac{2(1 - \cos\beta)}{\sqrt{2 + 2\cos\beta} - 4}\left(1 + \cos\beta + \sqrt{2 + 2\cos\beta}\right)\right\}. \quad (3.148)$$

The function f is shown in Figure 3.17. It is close to a straight line from -2 to $+2$ when β goes from 0 to 180 degrees. When β is larger than 92 degrees the function is positive, meaning that the cross-interaction tends to increase the wave number. In the particular case of a standing wave ($A_1 = A_2$, $\beta = \pi$) the cross-interaction overtakes the self-interaction: the wave number increases. This is for deep water; as the water depth decreases the self-interaction gradually becomes dominant. This is well known for sloshing in rectangular tanks, where there is a critical depth given by $h/L = 0.3368$ (L being the length of the tank), where both corrections cancel out for the first sloshing mode. In the case of sloshing, the wave length is imposed as twice the length of the tank, so it is the resonant frequency that varies with the amplitude of motion of the free surface. When h/L is lower than 0.3368, the system is "stiff," the resonant frequency increases with the response amplitude; when h/L is higher than 0.3368, the system is "soft," the resonant frequency decreases.

For plane waves propagating in open ocean, these mutual modifications of phase velocities are of lesser interest. However, similar phenomena occur in the case of incident waves interacting with the wave system reflected by a structure. When the reflected wave system is strong, for instance, in the case of ships or barges in beam seas, high runups may be observed amidships. These runups are due to the reflected wave system decreasing locally the wave lengths of the incoming waves (as a shoal would do in finite depth), inducing energy-focusing effects. These phenomena are further discussed in Chapter 8.

3.5.3   Multichromatic Seas

In the case of a multichromatic sea state, similar interactions as those presented above take place between all couples of wave components, leading to modifications of the wave lengths. The second-order correction to the wave number obtained in the bichromatic case can be extended to the case of an irregular sea state of spectrum $S(\omega, \beta)$ as

$$k_i^{(2)} = 2\,k_i \int_0^\infty \int_0^{2\pi} k^2(\omega)\, f(\omega_i, \omega, \beta - \beta_i, h)\, S(\omega, \beta)\, \mathrm{d}\beta\, \mathrm{d}\omega \qquad (3.149)$$

More interesting phenomena take place due to exchanges of energy between the wave components. The base case to consider is a quadruplet of wave components with frequencies $(\omega_1, \omega_2, \omega_3, \omega_4)$ and wave number vectors $(\vec{k_1}, \vec{k_2}, \vec{k_3}, \vec{k_4})$. When the following relationships are simultaneously fulfilled

$$\omega_1 + \omega_2 \simeq \omega_3 + \omega_4 \qquad (3.150)$$
$$\vec{k_1} + \vec{k_2} \simeq \vec{k_3} + \vec{k_4} \qquad (3.151)$$

where $\simeq$ means to the order of the steepness squared, then cross-modifications of the **amplitudes** can take place.

A particular, and historical, case, in infinite depth, is illustrated in Figure 3.18. There the component $(\omega_1, \vec{k_1})$ appears twice. When the wave number vector $\vec{k_3}$ covers the figure of eight, then $\omega_4 = 2\,\omega_1 - \omega_3$ and $\vec{k_4} = 2\,\vec{k_1} - \vec{k_3}$ obey the dispersion equation. Evidence of energy exchange was given experimentally by producing two regular waves at 90 degrees (in a wave tank): an oblique wave at frequency $2\,\omega_1 - \omega_3$, and with an amplitude increasing linearly down the tank, was observed (Longuet-Higgins & Smith, 1966; see also Bonnefoy *et al.*, 2016).

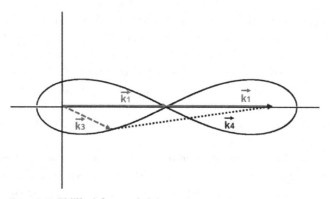

Figure 3.18 Phillips' figure of eight.

Another particular case is $\omega_1 = \omega_2 = \omega$, $\omega_3 = \omega - \Delta\omega$, $\omega_4 = \omega + \Delta\omega$, $\vec{k_1} = \vec{k_2} = \vec{k}$, $\vec{k_3} = \vec{k} - \vec{\Delta k}$, $\vec{k_4} = \vec{k} + \vec{\Delta k}$ where $\Delta\omega = O(kA)$. This combination leads to the Benjamin–Feir instability, considered in Section 3.7.2.

An important conclusion is that, from third-order, a sea state cannot be considered as stationary: the wave spectrum continuously evolves over space and time.

3.6   Regular Wave Theories to Order *N*

Extending the Stokes procedure to orders higher than 3 quickly becomes tedious and unpractical. Stokes gave a fifth-order solution in infinite depth. Finite depth solutions were proposed in the 1950s and 1960s, not always in complete agreement. In 1974, with the help of conformal mapping and computer resources, Schwartz solved the problem to arbitrary order N in infinite depth, up to the limiting steepness, when the crest has an inner angle of 120 degrees, and he gave limiting values for the steepness H/L.

3.6.1   The Stream Function Method

In 1965 a numerical method was proposed by Dean (1965) to compute highly non-linear regular waves. This method is known under different names: stream function method, Dean's model, Fourier model, Rienecker & Fenton (1981), etc.

The input data are the water depth h, the wave height H, and the wave length L. The problem is formulated in a coordinate system Oxz attached to a crest. In this coordinate system the flow is steady, and the free surface is a streamline. Hence it is more convenient to use the stream function $\Psi(x, z)$ to describe the flow than the velocity potential.

The stream function verifies the Laplace equation in the fluid domain. The boundary conditions are:

- $\Psi = 0$ on the sea floor $z = -h$.
- $\Psi = -Q$ at the free surface $z = \eta(x)$ (kinematic condition). Q is the flux through any vertical line.
- $\frac{1}{2}\left(\Psi_x^2 + \Psi_z^2\right) + g\,\eta = R$ at $z = \eta(x)$ (dynamic condition).

The values of the period T and phase velocity L/T depend on the choice of the "fixed" coordinate system. The intuitive choice of a coordinate system attached to the sea floor makes no sense since a free-slip condition is assumed there.

A first choice is a coordinate system such that the depth-averaged flux be zero: in this system the phase velocity is $C_{PL} = Q/h$ and the period is $T_L = L/C_{PL}$.

Another coordinate system is such that the time-averaged velocity, at any point always below the free surface, be nil. This choice leads to phase velocity C_{PE} and period T_E different from C_{PL} and T_L.

In the first coordinate system, the mean **Lagrangian** velocity is zero. In the second one the mean **Eulerian** velocity is zero.

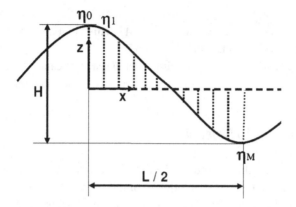

Figure 3.19 Geometry.

Another possibility is a coordinate system such that the wave coexist with some current. Then it must be made clear whether this current is defined in a Lagrangian or Eulerian sense.

The stream function is written as a Fourier series

$$\frac{k\,\Psi(x,z)}{C_{PE}} = -k\,(z+h) + \sum_{j=1}^{N} a_j\,\frac{\sinh j\,k\,(z+h)}{\cosh j\,k\,h}\,\cos j\,k\,x \tag{3.152}$$

where $k = 2\pi/L$.

The Laplace equation and the condition $\Psi(x,-h) = 0$ are fulfilled. It remains to verify the free-surface conditions, and to ensure that the wave height is H and that the mean elevation of the free surface is equal to zero.

The procedure is to discretize the interval $[0\ L/2]$ in $M + 1$ points with abscissas $x_i = i\,L/(2M)$, $i = 0, M$, and still unknown elevations η_i (see Figure 3.19).

There are $M+N+4$ unknowns: the constants Q and R in the free surface conditions, the phase velocity C_{PE}, $M + 1$ elevations η_i, and N coefficients a_j. In each point x_i there are two conditions, that is $2M + 2$ in total, to which two more must be added:

$$\eta_0 + \eta_M + 2\sum_{i=1}^{M-1} \eta_i = 0 \tag{3.153}$$

$$\eta_0 - \eta_M = H \tag{3.154}$$

expressing that the mean water level is zero and that the crest-to-trough value is H.

Taking $M = N$ leads to as many equations as unknowns. Another option is to take $M > N$ to obtain an over-determined system of equations. Different resolution methods have been used by different authors.

The wave period follows from the convention chosen to define the accompanying current U_C. If U_C is the Eulerian velocity, then the wave period is obtained as

$$T = \frac{L}{C_{PE} + U_C} \tag{3.155}$$

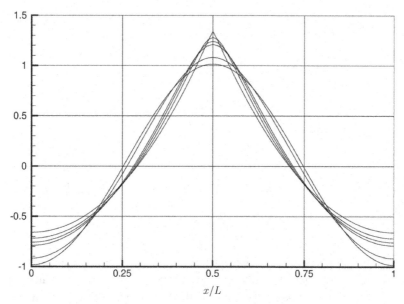

Figure 3.20 Normalized free surface profiles, in infinite depth, for $H/L = 0.01, 0.05, 0.10,$ $0.12, 0.13, 0.14$ (credits Pierre Ferrant).

When U_C is the Lagrangian velocity:

$$T = \frac{L}{C_{PL} + U_C} \tag{3.156}$$

where $C_{PL} = Q/h$.

Most frequently it is the period T, which is specified, not the wave length. Then either one of the equations above must be added to the $2M + 4$ previous equations, and included in the numerical resolution method.

The truncation order N in the Fourier series must be chosen according to the steepness: the steeper the wave the higher the truncation order. The water depth also matters. An advantage of the stream function model is that it applies to shallow as well as deep water, whatever the value of the Ursell number. It bridges the gap between the Stokes model and shallow water wave theories such as cnoidal waves.

Figure 3.20 shows normalized wave profiles computed with the stream function model, in deep water, at wave steepnesses H/L from 0.01 up to 0.14, the last value being very close to the limiting steepness 0.1412 where the crest becomes angular.

It is to be noted that, in finite depth, the limiting steepness is approximately given by

$$H/L_{\text{limit}} \simeq 0.14 \tanh kh \tag{3.157}$$

3.6.2   Fifth-Order Stokes Theory

As written earlier, several models of fifth-order Stokes theory have been proposed in the 1950s and early 1960s. Later Fenton (1985) published a model that is believed to be accurate. Fenton followed the stream function method, starting from

$$\frac{k \, \Psi(x,z)}{C_{PE}} = -k \, (z + h) + \sum_{i=1}^{5} \varepsilon^i \sum_{j=1}^{i} f_{ij} \, \frac{\sinh \, j \, k \, (z + h)}{\cosh \, j \, k \, h} \, \cos j \, k \, x \qquad (3.158)$$

$$k \, \eta(x) = \sum_{i=1}^{5} \varepsilon^i \sum_{j=1}^{i} b_{ij} \, \cos j \, k \, x \qquad (3.159)$$

$$C_{PE} \, \sqrt{\frac{k}{g}} = C_0 + \varepsilon^2 \, C_2 + \varepsilon^4 \, C_4 \qquad (3.160)$$

$$k \, Q \, \sqrt{\frac{k}{g}} = C_{PE} \, k \, h \, \sqrt{\frac{k}{g}} + \varepsilon^2 \, D_2 + \varepsilon^4 \, D_4 \qquad (3.161)$$

$$R \, \frac{k}{g} = \frac{1}{2} \, C_0^2 + k \, h + \varepsilon^2 \, E_2 + \varepsilon^4 \, E_4 \qquad (3.162)$$

The difference with the stream function model presented in the previous section is that the coefficients a_j of equation (3.152) are no longer obtained numerically, on a case-by-case basis, but analytically, following the Stokes development. Nowadays, formal calculation tools can be used and they would provide the exact solution with little effort. Fenton and his predecessors did everything by hand, repeating the developments many times until agreement.

As can be seen from equation (3.160), a further correction of the phase velocity occurs. When the wave frequency is kept the same, this results in a new modification of the wave length, an increase in deep water, and a decrease when kh is less than about 1.07.

Fenton (1985) gives the free surface elevation η and velocity potential Φ in the form

$$k \, \eta(x) = \varepsilon \, \cos k \, x + \varepsilon^2 \, B_{22} \, \cos 2 \, k \, x + \varepsilon^3 \, B_{31} \, (\cos k \, x - \cos 3 \, k \, x)$$
$$+ \, \varepsilon^4 \, (B_{42} \, \cos 2 \, k \, x + B_{44} \, \cos 4 \, k \, x)$$
$$+ \, \varepsilon^5 \, [-(B_{53} + B_{55}) \, \cos k \, x + B_{53} \, \cos 3 \, k \, x + B_{55} \, \cos 5 \, k \, x] \qquad (3.163)$$

$$\Phi(x,z) = -C_{PE} \, x + \frac{C_0}{k} \, \sqrt{\frac{g}{k}} \sum_{i=1}^{5} \varepsilon^i \sum_{j=1}^{i} A_{ij} \, \mathrm{ch} \, j \, k \, (z + h) \, \sin j \, k \, x \qquad (3.164)$$

where $\epsilon = kH/2$. All coefficients $A_{ij}, B_{ij}, C_i, D_i, E_i$ can be found in Fenton (1985, 1990).

When the wave height H, the water depth h, and the wave length L are given, the application of the equations given above is straightforward, and the wave period and phase velocity can easily be obtained. When the wave period T and a current velocity U_C are given, the wave number k is obtained from the relationship between velocities, period, and wave length

$$T = \frac{L}{C_{PE} + U_C} = \frac{2 \, \pi}{k \, (C_{PE} + U_C)} \qquad (3.165)$$

(the current being here defined in an Eulerian sense) and from equation (3.160).

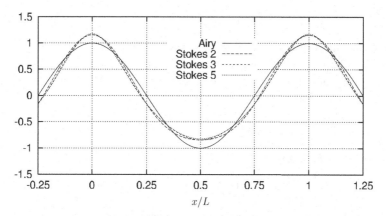

Figure 3.21 Normalized free surface profiles η/A vs x/L according to the Stokes model at orders 1 (Airy), 2, 3, and 5. Infinite depth. Steepness $kA = 0.1 \times \pi$.

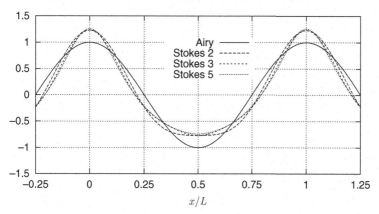

Figure 3.22 Normalized free surface profiles η/A vs x/L according to the Stokes model at orders 1 (Airy), 2, 3, and 5. Finite depth $kh = 1.5$. Steepness $kA = 0.1 \times \pi$.

When C_{PE} is eliminated, the following equation for k is obtained:

$$\left(U_C - \frac{2\pi}{kT}\right) \sqrt{\frac{k}{g}} + C_0(k\,h) + \left(\frac{k\,H}{2}\right)^2 C_2(k\,h) + \left(\frac{k\,H}{2}\right)^4 C_4(k\,h) = 0 \quad (3.166)$$

and needs to be solved through iterations.

Figures 3.21 and 3.22 show normalized free surface profiles η/A vs x/L, at a steepness $kA = 0.1\pi$, according to the Stokes theories of orders 1, 2, 3, and 5, in the cases of infinite depth (Figure 3.21) and for $kh = 1.5$ (Figure 3.22).

3.7 Nonlinear Wave Evolution and Wave Instabilities

3.7.1 The Nonlinear Schrödinger Equation

There is a wide range of theoretical models that apply to narrow-banded long-crested sea states, where the free surface elevation can be written as

$$\eta(x,t) = \Re\left\{E(x,t)\, e^{i(k_0 x - \omega_0 t)}\right\},$$ (3.167)

and the complex (pseudo-)envelope E varies slowly in space and time.

Within the scope of a linear theory, different orders of approximation can be derived for the time and space evolution of the complex envelope, in the following way. Writing η in the form

$$\eta(x,t) = \Re\left\{\int_{-\infty}^{\infty} \alpha(\omega)\, e^{i(k(\omega)x - \omega t)}\, d\omega\right\}$$ (3.168)

it follows that $E(x,t)$ verifies

$$E(x,t) = \int_{-\infty}^{\infty} \alpha(\omega)\, e^{i[(k(\omega)-k_0)x - (\omega-\omega_0)t]}\, d\omega$$ (3.169)

Through Taylor development of $k(\omega)$ around ω_0:

$$k(\omega) = k(\omega_0) + (\omega - \omega_0)\, k'(\omega_0) + \frac{1}{2}\,(\omega - \omega_0)^2\, k''(\omega_0) + \dots$$ (3.170)

one gets

$$E(x,t) = \int_{-\infty}^{\infty} \alpha(\omega)\, e^{i[(\omega-\omega_0)\,(k'(\omega_0)x - t) + \frac{1}{2}(\omega-\omega_0)^2\, k''(\omega_0)\, x + \dots]}\, d\omega$$ (3.171)

By derivations in x and t it is deduced that, to a term of order $(\omega - \omega_0)^2$, $E(x,t)$ satisfies the evolution equation:

$$k'(\omega_0)\, E_t + E_x = 0$$ (3.172)

meaning that the envelope travels at the group velocity $C_G = \partial\omega/\partial k = k'^{-1}$.

To the following order of approximation (with a $O(\omega - \omega_0)^3$ error) $E(x,t)$ satisfies:

$$i\,[E_x + k'(\omega_0)\, E_t] - \frac{1}{2}k''(\omega_0)\, E_{tt} = 0$$ (3.173)

where a dispersive term has now appeared. Physically this means that an initial envelope bump (a wave packet) will propagate at the group velocity while broadening and decreasing in height, the long-wave components traveling faster than the short-wave components.

This evolution equation is valid on distances of order $L\,(\omega_0/\Delta\omega)^2$ (or durations of order $T\,(\omega_0/\Delta\omega)^2$), where $\Delta\omega$ is the width of the wave spectrum.

When third-order nonlinearities are accounted for, a supplementary term in $E\,E^*E$ (with E^* the complex conjugate of E) appears. In infinite depth, the evolution equation takes the form

$$i\left[E_x + \frac{2\omega_0}{g}\, E_t\right] - \frac{1}{g}\, E_{tt} - k_0^3\, E\, E^*\, E = 0$$ (3.174)

known as the nonlinear Schrödinger equation.

In this theoretical model, it is assumed that the two small parameters $\Delta\omega/\omega_0$ and $k_0 A$ are comparable.

It is readily checked that the solution of this equation, for a regular wave of amplitude A is

$$E(x,t) = A\, e^{-ik_0^3 A^2 x}$$ (3.175)

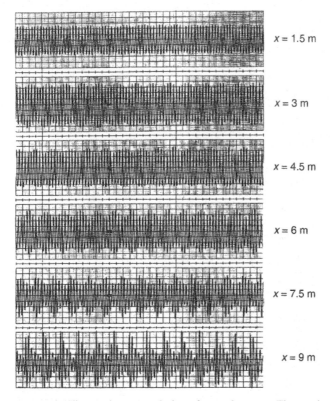

Figure 3.23 Time and space evolution of a regular wave. Time series of the free surface elevation along the tank (from *Lake et al.*, 1977) (Cambridge University Press, reprinted with permission).

in agreement with the third-order Stokes model:

$$\eta(x,t) = A \, \cos[(k_0 - k_0^3 A^2) \, x - \omega_0 t] \tag{3.176}$$

3.7.2   The Benjamin–Feir Instability

An application of the nonlinear Schrödinger equation is to show that a regular wave train is naturally unstable. This instability was first demonstrated by Benjamin and Feir (1967) using a different method.

To show that a regular wave is unstable, to $E(x,t)$ in (3.175), we superimpose two small perturbations, in the form

$$E = A \, \mathrm{e}^{-\mathrm{i}k_0^3 A^2 x} \left(1 + \alpha \, \mathrm{e}^{\mathrm{i}(Kx-\Omega t)} + \beta \, \mathrm{e}^{-\mathrm{i}(Kx-\Omega t)}\right) \tag{3.177}$$

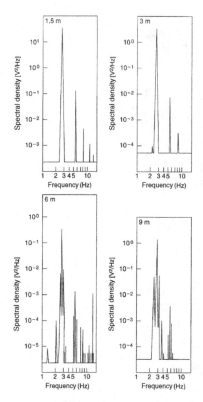

Figure 3.24 Time and space evolution of a regular wave. Power spectra of the measured free surface elevation along the tank (from *Lake et al.*, 1977) (Cambridge University Press, reprinted with permission).

When this expression is inserted into (3.174), and only terms linear in α and β are retained, the following system of equations is obtained:

$$\begin{pmatrix} -K + \frac{2\omega_0\Omega}{g} + \frac{\Omega^2}{g} - k_0^3 A^2 & -k_0^3 A^2 \\ -k_0^3 A^2 & K - \frac{2\omega_0\Omega}{g} + \frac{\Omega^2}{g} - k_0^3 A^2 \end{pmatrix} \begin{pmatrix} \alpha \\ \beta \end{pmatrix} = \begin{pmatrix} 0 \\ 0 \end{pmatrix} \tag{3.178}$$

To get a nontrivial solution, the determinant has to be nil:

$$\left(K - \frac{2\omega_0\Omega}{g} \right)^2 = \frac{\Omega^2}{g} \left(\frac{\Omega^2}{g} - 2k_0^3 A^2 \right) \tag{3.179}$$

giving a relationship between Ω and K. It is readily observed that when

$$\Omega^2 < 2\omega_0^2 k_0^2 A^2 \tag{3.180}$$

the right-hand side of (3.179) is negative. Then the imaginary part of $K - 2\omega_0\Omega/g$ is nonzero and one of the parasitic wave components grows exponentially. Its growth rate is maximum for

$$\Omega = k_0 A \omega_0 \tag{3.181}$$

when the imaginary part of K is $k_0^3 A^2$.

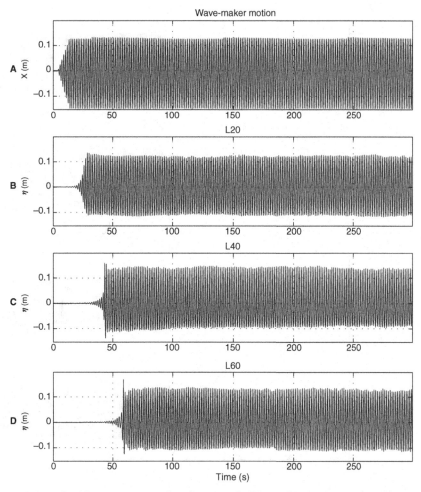

Figure 3.25 Regular wave propagation down a tank. Wavemaker motion (top) and free surface elevations measured at 20, 40, and 60 m. Wave period 1.5 s, amplitude 0.1 m.

When a regular wave is generated in a wave tank, it can be concluded that, if the wave paddle signal is a bit noisy, parasitic components will eventually appear at frequencies in between $\omega_0 (1 - \sqrt{2}\, k_0 A)$ and $\omega_0 (1 + \sqrt{2}\, k_0 A)$, and will attain maximum amplitudes at $\omega_0 (1 \pm k_0 A)$. Visually it means that the initially sinusoidal signal modulates in groups. The Schrödinger equation predicts recurrent cycles of modulations and demodulations, the initial sinusoidal shape being recovered at intervals. In practice, wave breaking is usually observed at peaks of the modulation, destroying the recurrent process.

Figure 3.23, taken from Lake *et al.* (1977), shows measurements of the free surface elevation along a canal for a regular wave of period 0.4 s and steepness $k_0 A = 0.16$. Figure 3.24 shows the associated power spectra: side-bands appear at frequencies $n\, \omega_0 (1 \pm k_0 A)$, as predicted by the nonlinear Schrödinger equation.

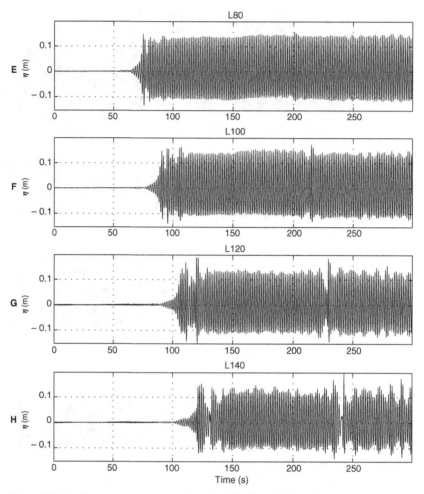

Figure 3.26 Regular wave propagation down a tank. Free surface elevations measured at 80, 100, 120, and 1400 m. Wave period 1.5 s, amplitude 0.1 m.

In the case of the experiments by Lake *et al.* (1977), because of the reduced length of the tank, the wave paddle signal was artificially contaminated at the critical frequencies. A more realistic case is provided in Figures 3.25 and 3.26, from experiments performed in a 150 m long tank at the former Bassin d'Essais des Carènes in Paris. The wave amplitude is about 0.1 m and the wave period is 1.5 s, meaning a wave length equal to 3.5 m and a steepness kA equal to 0.18. The top time trace in Figure 3.25 shows the motion of the wavemaker. Due to an improper control system, small kinks can be seen at times $t \simeq 140$ s and $t \simeq 245$ s. The following time traces show the free surface elevations measured down the tank every 20 m. At the first three gauges, the wave signal is quite regular, but from 80 m, a noticeable perturbation appears, which travels down the tank at the group velocity, eventually leading to wave breaking at the last gauge. When one tracks the perturbation back to the wavemaker at the group velocity, it turns out that it originates from the small kink at time 140 s. In the last time

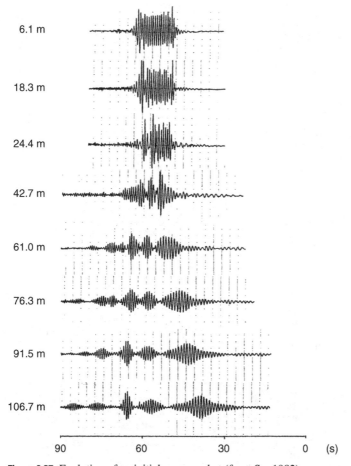

Figure 3.27 Evolution of an initial wave packet (from Su, 1982).

series strong modulations can also be seen, especially by the wave front, in groups of about 5 or 6 waves, as predicted by the Schrödinger equation.

This example shows that producing a regular wave in a tank is not as easy as it sounds, that efficient control of the wavemaker is important, and that it is safer to locate the model not too far from the wavemaker. In the case of towing tests in regular waves, over long distances, the Benjamin–Feir instability must be considered as a possible hindrance, and should be carefully monitored during wave calibration.

Another interesting phenomenon predicted by the nonlinear Schrödinger equation is the evolution of wave packets, obtained for example, by running the wavemaker for a limited time (and with a narrow frequency content). Linear theory predicts that the train of waves thus formed will spread out under dispersion. The Schrödinger equation predicts the splitting of the initial packet into several wave groups, each one propagating without modification of form (so-called envelope solitons). This second type of evolution is actually observed experimentally (see Figure 3.27), and also numerically

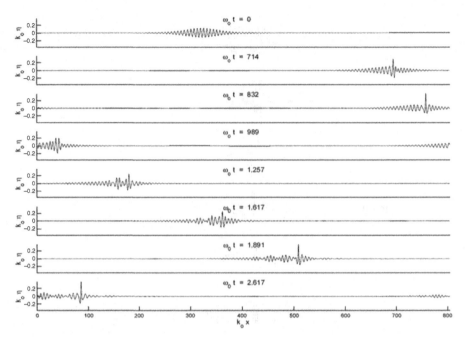

Figure 3.28 Application of a periodic numerical wavetank to the nonlinear evolution of an initial wave packet (from Clamond & Grue, 2001).

Figure 3.29 Crescent waves (from Su *et al.*, 1982) (Cambridge University Press, reprinted with permission).

when free surface nonlinearities are accounted for (see Figure 3.28). In the splitting process waves of abnormal heights can appear fleetingly.

It should not be forgotten, however, that the nonlinear Schrödinger equation provides only an approximate description of the phenomena at hand, and that it is based on fairly restrictive assumptions (narrow spectrum, limited evolution distance, only third-order nonlinearities taken into account). Furthermore, in finite depth, these results are somewhat modified, the Benjamin–Feir instability disappearing for $k\,h < 1.36$.

With different techniques, many authors have studied the stability of regular waves of finite amplitude. McLean (1982) obtains that, as soon as the steepness H/L exceeds about 10%, much stronger instabilities than the Benjamin–Feir instability come into play. Linked to nonlinearities of fourth order, these instabilities have a strong three-dimensional character. They quickly lead to wave-breaking, or to the bifurcation toward new three-dimensional states, in the shape of "horseshoes" or "crescents" (Figure 3.29). In wave tanks, these crescent shapes can sometimes be seen at the front of steep wave systems.

3.8 References

BENJAMIN T.B., FEIR J.E. 1967. The disintegration of wave trains on deep water. Part 1. Theory, *J. Fluid Mech.*, **27**, 417–430.

BONNEFOY F., HAUDIN F., MICHEL G., SEMIN B., HUMBERT T., AUMAÎTRE S., BERHANU M., FALCON E. 2016. Observation of resonant interactions among surface gravity waves, *J. Fluid Mech.*, **805**, R3. doi:10.1017/jfm.2016.576.

CLAMOND D., GRUE J. 2001. On a fast method for simulations of steep water waves, *Proc. 16th Int. Workshop on Water Waves & Floating Bodies*, Hiroshima, 25–28 (www.iwwwfb.org).

COCHE E. 1997. Analyse de la cinématique de la houle dans la crête des vagues, D.E.A. report (in French).

DEAN R.G. 1965. Stream function representation of nonlinear ocean waves, *J. Geophysical Res.*, **70**, 4561–4572.

FENTON J.D. 1985. A fifth-order Stokes theory for steady waves, *J. Waterway, Port, Coastal and Ocean Engineering*, **111**, 216–234.

FENTON J. D. 1990. Nonlinear wave theories, in *The Sea, Vol. 9: Ocean Engineering Science*, Eds. B. Le Méhauté and D.M. Hanes, Wiley, New York.

HOLTHUIJSEN L.H. 2007. *Waves in Oceanic and Coastal Waters*. Cambridge University Press.

LAKE B. M., YUEN H.C., RUNGALDIER H., FERGUSON W.E. 1977. Nonlinear deep-water waves: theory and experiment. Part 2. Evolution of a continuous wave train, *J. Fluid Mech.*, **83**, 49–74.

LONGUET-HIGGINS M.S. 1950. A theory of the origin of microseisms, *Philosophical Trans. Royal Society of London. Series A*, **243**, 1–35.

LONGUET-HIGGINS M.S., PHILLIPS O.M. 1962. Phase velocity effects in tertiary wave interactions, *J. Fluid Mech.*, **12**, 333–336.

LONGUET-HIGGINS M.S., SMITH N.D. 1966. An experiment on third-order resonant wave interactions, *J. Fluid Mech.*, **25**, 417–435.

MCALLISTER M.L., DRAYCOTT S., ADCOCK T.A.A., TAYLOR P.Y., VAN DEN BREMER T.S. 2019. Laboratory recreation of the Draupner wave and the role of breaking in crossing seas, *J. Fluid Mech.*, **860**, 767–786.

MCLEAN J.W. 1982. Instabilities of finite-amplitude water waves, *J. Fluid Mech.*, **114**, 315–330.

MOLIN B., FAUVEAU V. 1984. Effect of wave directionality on second-order loads induced by the set-down, *Applied Ocean Res.*, **6**, 66–72.

RIENECKER M.M., FENTON J.D. 1981. A Fourier approximation method for steady water waves, *J. Fluid Mech.*, **104**, 119–137.

RODENBUSCH G., FORRISTALL G.Z. 1986. An empirical model for random wave kinematics near the free surface, *Proc. Offshore Technology Conf.*, paper 5098.

SCHWARTZ L.W. 1974. Computer extension and analytic continuation of Stokes' expansion for gravity waves, *J. Fluid Mech.*, **62**, 553–578.

STANSBERG C.T., GUDMESTAD O.T., HAVER S.K. 2008. Kinematics under extreme waves, *J. Offshore Mech. Arct. Engineering*, **130**.

STASSEN Y. 1999. Simulation numérique d'un canal à houle bidimensionnel au troisième ordre d'approximation par une méthode intégrale, PhD thesis, Nantes University (in French).

SU M-Y. 1982. Evolution of groups of gravity waves with moderate to high steepness, *Phys. Fluids*, **25**, 2167–2174.

SU M-Y., BERGIN M., MARLER P., MYRICK R. 1982. Experiments on nonlinear instabilities and evolution of steep gravity-wave trains, *J. Fluid Mech.*, **124**, 45–72.

SUSBIELLES G., BRATU C. 1981. *Vagues et ouvrages pétroliers en mer*, Éditions Technip (in French).

WHEELER J.D. 1970. Method for prediction of forces produced by irregular waves, *J. Petrol. Techn.*, 359–367.

4 Wave and Current Loads on Slender Bodies

By definition, "small bodies" are small when related to the wave lengths. Slender bodies are small in two directions and elongated in the third direction. As a result of being small, diffraction effects are limited and the incoming waves remain quasi-unperturbed as they travel by. For a vertical cylinder piercing the free surface, one criterion is that the diameter be less than one-sixth of the wave length. A drawback of being small is that the K_C numbers are usually large: The flow separates massively, prohibiting potential flow theory. An advantage of transparency is that highly nonlinear wave models can be used to derive the local kinematics of the incoming flow.

In offshore engineering, the prototype of the small body is the jacket, composed of many tubular members with diameters of the order of one meter. Other slender bodies are all the connections between the floating structures and the seabed such as the risers and mooring lines. A recurrent geometry therefore is the circular cylinder.

As stated above, flow separation rules out numerical tools based on potential flow theory. Nowadays, although Computational Fluid Dynamics (CFD) has made great progress, CFD software is still a long way from being applicable to structures composed of hundreds of tubular elements. Semi-empirical approaches are still the only way.

Even though, in a jacket structure, the tubular elements are close to each other, proximity effects are ignored in the first approximation. Moreover, a strip theory approach is applied, and free surface effects are ignored. Finally, due to the assumption of small diameter related to the flow scale (the wave length), the incoming flow is idealized as **locally** uniform.

The elementary problem then is a two-dimensional cylinder, usually circular, in motion with velocity $(\dot{X}(t), \dot{Y}(t))$, in an incoming flow with velocity $(U(t), V(t))$ (Figure 4.1). First we solve this problem within the frame of potential flow theory, and we compare its findings with experimental results. This is done first in steady current. Then oscillatory flows are considered and the Morison equation is introduced. Its merits and deficiencies are illustrated in flows combining waves and current.

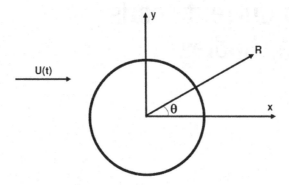

Figure 4.1 Geometry.

4.1   Potential Flow Theory

4.1.1   Fixed Cylinder in Uniform Flow

Without loss of generality, the incoming flow is taken along the Ox axis, with velocity $U(t)$.

The velocity potential of the incoming flow is then

$$\Phi_I = U(t)\, x = U(t)\, R\, \cos\theta \tag{4.1}$$

When the cylinder has been introduced, the potential becomes (see Appendix A)

$$\Phi(R,\theta,t) = U(t)\, a\, \left(\frac{R}{a} + \frac{a}{R}\right)\cos\theta \tag{4.2}$$

with a the radius.

The pressure follows from the Bernoulli equation, where gravity is neglected:

$$p = p_0(t) - \rho\,\Phi_t - \frac{1}{2}\,\rho\,(\nabla\Phi)^2 \tag{4.3}$$

At the cylinder wall it is

$$p = p_0(t) - 2\,\rho\,a\,\dot{U}(t)\,\cos\theta - 2\,\rho\,U^2(t)\,\sin^2\theta \tag{4.4}$$

When the pressure is integrated over the cylinder wall the following load is obtained

$$F_x = 2\rho\,\pi\,a^2\,\dot{U}(t) \tag{4.5}$$

As a result, when the flow is steady ($\dot{U} \equiv 0$), the force is zero (d'Alembert paradox).[1]

---

[1] Here we have assumed zero circulation. In the case when there is some circulation around the cylinder a lift force appears, perpendicular to the direction of the incoming flow.

4.1.2 Moving Cylinder in a Fluid at Rest

Still without loss of generality, we take the velocity $\dot{X}(t)$ of the cylinder along the Ox axis. In the cylinder-fitted coordinate system (R,θ), the velocity potential writes

$$\Phi(R,\theta,t) = -\dot{X}(t)\,\frac{a^2}{R}\,\cos\theta \tag{4.6}$$

while the pressure is obtained from

$$p = p_0(t) - \rho\,\Phi_t - \frac{1}{2}\,\rho(\nabla\Phi)^2 + \rho\,\dot{X}(t)\,\Phi_x \tag{4.7}$$

(the last term results from the fact that Φ is written in a moving coordinate system.)

Again it can be checked that only the unsteady term $-\rho\Phi_t$ contributes to the hydrodynamic load which writes

$$F_x = -\rho\,\pi a^2\,\ddot{X}(t) \tag{4.8}$$

4.1.3 Generalization

In the general case of a cylinder of arbitrary shape, but such that one of the two axes Ox or Oy is an axis of symmetry, the hydrodynamic loads are obtained as

$$F_x = (1 + C_{a11})\,\rho\,S\,\dot{U}(t) - C_{a11}\,\rho\,S\,\ddot{X}(t) \tag{4.9}$$
$$F_y = (1 + C_{a22})\,\rho\,S\,\dot{V}(t) - C_{a22}\,\rho\,S\,\ddot{Y}(t) \tag{4.10}$$

with S the sectional area.

C_{a11} and C_{a22} are the x and y added mass coefficients, equal to 1 in the case of a circular cylinder. Figure 4.2 shows the added mass coefficient $M_a/(\rho\pi a^2)$ of a rectangular cylinder of length $2b$ in the direction of motion and height $2a$, as a function of the ratio b/a.

For a body with no symmetry axis there appear supplementary terms such as $C_{a12}\,\dot{U}$, $C_{a12}\,\dot{V}$, $C_{a12}\,\ddot{X}$, and $C_{a12}\,\ddot{Y}$: a flow accelerated in the Ox direction produces loads in the transverse direction Oy, etc.

Here we have disregarded the rotational velocity, which gives rise to supplementary load components (see, e.g., ship maneuvering theory).

It should also be borne in mind that d'Alembert's paradox only applies to translation loads. Potential theory predicts a moment (known as Munk moment), given by

$$C_z = \rho\,(C_{a11} - C_{a22})\,S\,(U - \dot{X})\,(V - \dot{Y}) \tag{4.11}$$

This moment has a destabilizing effect. An illustration is that, in current, elongated bodies tend to orient themselves beamwise to the flow.

All these results are valid without any restriction on the motion amplitude of the cylinder. A remarkable feature is that, in the absences of a free surface and of viscosity, there are no memory effects: the loads at a given time only depend on the flow and cylinder kinematics at the same instant. This implies that, in the absence of incident flow, when the cylinder is suddenly stopped, the resulting fluid kinematics (and loads)

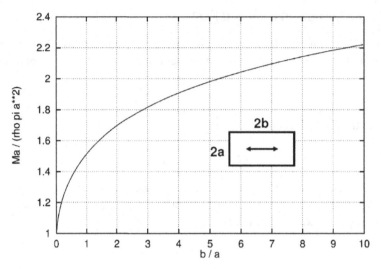

Figure 4.2 Added mass coefficient of a rectangular cylinder.

stop instantly as well. This is contrary to practical experience (when, for instance, stirring and stopping a spoon in a coffee cup).

In three dimensions, the equivalents to the cylinder and to the plate are the sphere and the disk, with added masses of $2 \rho \pi a^3 / 3$ for the sphere ($C_a = 1/2$) and $8 \rho a^3 / 3$ for the disk, a being the radius.

4.2   Circular Cylinder in Steady Flow

In this section, we consider the case of a circular cylinder in steady flow with velocity U, in the Ox direction, for instance, installed in a wind tunnel, from wall to wall. We have just seen that potential flow theory predicts an in-line load which is nil, and a pressure at the wall given by (still assuming no circulation):

$$p = p_\infty + \frac{1}{2} \rho U^2 - 2 \rho U^2 \sin^2 \theta \qquad (4.12)$$

where p_∞ is the ambient pressure, away from the cylinder.

One usually defines the pressure coefficient C_p as

$$C_p = \frac{p - p_\infty}{\frac{1}{2} \rho U^2} = 1 - 4 \sin^2 \theta \qquad (4.13)$$

Figure 4.3 shows experimental pressure coefficients (time-averaged), compared to the theoretical one given by (4.13). (In the figure the angle θ is counted from the upstream stagnation point.) It can be seen that the experimental values are in fair agreement with the theoretical ones only on the weather side, up to an angle of about 90 degrees. On the lee side the experimental pressure coefficients fail to increase up to the

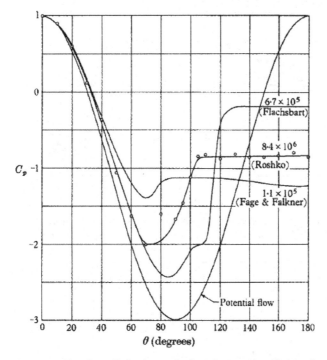

Figure 4.3 Circular cylinder in steady flow. Pressure coefficient C_p at the wall of a circular cylinder, from potential flow theory and from measurements (from Roshko, 1961) (Cambridge University Press, reprinted with permission).

potential flow values and level off to some plateau values that depend on the Reynolds number, defined as

$$\text{Re} = \frac{U\,D}{\nu} \tag{4.14}$$

with D the diameter and ν the kinematic viscosity ($\nu \sim 10^{-6}$ m^2 s^{-1} for water at 20 degrees; $\nu \sim 1.5 \cdot 10^{-5}$ m^2 s^{-1} for air at 20 degrees).

This pressure defect on the lee side is a sign that the flow has separated. As a consequence, a mean load has appeared, usually given in the form

$$F_D = \frac{1}{2}\rho\,C_D\,D\,U^2 \tag{4.15}$$

The drag coefficient C_D depends on the Reynolds number, on the roughness of the cylinder wall, and on the turbulence level of the incoming flow. Typical values of the Reynolds numbers that we are concerned with are from 10^4 up to 10^7. In this range the contribution of frictional stresses is negligible and most of the load is due to pressure. Figure 4.4 shows experimental values of the drag coefficient compiled by Cantwell & Coles (1983) from literature, for smooth cylinders. There is a notable scatter due to various experimental issues such as the turbulence level, vibration of the cylinder due to lack of stiffness of the setup, finite aspect ratio, confinement effects, etc.

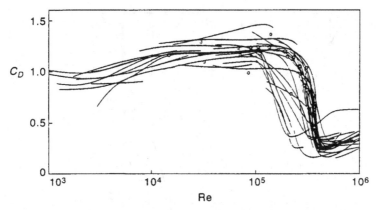

Figure 4.4 Circular cylinder in steady flow. Drag coefficient C_D vs Re (from Cantwell & Coles, 1983) (Cambridge University Press, reprinted with permission).

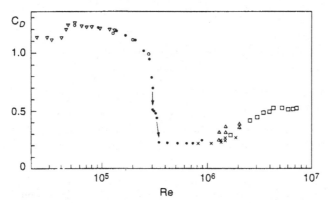

Figure 4.5 Circular cylinder in steady flow. Drag coefficient C_D vs Re (from Schewee, 1983) (Cambridge University Press, reprinted with permission).

Figure 4.5 shows experimental values obtained by Schewe (1983) in a wind tunnel, up to a Reynolds number close to 10^7, that are taken as reference.

From these figures, it appears that the drag coefficient is around 1.0–1.2 up to Re $\sim 2\,10^5$, where from there occurs a sharp drop down to very low values, around 0.2 or 0.3, and then, from Re $\sim 10^6$, a slow increase up to a plateau at about 0.5 or 0.6. These different Reynolds number ranges and different drag coefficients are associated with different flow regimes: In the **subcritical** regime, which extends from Re ~ 300 up to Re $\sim 2\,10^5$, the boundary layer is laminar up to the separation point and transition to turbulence occurs in the wake. As a result of being laminar, the boundary layer separates early, at an angle θ_S of about 82 degrees (counted from the upstream stagnation point). The other end, for Reynolds numbers higher than about 3.10^6, is the **transcritical** regime, where the boundary layer can be viewed as turbulent from the upstream stagnation point. As a result of being turbulent, the boundary layer is more stable than in the laminar case and the separation point is delayed to an angle of about 120 degrees, meaning a narrower wake and a smaller drag coefficient. In between is a peculiar **critical** regime, from Re $\sim 2.10^5$ to 5.10^5, where transition occurs right after the laminar separation, inducing reattachment and a second separation further downstream; this means a very narrow wake, hence the very low associated drag coefficients. (It has frequently been observed that the double separation may occur on only one side of the cylinder, inducing a transverse hydrodynamic load component.) The intermediate regime in between critical and transcritical (from $\text{Re} \sim 5.10^5$ to 3.10^6) is known as the **supercritical** regime.

Obviously, the boundaries between these different regimes are somewhat theoretical, and they are susceptible to variation, depending, as already stated, on the turbulence level, the rigidity of the experimental setup, etc. Figure 4.6, taken from Verley (1980), illustrates the different flow regimes.

4.2.1   Inclined Cylinders

When the cylinder is inclined with respect to the incoming flow with an angle α, the incoming velocity can be decomposed into a normal component U_n ($U_n = U \sin \alpha$) and a tangential component U_t ($U_t = U \cos \alpha$). See Figure 4.7. Similarly, the load consists in a normal component F_{Dn} and a tangential component F_t.

The normal load component can be related to the normal component of the velocity as

$$F_{Dn} = \frac{1}{2} \rho \, C_{Dn}(\alpha) \, D \, U_n^2 \qquad (4.16)$$

where the normal drag coefficient $C_{Dn}(\alpha)$ is usually taken as independent of the angle α and equal to C_D.

For smooth cylinders, the tangential load component results from friction and is thus quite small compared to the normal component. However for elongated cylindrical bodies in nearly tangential flow, for instance in the case of towing lines, it can be non-negligible. The tangential load is usually related to the total velocity U and written as

$$F_t = \frac{1}{2} \rho \, C_t(\alpha) \, D \, U^2 \qquad (4.17)$$

Different formulas have been proposed for $C_t(\alpha)$. For instance (Eames, 1968):

$$C_t(\alpha) = C_D \, (m + n \sin \alpha) \, \cos \alpha \qquad (4.18)$$

4.2.2   Effect of Roughness

Roughness is usually defined as a ratio k/D with D the diameter and k a typical size of the surface irregularities. For a pipe, roughness may result from corrosion (a typical value is then $k \sim 3$ mm) and/or from marine growth (with much larger k values).

An effect of roughness is to reduce the value of the critical Reynolds number, at the boundary between subcritical and critical ranges. According to Achenbach & Heinecke (1981) it is approximatively given by

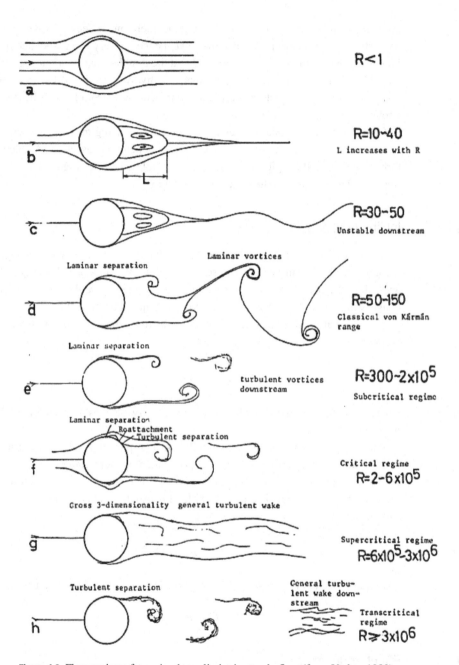

Figure 4.6 Flow regimes for a circular cylinder in steady flow (from Verley, 1980).

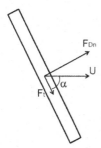

Figure 4.7 Inclined cylinder. Definition of normal and tangential forces.

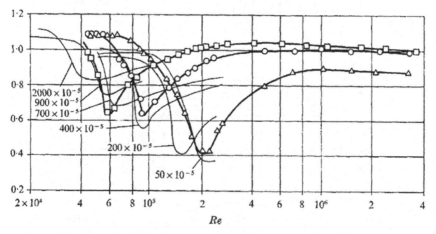

Figure 4.8 Drag coefficient for different roughnesses (from Achenbach, 1971).

$$\mathrm{Re}_C = 6000 \left(\frac{k}{D}\right)^{-1/2} \tag{4.19}$$

Another effect is that the drop in the drag coefficient, in the critical regime, is less pronounced. At very high roughness coefficients, the C_D value becomes independent of the Reynolds number.

Figure 4.8, taken from Achenbach (1971), shows the drag coefficient vs the Reynolds number for different k/D values. It may seem paradoxical that, for Reynolds numbers in the range $5 \cdot 10^4$–$2 \cdot 10^5$, the drag coefficient is less on a rough cylinder than on a smooth cylinder. This is also the case for spherical bodies, and it is the reason why golf balls are rough. It has also been observed that roughness may decrease the resistance of towing lines.

4.2.3 Noncircular Cylinders

Figure 4.9 shows the drag coefficient for a square cylinder vs the Reynolds number (here defined as $\mathrm{Re} = UD/\nu$ with D the side length), for three different roundnesses of the corners ($r/D = 0.021$; 0.167; 0.333). In the smallest r/D case, practically

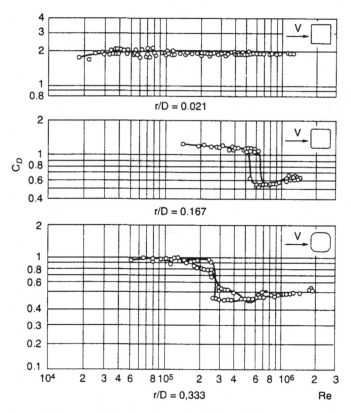

Figure 4.9 Drag coefficient for a square cylinder, with sharp and rounded corners (from Scruton & Rogers, 1971).

equivalent to a sharp corner, the drag coefficient is equal to 2, whatever the Reynolds number.

Conversely, in the rounded cases, the behavior is very similar to the circular cylinder, with a critical regime appearing at a somewhat higher Reynolds number, depending on the corner radius.

An important feature, illustrated here in the square case, is that the drag coefficient becomes more or less independent of the Reynolds number, as soon as the cylinder has sharp corners that impose the separation points of the boundary layer.

Figure 4.10 shows the drag coefficient for a rectangular cylinder with sharp corners vs the aspect ratio b/a with b the length of the side in the flow direction. It can be observed that, in between the plate and square cases, where $C_D \simeq 2$, there is a critical aspect ratio $b/a \simeq 0.6$ where the drag coefficient reaches almost 3. From then on there is a monotonous decrease of C_D with increasing b/a values.

4.2.4   Three-Dimensional Effects

The drag loads are directly related to the velocity increase of the outer flow in the neighborhood of the separation points of the boundary layers. These overspeeds are

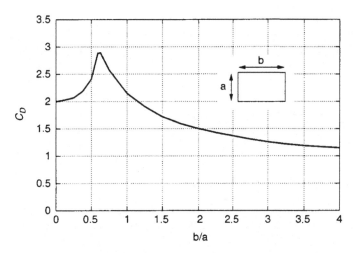

Figure 4.10 Drag coefficient for a sharp-angled rectangular cylinder (from Parkinson, 1974).

higher for plane flows than for their "equivalent" three-dimensional geometries. For instance, in the circular cylinder case, the potential flow velocity at ±90 degrees from the stagnation point is twice the incoming flow velocity. For a sphere it is 1.5 times. As a consequence, in the subcritical range, the drag coefficient of a sphere is about 0.5, when it is 1.2 for a circular cylinder.

Similarly, the drag coefficient for a cube is about 1, half that for a square cylinder.

For cylinders of finite length, correcting coefficients can be found in the literature as function of the flow regime and of the aspect ratio L/D (L the length and D a characteristic cross-sectional dimension).

4.2.5   The Lift Force

Again we consider the case of a two-dimensional circular cylinder in steady flow. In spite of the geometric symmetry, an oscillatory load appears in the transverse direction: the **lift force**. This lift force is due to the fact that the positions of the separation points of the boundary layer vary in time at the pace of the vortex shedding. As a result, an oscillatory circulation takes place around the cylinder, inducing, via Magnus effect, an oscillatory lift force.

In the subcritical and transcritical regimes, the lift force is quasi-sinusoidal in time and takes place at a frequency f_0 such that the **Strouhal number**, defined as $S_t = f_0 D/U$, is very close to 0.2 in the subcritical regime and somewhat higher in the transcritical case (see Figure 4.11). In the critical and supercritical regimes, the lift force is quite erratic with no dominant frequency.

Figures 4.12 and 4.13 show the measured lift force for a 20 cm diameter cylinder at Re = $8.8 \cdot 10^4$ (subcritical) and Re = $2.6 \cdot 10^5$ (critical). The high-frequency oscillations that appear are due to vibrations of the setup. In Figure 4.12 the lift force is quite periodic, at a frequency of about 0.5 Hz, while no dominant frequency appears in Figure 4.13. In this figure, a nonzero mean value is observed, probably due to the fact

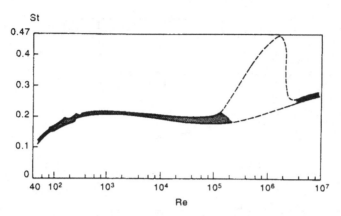

Figure 4.11 Strouhal number vs Reynolds number (from Blevins, 1990).

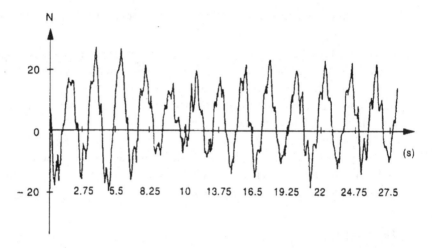

Figure 4.12 Time trace of the measured lift force on a cylinder at Re $= 8.8 \cdot 10^4$.

that the phenomenon of separation and re-attachment takes place at only one side of the cylinder.

Lift coefficients (C_L) are defined as the amplitude, or sometimes the standard deviation, of the lift force divided by $1/2 \, \rho \, D \, L \, U^2$ with L the length of the weighted part of the cylinder. A wide scatter is found in the literature. One reason is that it is not always clear whether C_L is based on the standard deviation or the amplitude of the lift force (and then how the amplitude is defined). Another reason is that the vortex shedding may not be synchronized over the whole length of the weighted part of the cylinder. This introduces an important parameter, the **correlation length**, which is the cylinder length over which the vortex shedding is synchronized. In the subcritical regime, the literature suggests correlation lengths in the range of 2–6 diameters.

In the time series of Figure 4.12, the weighted length is 5 diameters. The lift force amplitude is about 20 N. The diameter being 20 cm and the flow velocity 50 cm/s, this suggests $C_L \sim 0.8$.

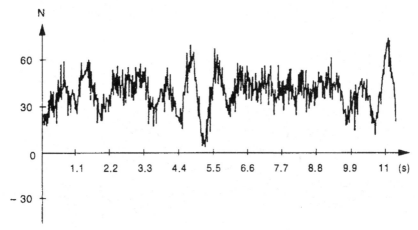

Figure 4.13 Time trace of the measured lift force on a cylinder at Re = $2.6 \cdot 10^5$.

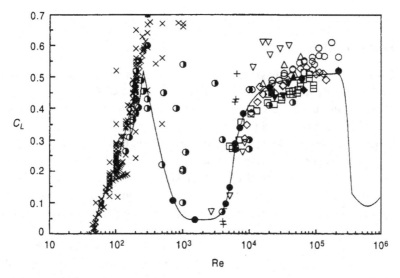

Figure 4.14 Lift coefficients compiled by Norberg (2000).

Figure 4.14 shows experimental values compiled by Norberg (2000), all for short-weighted length cylinders. The lift coefficient is defined from the standard deviation of the lift force. For Reynolds numbers in between 10^4 and 10^5, the figure suggests C_L values around 0.5, which would be 0.7 for the amplitude-based C_L, consistent with the value derived from Figure 4.12.

Lift forces are observed for most cylinder shapes, even though they have sharp corners. In the case of square cylinders, for instance, the Strouhal number drops to 0.13–0.15, and the lift force is much weaker than in the circular case.

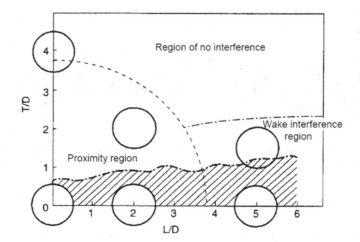

Figure 4.15 Hydrodynamic interaction zones for two cylinders in a current (from Zdravkovich, 1985).

4.2.6   Proximity Effects

Cylinders are frequently grouped together, for instance, as clusters of risers or tubular elements of jacket frames. Some cylinders may be located in the wakes of upstream cylinders and therefore subjected to currents of reduced velocity. This is known as the **shielding effect**. The upstream cylinders themselves are subjected to a reduced current, as compared to the case where they would be isolated, by the presence of the neighboring cylinders: this is the **blockage effect**. Drag loads may then be appreciably decreased as a result of blockage and shielding. For instance, Cuffe *et al.* (1990) relate that the mean current loads acting on the LENA-compliant tower were overestimated by 400% in the design! Another issue for risers is the risk of clashing when they are located too close together.

The case of two cylinders, of equal diameter D, in proximity, has been widely covered experimentally (for a survey see, for instance, Zdravkovich, 1979, 1985). When the spacing to diameter ratio L/D is less than some value, of the order 3–5 depending on the current direction, the two wakes merge into one (see Figure 4.15). When the cylinders are aligned with the current, the drag load on the upstream cylinder is decreased for relative spacing L/D up to 4 or 5. The current load on the lee cylinder is decreased much more to the point that it can be negative (the lee cylinder is "aspired" by the upstream cylinder) for L/D less than 2 or 3.

Velocity Defect in the Wake of a Cylinder
At high Reynolds numbers and at some distance from a cylinder, the flow becomes homogenized and the velocity defect takes a more or less Gaussian shape. The following profile has been proposed by Schlichting (1979) (*Boundary layer theory*, ch. XXIV, pp 742–743):

$$U(x,y) = U - 0.95\, U\, \sqrt{\frac{C_D\, D}{x}}\, \exp\left\{-11.3\, \frac{y^2}{C_D\, D\, x}\right\} \qquad (4.20)$$

The coefficients 0.95 and 11.3 are empirical but dependent since the loss of momentum in the wake has to be equal to the drag force on the cylinder.

With a C_D value of 1 one gets that the velocity is decreased by 30% at a distance of 10 diameters (in $y = 0$), so the drag load is decreased by 50% as compared to an isolated cylinder!

Huse (1991) has extended this model to the case of multiple cylinders. The principle is to start from the most upstream cylinders and to add up the velocity defects (actually the momentum flux defects) going downstream. Huse (1991) also proposes to account for blockage effects based on a potential flow approach. (As a result, the incoming flow velocity can be increased in some cases, for instance, for a row of cylinders in normal incidence.)

4.2.7   Cylinder Next to the Seabed

This is a configuration of interest for studies relating to pipelines, in particular, concerning their stability under the loads induced by the waves and current. Due to the irregularities of the seabed (the "free spans"), pipelines are not everywhere in contact with the bottom, which poses many other problems such as that of their vibrations induced by vortex shedding.

In a perfect fluid theory, as soon as the cylinder is detached from the sea floor, there is flow through the interstitial space, with the flow faster as the space is more reduced. There is therefore a pressure defect and the cylinder is drawn toward the seabed. Conversely, if there is contact, the flow can only pass over the cylinder, inducing an upward lift (with a lift coefficient C_L equal to 4.49 according to potential flow theory).

In a viscous fluid the situation is different. The incoming current presents a boundary layer profile, the thickness of which depends on the roughness of the seabed, the water depth, etc., and is usually much greater than the diameter of the cylinder. The viscosity limits the flow velocity in the interstitial space, with the result that, for small separations, the average lift is repulsive. When the cylinder lies on the bottom, or at a short distance from it, the alternate vortex shedding is inhibited: a recirculation zone is formed behind the cylinder, and the outer flow is reattached to the seabed some distance beyond.

The literature is rather confused as to the drag and lift coefficients to be applied, the main reason being that the thickness of the boundary layer (relative to the diameter of the cylinder) is a variable parameter from one experiment to the other, not always transcribed.

4.3   Cylinder in Sinusoidal Flow: The Morison Formula

Let us now address the case of a two-dimensional circular cylinder in uniform sinusoidal flow, with velocity

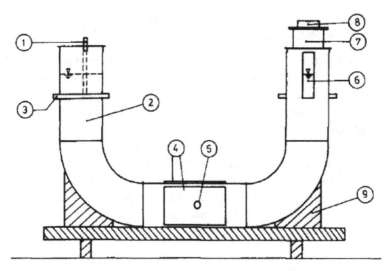

Figure 4.16 The U-tube facility at SINTEF Ocean (from Faltinsen & Sortland, 1987). 1, wave probe; 2, main body in aluminum; 3, wood stiffener; 4, Plexiglas window; 5, test cylinder; 6, Plexiglas window; 7, honeycomb section; 8, driving fan; 9, wood bed.

$$U(t) = A \omega \sin \omega t \qquad\qquad V = 0 \qquad\qquad (4.21)$$

This corresponds to the practical (and historical) case of a vertical cylinder in regular waves, with a wave length much greater than the diameter: in each horizontal strip the incoming flow is of this kind (when the vertical component of the flow is neglected).

An experimental apparatus, which allows genuine 2D periodic flows to be produced, is the U-tube (Figure 4.16). In a U-tube the frequency ω is imposed as the natural frequency of the U-shaped water column: $\omega \sim \sqrt{2\rho g\, S/M}$ with M the mass of water and S the cross-section area assumed to be nearly constant all throughout. It suffices to vary in time the air pressure over one of the two free surfaces (with a fan for instance) at the natural frequency to generate the oscillatory flow. Another experimental procedure is to impose forced sinusoidal motions to a cylinder in calm water, away from the free surface and from the tank walls. Kinematically, it is equivalent, but the hydrodynamic loads differ by the Froude–Krylov component ($\rho\, S\, L\, \dot{U}$). One may also perform tests on a vertical cylinder in regular waves, measuring the loads on a horizontal section of short height.

Besides the Reynolds number, here defined as Re $= A\,\omega\, D/\nu$, another dimensionless parameter comes into play, the Keulegan–Carpenter number[2], defined as

$$K_C = \frac{A\,\omega\, T}{D} = 2\,\pi\, \frac{A}{D} \qquad\qquad (4.22)$$

[2] from the paper (Keulegan & Carpenter, 1958).

In place of Re, the parameter $\beta = \text{Re}/\text{K}_C = D^2/\nu T$ is frequently used, $T = 2\pi/\omega$ being the period. β is known as the **Stokes parameter**. It can be given several physical meanings; for instance it is the square of the ratio of the diameter over the thickness of the boundary layer ($\delta \sim \sqrt{\nu T}$). In U-tube experiments, the β parameter remains constant for a given cylinder model.

For a smooth cylinder β (or Re) and K_C are the only two dimensionless parameters that characterize the flow and therefore the loads. It has been seen in the foregoing that potential flow theory predicts an inertia load proportional to acceleration and no effect of velocity, whereas in steady flow experiments a drag load, proportional to the velocity squared (more precisely to $U|U|$), is obtained.

The incoming flow being periodic in time, the hydrodynamic load is periodic as well and can be written as a Fourier series:

$$F_x(t) = F_{C1} \cos \omega t + F_{S1} \sin \omega t + F_{C2} \cos 2\omega t + F_{S2} \sin 2\omega t + \ldots \quad (4.23)$$

The first term $F_{C1} \cos \omega t$ can be interpreted as an inertia load (in phase with acceleration) and the second one $F_{S1} \sin \omega t$ as a drag force (in phase with velocity). The remaining terms at higher harmonics (2ω, 3ω, etc.) being neglected, the total load is written as

$$F_x = \rho\, C_M\, S\, \dot{U} + \frac{1}{2} \rho\, C_D\, D\, U\, |U| \quad (4.24)$$

where the **inertia** coefficient $C_M = 1 + C_a$ and the drag coefficient C_D only depend on K_C and Re. They can be derived through harmonic analysis of the experimental measurements (another technique is to minimize in time the difference between the measured load and (4.24); this leads to the same C_M and to a slightly different C_D).

The formula (4.24) is known as the **Morison equation** (from Morison et al., 1950). Its main virtue is to produce two terms, in $\cos \omega t$ and $\sin \omega t$,[3] when a periodic load at frequency ω is expected and higher harmonics can be neglected.

It matters to compare the relative importance of the inertia and drag loads. The ratio drag / inertia is obtained as

$$\frac{F_{\text{drag}}}{F_{\text{inertia}}} = \frac{8}{3\pi^3} \frac{C_D}{C_M} \text{K}_C \quad (4.25)$$

(here $\sin \omega t\, |\sin \omega t|$ has been approximated as $8/3\pi \sin \omega t$).

When $C_D \simeq 1$ and $C_M \simeq 2$, this gives:

$$\frac{F_{\text{drag}}}{F_{\text{inertia}}} \simeq 0.043\, \text{K}_C \quad (4.26)$$

---

[3] $\sin \omega t\, |\sin \omega t|$ is very similar to $\sin \omega t$. More precisely:

$$\sin \omega t\, |\sin \omega t| = \frac{8}{3\pi} \sin \omega t - \sum_{n=1}^{\infty} \frac{8}{(2n-1)(2n+1)(2n+3)\pi} \sin(2n+1)\omega t$$

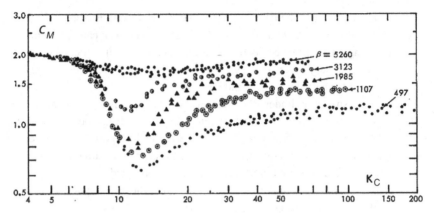

Figure 4.17 Circular cylinder in oscillatory flow. Inertia coefficients ($C_M = 1 + C_a$) obtained by Sarpkaya vs K_C for different β values (from Sarpkaya, 2010) (Cambridge University Press, reprinted with permission).

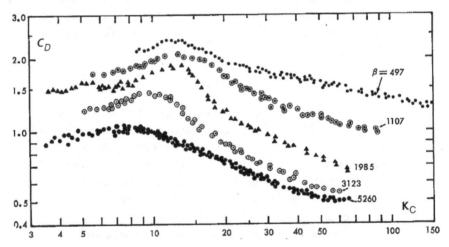

Figure 4.18 Circular cylinder in oscillatory flow. Drag coefficients (C_D) obtained by Sarpkaya vs K_C for different β values (from Sarpkaya, 2010) (Cambridge University Press, reprinted with permission).

The relative contribution of the drag and inertia components depends on the Keulegan–Carpenter number. When K_C is low, inertia is dominant when K_C is high, drag is dominant. At $K_C \sim 20$ the two components are comparable.

The cases $K_C \to 0$ and $K_C \to \infty$ are also the cases when the Morison equation can be given some physical meaning: when K_C is low (less than some value in between 2 and 5 depending on β) the boundary layer remains attached to the cylinder and potential flow theory becomes applicable. When K_C is high, many vortices are shed in each half cycle and the wake is convected far away: asymptotically the case of a cylinder in steady flow is attained.

There have been numerous experiments to determine the C_M and C_D values in oscillatory flows. Figures 4.17 and 4.18 show values obtained by Sarpkaya through

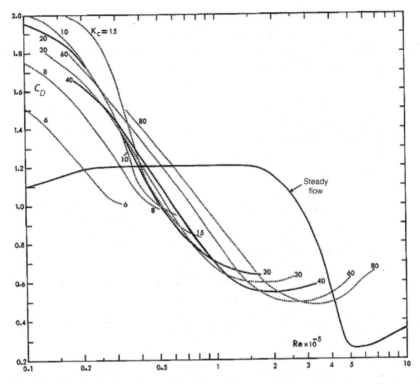

Figure 4.19 Circular cylinder in oscillatory flow. Drag coefficients (C_D) obtained by Sarpkaya vs Re = $\beta \times K_C$ (from Sarpkaya, 2010) (Cambridge University Press, reprinted with permission).

U-tube experiments. Different β values in the figures correspond to different cylinder diameters, from 5 cm up to 16.5 cm (the U-tube cross-section being 30 cm × 30 cm).

Considering first the inertia coefficient $C_M = 1 + C_a$, it can be checked that the theoretical potential flow value 2 is actually attained at low K_C values. When K_C increases from zero, C_M decreases slowly[4] and drops to a minimum value that depends on β. It is striking that, for $\beta < 2000$, this minimum value is below 1, which means that the added mass coefficient C_a is negative! It is comforting to see that, at the highest β values, C_M is close to 2, over the whole K_C range.

Figure 4.19 shows the same experimental C_D values as Figure 4.18, plotted vs the Reynolds number Re = β K_C, for different K_C values. When $K_C \to \infty$, one would expect to recover the drag coefficient $C_D(\text{Re})$ in the steady flow case, as argued above. Sarpkaya's results do not go beyond $K_C = 100$, but they do not seem to converge toward the $C_D(\text{Re})$ plot in Figure 4.5. It is noteworthy that the critical zone is shifted to lower Reynolds numbers, and that it is less pronounced, as in the case of a rough cylinder (see Figure 4.8), or of a smooth cylinder in highly turbulent flow. This is due to the fact that the cylinder reencounters its own wake at each flow cycle.

---

[4] Actually there is first a very slight increase, due to viscous effects.

Moving back to Figure 4.18 and considering the C_D values from right to left, for decreasing K_C values, one notices that the drag coefficient increases to some maximum for $K_C \sim 10$, and then decreases. An interpretation is that the "effective" flow velocity is higher than the free stream velocity, due to the additional velocity of the wake shed at the previous half-cycle (this wake effect is better seen if one considers the cylinder oscillating in a fluid at rest). When K_C is less than 10, the vortex shedding decreases until the boundary layer remains attached, with the result that C_D decreases.

To account for this wake effect, DNVGL, for instance, relates the oscillatory flow C_D to the steady flow C_{DS} through

$$C_D = C_{DS} \, \psi(K_C) \tag{4.27}$$

with ψ the "wake amplification factor" (e.g., see DNVGL-RP-C205).

4.3.1   Wave Load on a Vertical Pile

Consider a bottom-mounted vertical cylinder, of diameter D, with its vertical axis in $x = y = 0$, submitted to a regular wave.

From the velocity potential

$$\Phi(x,z,t) = \frac{A\,g}{\omega} \frac{\cosh k(z + h)}{\cosh kh} \, \sin(k\,x - \omega\,t) \tag{4.28}$$

the horizontal velocity and acceleration, in $x = 0$, are obtained as

$$U(z,t) = \Phi_x(0,z,t) = \frac{A\,g\,k}{\omega} \frac{\cosh k(z + h)}{\cosh kh} \, \cos \omega t \tag{4.29}$$

$$\dot{U}(z,t) = -A\,g\,k \, \frac{\cosh k(z + h)}{\cosh kh} \, \sin \omega t \tag{4.30}$$

with h the water depth.

The inertia (F_I) and drag (F_D) loads are obtained through integration over the water height:

$$F_I(t) = \rho\,C_M \, \frac{\pi D^2}{4} \int_{-h}^{0} \dot{U}(z,t) \, \mathrm{d}z \tag{4.31}$$

giving

$$F_I(t) = -\rho\,C_M \, \frac{\pi D^2}{4} \, A\,g \, \tanh kh \, \sin \omega t \tag{4.32}$$

for the inertia load, and

$$\begin{aligned}
F_D &= \frac{1}{2} \rho\,C_D\,D \int_{-h}^{0} U(z,t) \, |U(z,t)| \, \mathrm{d}z \\
&= \frac{1}{4} \rho\,C_D\,D\,A^2\,g \left(1 + \frac{2kh}{\sinh 2kh} \right) \cos \omega t \, |\cos \omega t| \tag{4.33}
\end{aligned}$$

for the drag load.

The overturning moment, with respect to the sea floor, may similarly be derived, by adding a factor $(z + h)$ in the integrands. In deep water $(kh > 3)$ one obtains that the inertia force is applied at a distance $1/k = L/(2\pi)$ below the free surface, and the drag force at the half distance $1/(2k)$.

In Chapter 6 it will be seen that equation (4.32) is a good approximation of the wave load on a vertical cylinder when its diameter is less than one-sixth of the wave length. For larger diameters diffraction effects cannot be neglected (the coefficient C_M shifts from its reference value 2 and a phase lag appears between horizontal acceleration and load).

To derive equations (4.32) and (4.33) the Airy wave model has been used and the z integration has been carried out up to the still water level. Nonlinear effects can be introduced by choosing a nonlinear wave model (Stokes fifth order or stream function for instance) and integrating up to the instantaneous free surface.

4.3.2   Drag Coefficient at Low K_C Values

It has been observed that at low K_C values inertia loads are dominant (and well predicted by a potential flow approach), so there seems to be no need to worry about C_D values.

This is true for fixed bodies. However, in the case of bodies free to move in waves, there are many cases where the damping effect due to viscous forces needs to be taken into account, for instance, when calculating the slow-drift motion of moored TLPs or semis.

When the flow is attached and laminar, the damping force is for half due to friction and for the other half due to pressure forces (Stokes, 1851). This is only for a circular shape, in the general case the two components are not equal (Molin & Etienne, 2000). The equivalent drag coefficient is given by:

$$C_D = \frac{3\pi^3}{2\,K_C} \left(\frac{1}{\sqrt{\pi\,\beta}} + O\left(\frac{1}{\beta}\right) \right) \tag{4.34}$$

C_D is inversely proportional to K_C: the viscous damping force is no longer quadratic with the velocity, but linear.

As Figures 4.20 and 4.21 show, the domain of validity of equation (4.34) (a straight line in the figures due to the logarithmic scale) is quite limited. When β and/or K_C increase, three-dimensional instabilities appear in the boundary layer and the flow becomes turbulent. Equation (4.34) remains more or less applicable when the molecular viscosity ν is replaced with a turbulent viscosity ν_e, 5–50 times larger.

According to Sarpkaya (1986), the minimum value of the drag coefficient coincides with the onset of separation. From Figures 4.20 and 4.21, it appears that the K_C value where this minimum occurs moves from about 1.5 at $\beta = 1380$ up to about 4 at $\beta = 11240$, while the associated C_D value decreases from 1 to 0.5.

In the case of the slow-drift motion of semi-submersible platforms or TLPs, the Keulegan-Carpenter numbers (referenced to the diameters of the columns) lie in between 0 and 5, and the β parameters are very high, in the range 10^6 to 10^7. The choice of the drag coefficient is not easy! Figure 4.22 shows that minimum C_D values

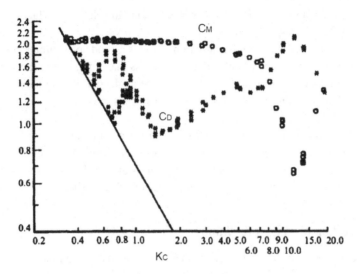

Figure 4.20 Circular cylinder in low K_C oscillatory flow. Coefficients C_M and C_D at $\beta = 1380$ (from Sarpkaya, 1986) (Cambridge University Press, reprinted with permission).

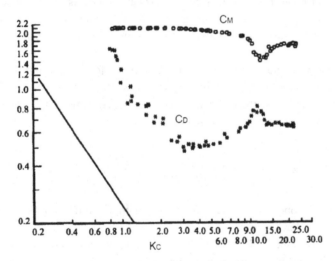

Figure 4.21 Circular cylinder in low K_C oscillatory flow. Coefficients C_M and C_D at $\beta = 11240$ (from Sarpkaya, 1986) (Cambridge University Press, reprinted with permission).

very close to zero have been obtained experimentally for β values around 10^5. In this figure, it can also be noticed that the drag coefficient is quite sensitive to roughness.

Sharp-Edged Cylinders

Semi-submersible platforms and TLPs usually have rectangular pontoons, with sharp or chamfered corners. The same issue arises as with the circular columns, that is the choice of the drag coefficient at low K_C values.

Sharp corners mean that the flow always separates, whatever the K_C value. In the square cylinder case, an asymptotic analysis, due to Bearman *et al.*, (1985), gives a

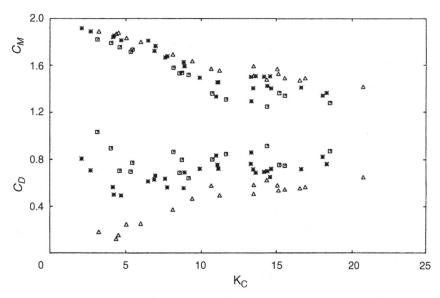

Figure 4.22 Vertical cylinder in regular waves. Inertia coefficient C_M and drag coefficient C_D at large Reynolds numbers, for smooth ($\triangle$) and rough ($\square$ $*$) surfaces (from Bearman, 1988).

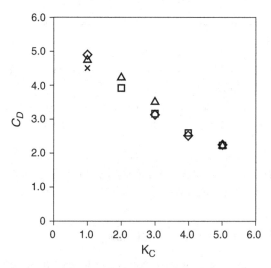

Figure 4.23 Square cylinder in low K_C oscillatory flow. Drag coefficient C_D at β numbers from $2.5 \cdot 10^4$ (triangles) to $7 \cdot 10^4$ (squares).

C_D value equal to 4.2 when K_C goes to zero. This is for an infinite β value, omitting the contribution of viscous friction. Figure 4.23 shows experimental values of the drag coefficient for a 35 cm × 35 cm square cylinder at β numbers around $5 \cdot 10^4$. At variance with the circular case, there is no minimum: the drag coefficient decreases monotonously when K_C increases, down to its steady flow value of 2. Inversely, the C_D increase from 2 to about 4 or 5 as $K_C \to 0$ is reasonably well predicted by wake models (Huse & Muren, 1987).

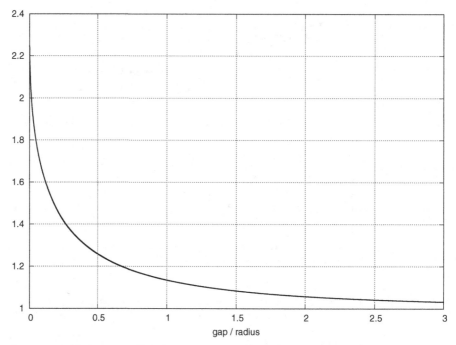

Figure 4.24 Added mass coefficient of a circular cylinder close to the sea floor.

When the edges are rounded or chamfered a similar behavior as for the circular cylinder at low K_C is found, with a drop of C_D.

In the case of a flat plate in normal flow, still according to Bearman *et al.*, (1985), the drag coefficient behaves as 7.8 $K_C^{-1/3}$ when K_C goes to zero, K_C being defined as $2\pi A/b$, with b the plate width.

Pontoons with sharp corners are thus preferable to pontoons with rounded or chamfered corners when one is concerned with damping the low-frequency motion of a TLP or semi. On the other hand, the mean current loads may be increased.

4.3.3   Cylinder Next to the Sea Floor in Sinusoidal Flow

In a perfect fluid theory, the proximity of the seabed leads to an increase of the added mass coefficient, which reaches the theoretical value of 2.29 for a cylinder placed on the bottom (i.e. $C_M = 1 + C_a = 3.29$). Figure 4.24 shows the added mass coefficient C_a vs the distance to the seabed divided by the diameter. (For this particular geometry the horizontal and vertical added mass coefficients are equal.)

Experimental investigations carried out on this configuration have confirmed this value, at least for low Keulegan–Carpenter numbers. They also showed a significant increase in the drag coefficient C_D when the cylinder approaches the bottom, values of the order of twice those in unbounded fluid being then reached.

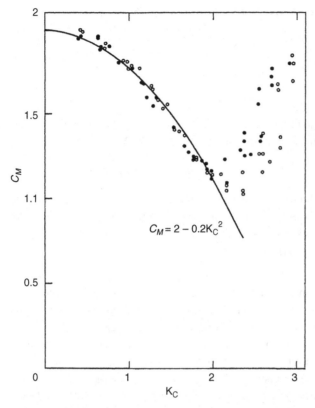

Figure 4.25 Submerged horizontal cylinder under regular waves. Measured inertia coefficient C_M at low K_C numbers (from Chaplin, 1984) (Cambridge University Press, reprinted with permission).

4.3.4   Cylinder in Orbital Flow

A submerged horizontal cylinder under regular waves is subjected to orbital flow:

$$U(t) = A\,\omega\,\sin\omega t \qquad\qquad V(t) = B\,\omega\,\cos\omega t \qquad (4.35)$$

A peculiar phenomenon, observed in wave tank experiments (Chaplin, 1984), is a rapid decrease of the inertia coefficient C_M at low K_C values (Figure 4.25). This decrease is due to viscosity: a steady streaming effect takes place around the cylinder, meaning a nonzero circulation. (Similar steady streaming occurs at the sea floor, at the outer end of the oscillatory boundary layer – see Longuet-Higgins, 1953). Through Magnus effect, a lift force appears that subtracts from the inertia component.

This phenomenon is not restricted to circular cylinders and is also observed for rectangular ones, whether the corners be sharp or rounded! Implications for full-scale bodies (pontoons of semi-submersible platforms) are unclear.

4.4   Circular Cylinder in Steady Plus Oscillatory Flow

4.4.1   In-Line Case

Let the incoming flow velocity be given as

$$U = U_0 + A\omega \sin \omega t \qquad\qquad V = 0 \qquad\qquad (4.36)$$

This applies to the practical case of a vertical cylinder in regular waves and current. It seems to make sense to continue applying the Morison equation:

$$F_x = \rho\, C_M\, S\, \dot{U} + \frac{1}{2}\rho\, C_D\, D\, U\, |U| \qquad\qquad (4.37)$$

and to adjust, through a best fit with the experimental measurements, the hydrodynamic coefficients C_M and C_D that depend on the nondimensional flow parameters, which are now one more than in the sinusoidal flow case, for instance:

$$\beta = \frac{D^2}{\nu T} \qquad\qquad K_C = \frac{2\pi A}{D} \qquad\qquad r = \frac{U_0}{A\omega} \qquad (4.38)$$

The third parameter r must be expected to play an important role. When r is larger than one, the flow never reverses; this is a situation somewhat similar to the current only case where the flow separates whatever the K_C value is. When r is less than one, the flow reverses periodically; if the Keulegan–Carpenter number is sufficiently low, the flow does not separate, even though the current velocity is not zero.

The actual problem is the information that one is looking for. It can be the maximum value of the hydrodynamic load, over one cycle. It can also be the time-averaged value, of interest for designing anchoring systems; or the load component opposing the fluctuating velocity, to derive the hydrodynamic damping. The problem is that one has only two coefficients to adjust.

Verley (1980) performed extensive experiments on an oscillating cylinder in a circulation channel. The cylinder was mounted as a pendulum, given an initial displacement and released. From the slowly varying mean offset, the first estimate of the drag coefficient was obtained. From the rate of decay a second value was obtained, characterizing the hydrodynamic load opposing the time-varying velocity. Figures 4.26 and 4.27 show the obtained drag coefficients, as presented in Verley's thesis; they are given vs the ratio A/D, that is $K_C/2\pi$, for different values of the reduced velocity $U_0 T/D = K_C\, r$, T being the period. The parameter β is in between 200 and 500.

Comparing the two figures, it can be observed that the curves do not look the same, except for $U_0 T/D > 20$ where they converge toward the horizontal line $C_D \simeq 1$. A high value of the reduced velocity $U_0 T/D$ means that at least one of the two parameters K_C or r is high, and that drag is dominating the loading. A conclusion from this comparison is that the Morison formula lacks physical ground for values of the reduced velocity less than 10 or 20.

In Figures 4.26 and 4.27 it may be noticed that the drag coefficients take negative values for some combinations of the parameters A/D and $U_0 T/D$. As for the fluctuating load component (Figure 4.26), C_D is negative for small values of A/D (less than 0.2) and for $U_0 T/D$ around 2 or 3. The period T is then about half the vortex shedding

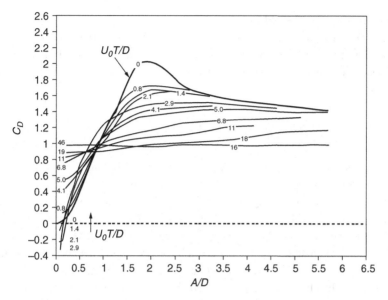

Figure 4.26 Circular cylinder in current + oscillatory flow. Drag coefficient derived from the load component opposing the fluctuating velocity (from Verley, 1980).

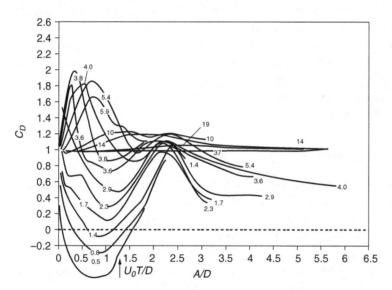

Figure 4.27 Circular cylinder in current + oscillatory flow. Drag coefficient derived from the time-averaged load (from Verley, 1980).

period and therefore equal to the resulting fluctuating in-line force. This suggests that an elastically mounted cylinder may naturally undergo in-line oscillations in current, which is actually the case (see paragraph 5.1.6).

In Figure 4.27 negative values of the mean C_D coefficient are also seen, for reduced velocities $U_0\,T/D$ less than one, and amplitudes A around one diameter. This is counterintuitive.

Independent Flow Field Formulation

An alternative to (4.37) is to separate the drag load into two components, one associated with the current, taking care of the time-averaged load, and the other one associated with the oscillating velocity, taking care of the load opposing the fluctuating velocity:

$$F_x = \rho \, C_M \, S \, \dot{U} + \frac{1}{2} \rho \, C_{D0} \, D \, U_0 \, |U_0| + \frac{1}{2} \rho \, C_{D1} \, D \, A^2 \, \omega^2 \, \sin \omega t \, |\sin \omega t| \quad (4.39)$$

This is known as the independent flow field formulation.

When they are derived from Verley's experiments, the C_{D1} and C_{D0} coefficients are hardly more convincing than the C_D coefficients shown in Figures 4.26 and 4.27: They also show a lot of scatter at the low values of the reduced velocity $U_0 T/D$. Neither formulation seems to be applicable in the general case.

The hydrodynamic coefficients given by Verley were obtained for quite low β values and they cannot be extrapolated to large scale. However, they show that neither the dependent flow field nor the independent flow field formulations apply at low values of the reduced velocity $U_0 T/D$. When K_C is low, the loading is dominated by inertia, and the choice of the formulation and of the associated C_D values will mostly matter for the time-averaged loads. From Figure 4.27 it can be seen that, when using the dependent flow field formulation, taking a C_D equal to about one will overpredict the mean load when K_C is less than about 5 or 6. In the case of slow-drift response, where the apparent current (the opposite of the slow-drift velocity) varies slowly in time, the damping effect will be overestimated. This is the reason why the independent flow field formulation is preferably used to model the viscous damping of slow-drift motions. This is further discussed in Chapter 7.

4.4.2   On-Bottom Stability of Pipelines and Wake Models

Another example of the deficiencies of the Morison formula is provided by studies of the stability of pipelines laid on the seabed, under the action of waves and current.

The principle of the stability analysis is to verify that the horizontal resistance of the seabed is greater than the maximum value of the horizontal component of the hydrodynamic force. This seabed resistance is generally expressed as the apparent weight multiplied by a friction coefficient. One therefore ends up with the following inequality, which must be verified at all times:

$$[P - F_z(t)] \, f_c > k \, F_x(t) \quad (4.40)$$

with P the weight in water, $F_z(t)$ the lift force, f_c the friction coefficient, k a safety coefficient, and $F_x(t)$ the horizontal load.

When the Morison equation is applied, the horizontal load is given by

$$F_x(t) = \frac{1}{2} \rho \, C_D \, D \, U \, |U| + \rho \, C_M \, \pi \, \frac{D^2}{4} \, \dot{U} \quad (4.41)$$

with, in collinear regular waves plus current

$$U(t) = U_0 + \frac{A\,\omega}{\sinh kh}\,\cos\omega t \qquad (4.42)$$

The vertical load is assumed to consist only in the lift force, given by

$$F_z(t) = \frac{1}{2}\,\rho\,C_L\,D\,U^2(t) \qquad (4.43)$$

It has already been mentioned that, in perfect fluid, the theoretical C_M and C_L values are, respectively, 3.29 and 4.49, and that experimental measurements, in purely sinusoidal flow, suggest C_D coefficients approximately twice their values in unbounded fluid. With regard to the lift coefficient in sinusoidal flow, Sarpkaya (1977) obtains experimental values even higher than 4.49, at low Keulegan–Carpenter numbers.

It has been realized that taking that high drag and lift coefficients in mixed wave + current flow leads to overestimating the hydrodynamic loads and, consequently, to oversizing the coatings of pipelines (to make them heavier). An example of discrepancies between forces predicted numerically and measured experimentally (in irregular waves and current) is provided by Figure 4.28, taken from Verley *et al.* (1987): the negative peaks of the horizontal force are overestimated, as are the peaks of the lift force.

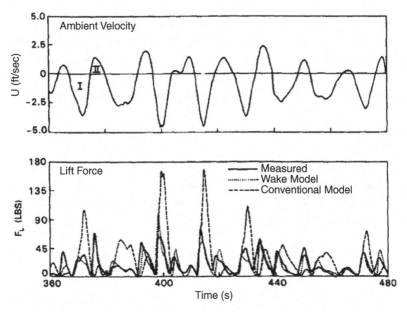

Figure 4.28 Comparison of measured and calculated lift force in irregular waves and steady current (from Lambrakos *et al.*, 1987).

Therefore one finds, in recommended procedures for the calculation of pipe stability, much lower hydrodynamic coefficients, for example $C_L = 0.9$ and $C_D = 1.0$.[5]

Some researchers have attempted, with some success, to improve the formulation of the hydrodynamic loads by accounting for the past history of the flow, via so-called **wake models**. These methods assume that a wake is formed at each half-cycle, assuming that the K_C number is sufficiently high. They are based on the physical observation that the fluid velocity encountered by the cylinder during each half-cycle is affected by the wakes emitted during the previous half-cycles. It has already been mentioned that this wake effect is at the origin of the increase in the drag coefficient C_D, in oscillatory flow, compared to its value in current only.

In paragraph 4.2.6, devoted to proximity effects in current alone, the Huse method has been presented, which allows, by considerations and summations on the wakes emitted by the different cylinders, to account for shielding effects. The same idea is applied here, the additional difficulty being that the incident flow varies in time and reverses. The spatio-temporal monitoring of the wakes emitted therefore raises some difficulties, resolved in different ways by the researchers who addressed the problem. Most often only the wake emitted during the half-cycle preceding that in progress is taken into account. The reader is referred to the works in reference (Lambrakos *et al.*, 1987; Verley *et al.*, 1987 – for cases of irregular waves + current – Huse & Muren, 1987; Faltinsen, 1990, pp 244–248 – for the simpler cases of sinusoidal flow without current). Figures 4.28 and 4.29, taken from Lambrakos *et al.* (1987), show favorable agreement with experimental results.

4.4.3   Perpendicular Case

This is a case of practical importance to address the problems of Vortex-Induced Vibrations (VIVs) and Vortex-Induced Motions (VIMs), considered in the following Chapter. A circular cylinder undergoes forced periodic oscillations, of amplitude A and frequency ω, in the direction perpendicular to a uniform current of velocity U_0:

$$U = U_0 \qquad\qquad Y(t) = A \cos \omega t \qquad\qquad (4.44)$$

Numerous experimental campaigns have been performed to determine the added mass and drag coefficients of the transverse motion, vs the flow non-dimensional parameters. As in the in-line case, besides the Reynolds (or β) number, there are two more parameters. For instance:

$$K_C = 2\pi \frac{A}{D} \qquad\qquad r = \frac{U_0}{A\omega} \qquad\qquad (4.45)$$

---

[5] DNVGL (2017) provides tables of drag and lift coefficients vs the Keulegan–Carpenter number and the ratio of the steady to oscillatory current velocity.

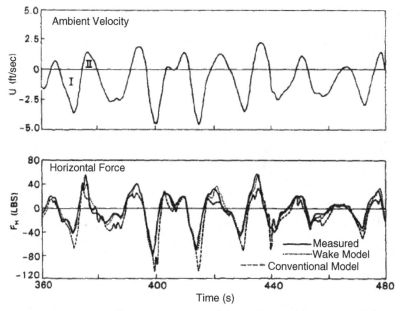

Figure 4.29 Comparison of measured and calculated horizontal force in irregular waves and steady current (from Lambrakos *et al.*, 1987).

In place of the velocity ratio $U_0/A\omega$, it is customary to use the **reduced velocity** U_R, defined by

$$U_R = \frac{U_0 T}{D} = r \, K_C \qquad (4.46)$$

It is the distance traveled by the free stream over one period, divided by the diameter.

The added mass and drag coefficients are obtained through Fourier analysis (or best fit) of the transverse hydrodynamic load, written as:

$$F_y(t) = -\rho \, C_{ay} \, S \, \ddot{Y}(t) - \frac{1}{2}\rho \, C_{Dy} \, D \, \dot{Y}(t) \, |\dot{Y}(t)| \qquad (4.47)$$

$$F_y(t) = \rho \, C_{ay} \, S \, A \, \omega^2 \, \cos \omega t + \frac{1}{2} \, \rho \, C_{Dy} \, D \, A^2 \, \omega^2 \, \sin \omega t \, |\sin \omega t| \qquad (4.48)$$

Figures 4.30 and 4.31 show the coefficients C_{ay} and C_{Dy} obtained experimentally by Sarpkaya (1978), vs the reduced velocity U_R for different A/D ratios. The Reynolds number is in between 5000 and 25000: the flow is subcritical. It can be seen that the added mass coefficient, from its reference value 1 at low U_R values, first increases to about 2, then drops down sharply to negative values at $U_R \sim 5$, and remains negative. As for the drag coefficient C_{Dy}, it also takes negative values for some combinations of U_R and A/D, in particular, for reduced velocities in between 5 and 6.

In the subcritical regime, the Strouhal number is equal to 0.20. A reduced velocity equal to 5 means that the oscillation frequency is equal to the vortex shedding frequency for a fixed cylinder.

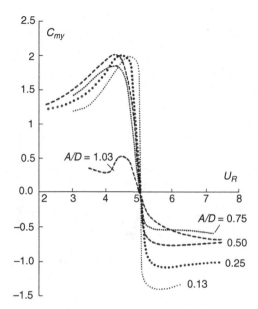

Figure 4.30 Circular cylinder oscillating transversally to a current. Added mass coefficient C_{ay} vs the reduced velocity U_R, for different ratios A/D (from Sarpkaya, 1978) (with permission from ASCE).

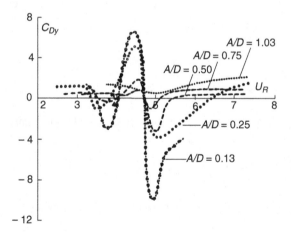

Figure 4.31 Circular cylinder oscillating transversally to a current. Drag coefficient C_{Dy} vs the reduced velocity U_R, for different ratios A/D (from Sarpkaya, 1978) (with permission from ASCE).

When the cylinder is undergoing forced motion, the vortex shedding frequency adjusts, over some U_R range around 5, to the oscillation frequency. This is known as **lock-in**. The intensity of the shed vortices is increased and the phase shift of the lift force, with respect to the imposed motion, varies rapidly and leads to C_{ay} and C_{Dy} varying quickly with U_R. When the ratio A/D increases, the vortex shedding becomes disorganized, with the result that the drag coefficient becomes positive beyond some A/D value.

That the drag coefficient is negative for some combinations of U_R and A/D suggests that, the cylinder extracting energy from the fluid, unstable behaviors may be observed for an elastically supported cylinder. These Vortex-Induced Vibrations are studied in Section 5.1.

Similar forced motion experiments have been performed by many other investigators (Gopalkrishnan, 1993; Morse & Williamson, 2009; etc.), also in the subcritical regime, but for more extensive and more refined combinations of U_R and A/D. For instance Morse & Williamson present maps of the hydrodynamic coefficients from (two times) 5,680 individual tests, with A/D from 0 to 1.6 and U_R from 2 to 16, at Reynolds numbers 4,000 and 12,000.

It is to be noted that, in these papers, the hydrodynamic coefficients are defined differently from (4.48), from the usual expression of the lift force on a fixed cylinder:

$$F_L = \frac{1}{2} \rho\, C_L\, D\, U_0^2\, \sin 2\pi\, f_0\, t \qquad (4.49)$$

The transverse load is then written as

$$F_y = \frac{1}{2} \rho\, D\, U_0^2\, (C_{La}\, \cos \omega t + C_{LD}\, \sin \omega t) \qquad (4.50)$$

These coefficients relate to the previous ones via

$$C_{La} = \pi^2\, C_{ay}\, K_C\, U_R^{-2} \qquad\qquad C_{LD} = \frac{8}{3\,\pi}\, C_{Dy}\, K_C^2\, U_R^{-2} \qquad (4.51)$$

The coefficients given in Morse & Williamson (2009), for instance, are C_{ay} and $-C_{LD}$.

There is a need for similar test results at higher Reynolds numbers. It has been mentioned that, in the critical regime (Reynolds numbers higher than $2 \cdot 10^5$ and lower than $6 \cdot 10^5$), the lift force is very erratic, with no dominant frequency. On the other hand, forced transverse motion has been acknowledged to synchronize the vortex shedding along the length of the cylinder. For risers used in the offshore oil industry, the Reynolds numbers are typically in the critical regime, and their fatigue under VIVs is a major issue. Large-scale experiments have been carried out by oil companies but not all the results have been published. For instance, Exxon performed extensive series of tests in the high-speed tank of the David Taylor Model Basin, investigating the effects of Reynolds numbers and roughness, but they published only one incomplete (and somewhat confusing) paper (Ding et al., 2004).

Tests performed within a French JIP (Cinello et al., 2013) with smooth cylinders of large diameters (from 150 mm up to 350 mm), at Reynolds numbers up to $2.3 \cdot 10^5$ (in a current of about 3% turbulence intensity), have given negative values of the C_{Dy} coefficients over an even wider U_R range (as compared to the subcritical case). Also noteworthy is that, at the highest Reynolds number reached, the added mass coefficient behaves differently, slowly decreasing from about one to nearly zero (remaining positive) as U_R increases up to 15. On the other hand, Oakley & Spencer (2004), at even higher Reynolds numbers (up to 10^6), find negative added mass coefficients at high U_R values. Roughness and turbulence intensity have also been found to play an important role in the critical regime, so in the end, the literature is somewhat confusing and incomplete.

Increase of the Mean In-line Force

An important effect of the cross-flow motion is that the mean in-line force increases, compared to the fixed cylinder case. An intuitive interpretation is that the effective frontal area increases. Some formulations have been proposed for the increase in the drag coefficient, such as

$$C_D(A/D) = C_D(0)\left(1 + 2\frac{A}{D}\right)$$
(4.52)

or (Skop *et al.*, 1977):

$$C_D(A/D) = C_D(0)\left[1 + 1.16\,(W_R - 1)^{0.65}\right]$$
(4.53)

where W_R is defined as

$$W_R = \left(1 + 2\frac{A}{D}\right)\frac{f}{f_0}$$
(4.54)

f being the oscillation frequency and f_0 the Strouhal frequency.

Figure 4.32, taken from Cinello *et al.* (2013), shows a comparison between the formula (4.53) and experimental values obtained at a Reynolds number of 10^5, with a fair agreement.

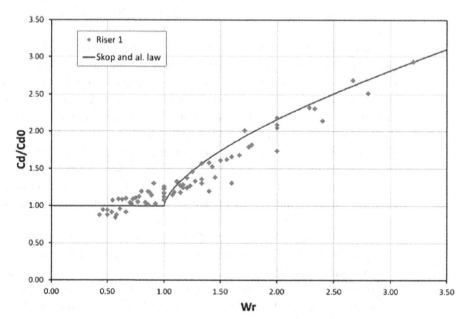

Figure 4.32 Circular cylinder oscillating in the cross-flow direction to a current. Increase of the mean in-line drag coefficient, from (4.53), and from experiments (from Cinello *et al.*, 2013).

4.5   Generalized Morison Formula

The preceding sections have enabled the reader to realize that the Morison equation expresses physical reality very imperfectly. This is a point that one quickly tends to forget in practical applications.

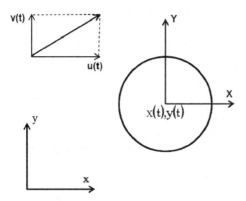

Figure 4.33 Geometry.

For tubular elements such as jacket bars or risers, the Morison formula is applied via a strip theory, where the axial component of the flow is neglected.

The first step is to determine the components (u, v, w) of the (local) velocity of the incident flow (induced by waves and current), the associated Eulerian acceleration $(\dot{u}, \dot{v}, \dot{w})$, and, likewise, the components $(\dot{x}, \dot{y}, \dot{z})$ of the (local) velocity of the tubular element, and its acceleration $(\ddot{x}, \ddot{y}, \ddot{z})$. One then projects these vectors on the axes XYZ of a local coordinate system linked to the bar element, Z coinciding with the axial direction. The components of the incident velocity are then (U, V, W), those of the own velocity of the bar element $(\dot{X}, \dot{Y}, \dot{Z})$, etc. See Figure 4.33.

The load on the bar element are then obtained as

$$\begin{pmatrix} dF_X \\ dF_Y \end{pmatrix} = \frac{1}{2}\rho C_D\, D \begin{pmatrix} U - \dot{X} \\ V - \dot{Y} \end{pmatrix} \sqrt{(U - \dot{X})^2 + (V - \dot{Y})^2}\; dL$$

$$+ \left\{ \rho\,(1 + C_a)\,\pi\,\frac{D^2}{4} \begin{pmatrix} \dot{U} \\ \dot{V} \end{pmatrix} - \rho\,C_a\,\pi\,\frac{D^2}{4} \begin{pmatrix} \ddot{X} \\ \ddot{Y} \end{pmatrix} \right\}\, dL \qquad (4.55)$$

One may then go back to the fixed reference to express the components (df_x, df_y, df_z) of the hydrodynamic load.

Equation (4.55) assumes that the drag component is in line with the instantaneous relative flow velocity. This is justified when the K_C number is high and/or the current velocity is dominant.

Linearization of the Drag Component

The drag term, in $U\,|U|$, does not lend itself well to certain types of computations, for example in the frequency domain.

In the case where U is periodic: $U = u \sin \omega t$, one may apply

$$\sin \omega t \,|\sin \omega t| = \frac{8}{3\pi} \sin \omega t - \sum_{n=1}^{\infty} \frac{8}{(2n-1)(2n+1)(2n+3)\,\pi} \sin(2n+1)\omega t \quad (4.56)$$

and retain only the first term.

Therefore, when incident velocity $U(t)$ and response $X(t)$ are written as

$$U(t) = \Re\left\{u\, e^{-i\omega t}\right\} \qquad\qquad X(t) = \Re\left\{x\, e^{-i\omega t}\right\} \quad (4.57)$$

where u and x are complex, the product $(U - \dot{X})\,|U - \dot{X}|$ is written as

$$(U - \dot{X})\,|U - \dot{X}| \simeq \frac{8}{3\pi} \|u + i\omega x\|\, \Re\left\{(u + i\omega x)\, e^{-i\omega t}\right\} \quad (4.58)$$

where $\|\ \|$ designates the modulus of the complex number.

When the response $X(t)$ is unknown, an iterative procedure may be followed, by writing

$$(U - \dot{X})\,|U - \dot{X}|^{(n)} \simeq \frac{8}{3\pi} \|u + i\omega\, x^{(n-1)}\|\, \Re\left\{(u + i\omega\, x^{(n)})\, e^{-i\omega t}\right\} \quad (4.59)$$

yielding, at each iteration, a linear system. Convergence is usually obtained with a few iterations.

When $U(t)$ is a random variable, of zero mean value, an equivalent linear form may be looked for, in the form

$$U\,|U| \simeq \lambda\, \sigma_U\, U \quad (4.60)$$

with σ_U the standard deviation.

When $U(t)$ follows a Gaussian distribution, by minimizing the difference between both sides, one obtains $\lambda = \sqrt{8/\pi}$. This is also the value that ensures the same energy dissipation.

With this linearization, known as **stochastic linearization**, it is possible to work in the frequency domain and to calculate the spectrum of the loads applied on a bar element or on a complete structure. If the motion is to be determined, one can follow an iterative scheme equivalent to that proposed above

$$(U - \dot{X})\,|U - \dot{X}|^{(n)} \simeq \sqrt{\frac{8}{\pi}}\, \sigma_{U-\dot{X}}^{(n-1)}\, (U - \dot{X}^{(n)}) \quad (4.61)$$

the standard deviation $\sigma_{U-\dot{X}}^{(n-1)}$ being obtained from the wave spectrum and from the RAOs of the process $U(t) - \dot{X}^{(n-1)}(t)$.

One should be aware, however, that applying this type of approximation is equivalent to representing a non-Gaussian signal $(U\,|U|)$ by a Gaussian one. As shown in Figure 4.34, the extrema of the two processes are quite different. This procedure is therefore not applicable, for example, to predict extreme loads on a fixed cylinder. It is more justifiable when calculating the wave response, where the drag term accounts for the dissipation of energy which comes to limit the resonance.

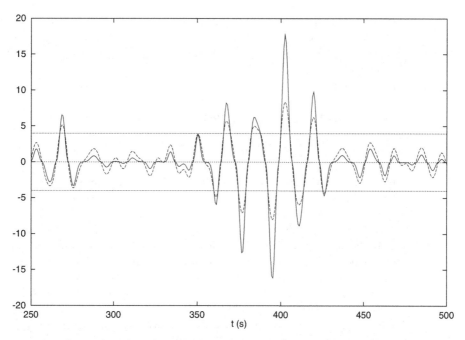

Figure 4.34 Comparison, for a Gaussian variable $U(t)$, of the processes $U|U|$ (solid line) and $\sqrt{8/\pi}\,\sigma_U\,U$ (dash line). The two dotted horizontal lines show $\pm\sqrt{8/\pi}\,\sigma_U$.

4.6 References

ACHENBACH E. 1971. Influence of surface roughness on the cross-flow around a circular cylinder, *J. Fluid Mech.*, **46**, 321–335.

ACHENBACH E., HEINECKE E. 1981. On vortex shedding from smooth and rough cylinders in the range of Reynolds numbers 6 x 10^3 to 5 x 10^6, *J. Fluid Mech.*, **109**, 239–251.

BEARMAN P.W. 1988. Wave loading experiments on circular cylinders at large scale, *Proc. 5th Int. Conf. on Behaviour of Offshore Structures, BOSS'88*.

BEARMAN P.W., DOWNIE M.J., GRAHAM J.M.R., OBASAJU E.D. 1985. Forces on cylinders in viscous oscillatory flow at low Keulegan-Carpenter numbers, *J. Fluid Mech.*, **154**, 337–356.

BLEVINS R.D. 1990. *Flow-Induced Vibration*, 2nd edition. Krieger Publishing Company.

CANTWELL B., COLES D. 1983. An experimental study of entrainment and transport in the turbulent near wake of a circular cylinder, *J. Fluid Mech.*, **136**, 321–374.

CHAPLIN J.R. 1984. Non-linear forces on a horizontal cylinder beneath waves, *J. Fluid Mech.*, **147**, 449–464.

CINELLO A., PÉTRIÉ F., RIPPOL T., MOLIN B., LE CUNFF C. 2013. Experimental investigations of VIV at high Reynolds numbers for smooth circular cylinders in single and tandem arrangements, in *Proc. ASME 2013 32nd International Conference on Ocean, Offshore and Arctic Engineering*, paper OMAE2013-10638.

CUFFE P.D., FINN L.D., LAMBRAKOS K.F. 1990. Compliant tower loading and response measurements, *Proc. 22nd Offshore Techn. Conf.*, paper 6313.

DNVGL 2017. On-bottom stability design of submarine pipelines, Recommended Practice DNVGL-RP-F109.

DNVGL 2017. Environmental conditions and environmental loads, Recommended Practice DNV-RP-C205.

EAMES M.C. 1968. Steady state theory of towing cables, *Trans. Royal Inst. Naval Architects*, **10**.

FALTINSEN O.M. 1990. *Sea Loads on Ships and Offshore Structures*. Cambridge University Press.

FALTINSEN O.M., SORTLAND B. 1987. Slow drift eddy making damping of a ship, *Applied Ocean Res.*, **9**, 37–46.

GOPALKRISHNAN R. 1993. Vortex-induced forces on oscillating bluff cylinders. PhD thesis, Massachusetts Institute of Technology, Cambridge, MA.

HUSE E. 1991. Current force on individual elements of risers arrays, Marintek Report 513902.

HUSE E., MUREN P. 1987. Drag in oscillatory flow interpreted from wake consideration, *Proc. 19th Offshore Techn. Conf.*, paper 5370.

KEULEGAN G.H., CARPENTER L.H. 1958. Forces on cylinders and plates in an oscillating fluid, *Journal of Research of the National Bureau of Standards*, **60**, research paper 2857.

LAMBRAKOS K.F., CHAO J.C., BECKMANN H., BRANNON H.R. 1987. Wake model of hydrodynamic forces on pipelines, *Ocean Engineering*, **14**, 117–136.

LONGUET-HIGGINS M.S. 1953. Mass transport in water waves, *Phil. Trans. Royal Soc. A*, **245**, Issue 903, 535–581.

MOLIN B., ETIENNE S. 2000. On viscous forces on non-circular cylinders in low KC oscillatory flows, *Eur. J. Mech. B-Fluids*, **19**, 453–457.

MORISON J.R., O'BRIEN M.P., JOHNSON J.W., SCHAAF S.A. 1950. The force exerted by surface waves on piles, *Petrol. Trans.*, **189**, 149–154.

MORSE T.L., WILLIAMSON C.H.K. 2009. Prediction of vortex-induced vibration response by employing controlled motion, *J. Fluid Mech.*, **634**, 5–39.

NORBERG C. 2000. Flow around a circular cylinder: aspects of fluctuating lift, *J. Fluids and Structures*, **15**, 459–469.

OAKLEY O.H., SPENCER D. 2004. Deepstar VIV experiments with a cylinder at high Reynolds numbers, in *Proc. Deep Offshore Technology Conf.*, DOT04.

PARKINSON G.V. 1974. Mathematical models of flow-induced vibrations of bluff bodies, in *Flow Induced Structural Vibrations*, ed. E. Naudascher, 81–127, Berlin: Springer.

ROSHKO A. 1961. Experiments on the flow past a circular cylinder at very high Reynolds number, *J. Fluid Mech.*, **10**, 345–356.

SARPKAYA T. 1977. In-line and transverse forces on cylinders near a wall in oscillatory flow at high Reynolds numbers, *Proc. 9th Offshore Techn. Conf.*, paper 2898.

SARPKAYA T. 1978. Fluid forces on oscillating cylinders, *J. of Waterway, Port, Coastal and Ocean Division, ASCE*, **104**, 275–290.

SARPKAYA T. 1986. Force on a circular cylinder in viscous oscillatory flow at low Keulegan-Carpenter numbers, *J. Fluid Mech.*, **165**, 61–71.

SARPKAYA T. 2010. *Wave Forces on Offshore Structures*, Cambridge University Press.

SCHEWE G. 1983. On the force fluctuations acting on a circular cylinder in cross flow from subcritical up to transcritical Reynolds numbers, *J. Fluid Mech.*, **133**, 265–285.

SCHLICHTING H. 1979. *Boundary Layer Theory*. McGraw-Hill Book Company.

SCRUTON C., ROGERS E.W.E. 1971. Steady and unsteady wind loading of buildings and structures, *Phil. Trans. Roy. Soc. Lond., A*, **269**, 353–383.

STOKES G.G. 1851. On the effect of the internal friction of fluids on the motion of pendulums, *T. Camb. Phil. Soc.*, **9**, 8–106.

VERLEY R.L.P. 1980. Oscillations of cylinders in waves and currents, PhD thesis, Loughborough University.

VERLEY R.L.P., LAMBRAKOS K.F., REED K. 1987. Prediction of hydrodynamic forces on seabed pipelines, *Proc. 19th Offshore Techn. Conf.*, paper 5503.

ZDRAVKOVICH M.M. 1977. Review of flow interference between two cylinders in various arrangements, *ASME Journal of Fluids Engineering*, **99**, 618–633.

ZDRAVKOVICH M.M. 1985. Flow induced oscillations of two interfering cylinders, *J. Sound & Vibration*, **101**, 511–521.

5   Flow-Induced Instabilities

5.1   Vortex-Induced Vibrations

The experimental results presented in Section 4.4.3 suggest that an elastically supported cylinder, in a current, can naturally oscillate in the cross-flow direction if its natural frequency is close to the vortex-shedding frequency. This is actually what happens. Examples are numerous in offshore, from towing cables to risers to Spar platforms. In the case of large rigid bodies, such as Spars or semis, the coinage VIM, for Vortex-Induced Motion, is frequently used.

Vortex-induced vibrations (VIVs) are only one kind of flow-induced instabilities. There are many others, in particular, in the case of lifting bodies such as wings of planes or bridge decks, known as "flutter" or "galloping," well known in aerodynamics. In the case of cylinder arrays, there are so-called wake-induced oscillations (WIOs), when lee cylinders move in and out of the wakes of upstream cylinders. Galloping and flutter are considered in Section 5.2, and wake-induced oscillations in Section 5.3.

Isolated circular cylinders are susceptible to only one kind of instability, induced by vortex shedding. Isolated noncircular cylinders are also prone to VIVs, but they may moreover undergo galloping.

Description of the Phenomenon

Consider the simple dynamic system consisting in a circular cylinder of mass M, in between two springs of combined stiffness K, in a uniform current of velocity U. Depending on the experimental setup, the in-line response of the cylinder may be restrained, or allowed with the same stiffness K or with a different stiffness. Figure 5.1 shows a pendulum setup used to study VIVs at high Reynolds numbers (Molin *et al.* 2010; Cinello *et al.*, 2013), where the in-line and cross-flow stiffnesses are the same.

The cross-flow motion obeys the simple massspring equation

$$(M + M_a)\ddot{Y} + B\dot{Y} + KY = F_Y(t) \tag{5.1}$$

with M_a the added mass, B some damping (the non-hydrodynamic part, due to mechanical friction in the experimental setup for instance), K the stiffness and F_Y the transverse hydrodynamic load.

In still water (no incoming current), the natural period of the cross-flow motion is given by

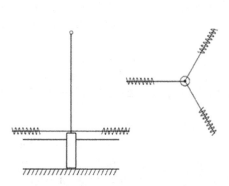 

Figure 5.1 Pendulum setup (from Molin *et al.* 2010).

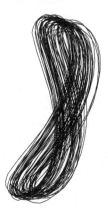

Figure 5.2 Example of trajectory under VIV.

$$T_n = 2\pi \sqrt{\frac{M + M_a}{K}} = \frac{1}{f_n}$$ (5.2)

with $M_a = \rho \pi D^2/4$ the added mass per unit length.

When the current is turned on, spontaneously, the cylinder undergoes nearly periodic cross-flow motion, if the reduced velocity $U_{Rn} = U T_n/D$ is comprised within some interval, typically [5 8] in air, but usually much greater in water. When the in-line motion is allowed, trajectories take the shape of elongated eights or bananas (Figure 5.2).

No matter how low the mechanical damping B, the amplitude A of the cross-flow motion hardly exceeds one or one and a half diameter. The reason is that, beyond some A/D value, the vortex shedding process gets disorganized and the C_{Dy} coefficient becomes positive again (see Figure 4.31). VIVs are said to be self-limited, at variance with other hydroelastic instabilities, such as galloping, which are unbounded.

One reason why VIVs take place in some U_{Rn} range, and not just at the inverse of the Strouhal number, is that, as observed previously, the C_{Dy} coefficient is negative over some U_R interval where the vortex shedding frequency locks to the oscillation frequency. Another reason, which is specific to VIVs occurring in water, is that the added mass varies quite a lot with the reduced velocity and amplitude (see Figure 4.30). In air this is of no effect since the added mass is negligible compared to the actual mass, but in water it is appreciable. An important parameter therefore coming into play is the mass ratio m^*, usually defined as $M/(\rho \pi D^2/4)$ (M being the mass per unit length). Typical values of the mass ratio for risers are between 1.5 and 2. For cylindrical spar platforms the mass ratio is just equal to one; it is less than one in the case of buoyant cylinders, for instance, the cylindrical cans tensioning the riser-towers at Girassol.

This leads us to distinguish the actual reduced velocity $U_R = UT/D$ based on the actual oscillation period from the reduced velocity U_{Rn} based on the natural period in still water. If one considers that, in air, VIVs take place in the U_R range [5 8] and that the added mass coefficient varies from a maximum value of 2 for $U_R \simeq 5$ down to $-1/2$ at the end of the locking region (from Figure 4.30), one may deduce that in water VIVs take place in the following U_{Rn} range

$$5 \sqrt{\frac{m^* + 1}{m^* + 2}} \leq U_{Rn} \leq 8 \sqrt{\frac{m^* + 1}{m^* - 1/2}} \tag{5.3}$$

There have in fact been observations of VIVs up to U_{Rn} values around 15 or 20.

Figure 5.3, taken from Khalak & Williamson (1997), shows the non-dimensional cross-flow amplitude A/D vs the reduced velocity U_{Rn} at a Reynolds number of about 6000 and for a mass ratio equal to 2.4. It is compared with old wind tunnel results by Feng (1968) at a similar Reynolds number. It can be checked that, in water, the U_{Rn} range where VIVs take place is much larger than in air, and agrees fairly well with equation (5.3).

It can be seen that the VIV amplitude is much greater than obtained by Feng, in air, presumably due to additional structural damping in Feng's experiments. Noteworthy are jumps in the VIV amplitudes, due to different vortex-shedding regimes taking place; and also overlaps due to hysteresis effects (different responses being obtained depending on whether the current velocity is gradually increased or decreased).

The bottom plot in Figure 5.3 shows that, in the high U_{Rn} range, the oscillation frequency locks to a value about 30%–40% higher than in still water, confirming that the added mass has become negative, close to $-1/2$ (referring to equation (5.3), $\sqrt{(2.4 + 1)/(2.4 - 0.5)} \simeq 1.33$).

Figure 5.4 shows the response amplitude of different cylinders, all 31.5 cm in diameter, tested at BGO-FIRST with the setup shown in Figure 5.1. What is plotted is the standard deviation of the cross-flow motion normalized by the diameter. The mass ratio is about 1.5 and the Reynolds number ranges from $5 \cdot 10^4$ up to $3 \cdot 10^5$. Two of the cylinders are rough, with roughnesses k/D equal to 0.01 and 0.03. In the "smooth3" case, the in-line motion is hindered. Compared to the experimental results of Khalak & Williamson shown in Figure 5.3, it can be observed that the cross-flow amplitude is about 50% higher, and that no jumps in amplitude seem to take place as the reduced velocity is varied.

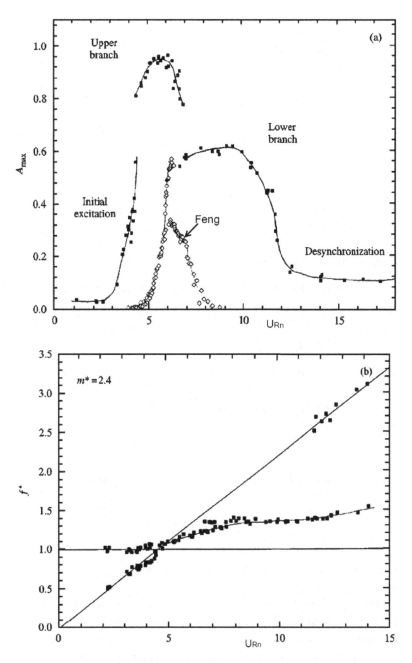

Figure 5.3 Response amplitude A/D (top) and frequency (normalized by the still water frequency) at low mass ratio and damping, compared to Feng's results (diamond symbols). Reynolds number around 6,000 (from Khalak & Williamson, 1997).

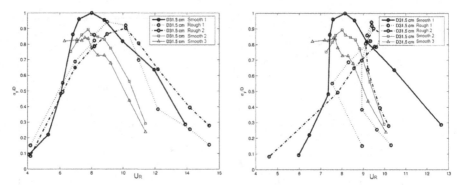

Figure 5.4 Response amplitude σ_Y/D of different cylinders of diameter 0.315 m vs reduced velocity based on still water natural period (left) or oscillation period under VIV (right) (from Molin *et al.* 2010).

5.1.1   Calculating the VIV Amplitude

Still considering the two-dimensional case of an elastically supported cylinder, its transverse equation of motion can schematically be written as

$$(M + M_a)\ddot{Y} + B\dot{Y} + KY = \frac{1}{2}\rho\, C_L(U_R, A/D)\, U^2\, D\, \sin 2\pi\, f_0 t \qquad (5.4)$$

with f_0 the shedding frequency, B the structural damping and C_L the lift force coefficient.

At resonance the response amplitude is given by

$$A = \frac{1/2\,\rho\, C_L\, U^2\, D}{2\,\pi\, f_0\, B} \qquad (5.5)$$

or, introducing the damping ratio $\zeta = B/B_C$ referred to the critical damping $B_C = 4\pi\, f_0\,(M + M_a)$:

$$\frac{A}{D} = \frac{C_L(U_R, A/D)}{8\,\pi^2\, S_t^2\, K_S} \qquad (5.6)$$

where $S_t = f_0 D/U$ is the Strouhal number and $K_S = 2\zeta\,(M + M_a)/(\rho D^2)$ is a stability parameter known as the Scruton number.

The dependence of the lift coefficient C_L on the reduced velocity U_R and on the reduced amplitude A/D must be taken into account in the resolution of this equation. For this it is necessary to resort to information derived from experimentation (CFD being an alternative) such as that described in paragraph 4.4.3.

When the reduced velocity is close to 5 (and in the subcritical regime), the following simple formula, obtained by regression, has been proposed (Blevins, 1977):

$$C_L(A/D) = a + b\,\frac{A}{D} + c\left(\frac{A}{D}\right)^2 \qquad (5.7)$$

with $a = 0.35$, $b = 0.60$, $c = -0.93$.

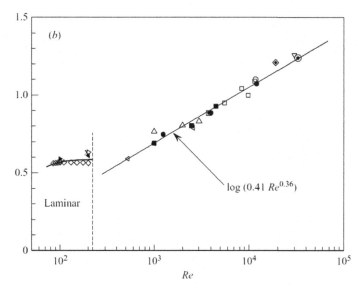

Figure 5.5 Effect of Reynolds number on maximum A/D (from Govardhan & Williamson, 2006) (Cambridge University Press, reprinted with permission).

The reduced amplitude A/D can then be obtained by solving a second-degree equation.

This procedure applies well to cylinders undergoing VIV in air, where the structural damping force is comparable to the wind excitation. In water the stability parameter K_S is usually very low and the response amplitude closely satisfies $C_L(A/D) \simeq 0$, that is, $A/D \simeq 1$ with the polynomial expression given above.

Blevins formula (5.7) makes no reference to the Reynolds number. In fact, in the subcritical regime, the VIV amplitude steadily increases with the Reynolds number. Based on their own experimental results and other data from literature, Govardhan & Williamson (2006) have found that the nondimensional VIV amplitude A/D (at the apex of the curve) is reasonably well predicted by

$$\frac{A}{D}\text{max} = \left(1 - 1.12\,\alpha + 0.30\,\alpha^2\right)\,\log_{10}\left(0.41\,\text{Re}^{0.36}\right) \tag{5.8}$$

where α is another stability parameter $\alpha = (1 + m^*)\,\zeta$.

Figure 5.5 shows that the proposed equation (5.8) (with $\alpha = 0$) fits reasonably well with the measurements. Figure 5.6 shows the effect of the stability parameter α, usually close to zero in water.

One issue is whether the VIV amplitude depends on the stiffness of the in-line response. In most experiments the in-line motion is hindered while in reality, for risers for instance, the in-line and cross-flow stiffnesses are the same.

Another issue is related to the aspect ratio, that is the cylinder length compared to its diameter. Beyond a certain length, it is known that the eddy shedding is not synchronized over the whole of the cylinder wall. This is for a fixed cylinder, where it is known that the correlation length is only a few diameters in the subcritical regime

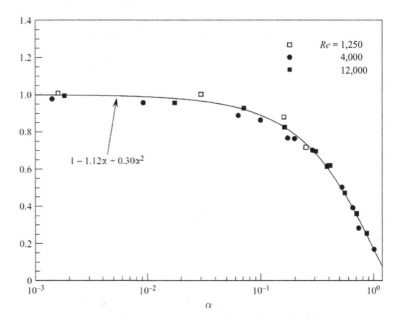

Figure 5.6 Effect of stability parameter α on maximum A/D (from Govardhan & Williamson, 2006) (Cambridge University Press, reprinted with permission).

and much less in the critical regime. As the cylinder undergoes VIVs, the vortex shedding tends to synchronize, increasing the fluctuating lift load and therefore the VIV response.

5.1.2   Wake Oscillator Models

In 1970 Hartlen & Currie proposed modeling the fluctuating lift force as a Van der Pol oscillator. This idea has been revisited by many investigators (e.g., see Skop & Griffin, 1973). Facchinetti *et al.* (2004) have proposed a model that is now implemented in some software dedicated to the VIV analysis of risers.

The model is as follows:

$$(M + M_a)\ddot{Y} + \left(B + \frac{1}{2}\rho\,U\,D\,\mathrm{S_t}\,C_D\right)\dot{Y} + K\,Y = \frac{1}{4}\rho\,U^2\,D\,L\,C_L^0\,q \qquad (5.9)$$

$$\ddot{q} + 2\pi\,\mathrm{S_t}\frac{U}{D}\,\varepsilon\,(q^2 - 1)\,\dot{q} + \left(2\pi\,\mathrm{S_t}\frac{U}{D}\right)^2 q = \frac{A}{D}\ddot{Y} \qquad (5.10)$$

Here $q(t)$ is a dimensionless lift coefficient $q(t) = 2\,C_L(t)/C_L^0$. B is the structural damping, C_L^0 a reference lift coefficient (typically 0.3), C_D a drag coefficient, and ε and A two dimensionless coefficients to be adjusted, for which Facchinetti *et al.* (2004) have proposed $\varepsilon = 0.3$ and $A = 12$.

When they are linearized and damping terms are dropped, equations (5.9) and (5.10) have the form

$$\ddot{y} + y = M \Omega^2 q \tag{5.11}$$

$$\ddot{q} + \Omega^2 q = A \ddot{y} \tag{5.12}$$

where M is a dimensionless mass ratio

$$M = \frac{C_L^0}{2} \frac{1}{8 \pi^2 S_t^2 \mu} \tag{5.13}$$

with $\mu = \pi (m^* + 1)/4$ and $\Omega = S_t U_R$. S_t is the Strouhal number and U_R the reduced velocity. Simple analysis shows that these two coupled oscillators have an instability zone for $1/(1 + \sqrt{A M}) < \Omega < 1/(1 - \sqrt{A M})$. (This is very similar to the coupled-mode flutter of lifting bodies considered in paragraph 5.2.2.)

M being given from (5.13), the coefficient A can be adjusted so that the instability zone corresponds to the lock-in region.

5.1.3 VIV Response of Cables and Risers

In the case of elongated cables and risers two new problems arise:

- there is no longer **one** resonant frequency of the cross-flow response, but several, associated with the natural modes of deformation;
- in the case of risers, the current velocity usually varies from the sea floor up to the free surface. As a result, several modes may participate, each mode being excited in some part of the water column while elsewhere it is damped by drag forces.

The first step is to determine the natural modes and associated frequencies. For the sake of simplicity, consider a weightless (in water) string of length l, under tension. Its transverse response $y(z,t)$ obeys the equation

$$m \, y_{tt} + b \, y_t - T_0 \, y_{zz} = f_L(z,t) \tag{5.14}$$

with m the mass plus added mass per unit length, T_0 the tension, and b the structural damping.

When the extremities are fixed, the natural modes are simply $Y_n(z) = \sin \lambda_n z$ with $\lambda_n = n \pi / l$. The natural frequencies are given by

$$\omega_n = \frac{n \pi}{l} \sqrt{\frac{T_0}{m}} \tag{5.15}$$

The VIV response can be derived by decomposing $y(z,t)$ on the modal base:

$$y(z,t) = \sum_{n=1}^{N} A_n(t) \, Y_n(z) = \sum_{n=1}^{N} A_n(t) \, \sin \lambda_n z \tag{5.16}$$

giving

$$\sum_{n=1}^{N} (m \, \ddot{A}_n + b \, \dot{A}_n + m \, \omega_n^2 \, A_n) \, \sin \lambda_n z = f_L(z,t) \tag{5.17}$$

or, taking advantage of the orthogonality of the modal set:

$$m \ddot{A}_n + b \dot{A}_n + m \omega_n^2 A_n = \frac{2}{l} \int_0^l f_L(z,t) \sin \lambda_n z \, dz \qquad (5.18)$$

Evolution equations are thus derived for the modal amplitudes $A_n(t)$.

The simplest case to address is when the current is constant over the depth and only one mode is capable of responding: Taking a length l equal to 100 m, a tension T_0 equal to 1000 kN, and a lineic mass m equal to 100 kg/m, one obtains that the natural periods (in seconds) are approximately given by $T_n \simeq 2/n$ with n the mode number.

With a current velocity of 1 m/s and a diameter of 30 cm, the modal reduced velocities are obtained as $U_{Rn} = U T_n/D \simeq 6.67/n$. Only the first mode ($n = 1$) has the capacity to resonate. With a diameter of 20 cm only the second mode is concerned.

Considering that the only resonating mode is the mode n, the VIV response can be looked for as

$$y(z,t) = A_n(t) \sin \lambda_n z = A_{0n} \cos \omega_n t \sin \lambda_n z \qquad (5.19)$$

The response amplitude $A_{0n} \sin \lambda_n z$ varies along the riser. At some places it is less than what it would be in the two-dimensional case considered in the previous section: energy is transferred from the fluid to the riser. At the antinodes the response amplitude is higher: energy is transferred from the riser back to the fluid. The total energy budget is equal to the energy dissipated by the structural damping. When this structural energy loss is assumed to be negligible, the modal amplitude A_{0n} is obtained by setting

$$\int_0^T \sin \omega_n t \left[\int_0^l f_L(z,t) \sin \lambda_n z \, dz \right] dt \equiv 0 \qquad (5.20)$$

where f_L is for instance given by

$$f_L(z,t) = \frac{1}{2} \rho \, C_L \left(\frac{A_{0n} \, |\sin \lambda_n z|}{D} \right) D \, U^2 \, \sin \omega_n t \qquad (5.21)$$

If we assume that the lift coefficient is given by

$$C_L(A/D) = 0.35 + 0.60 \, \frac{A}{D} - 0.93 \left(\frac{A}{D} \right)^2 \qquad (5.22)$$

then we get $A_{0n} = 1.2 \, D$: The VIV amplitude, at the antinodes, is 20% higher than in the 2D case.

This result supposes that the polynomial form (5.22) represents the physical reality both when C_L is negative (energy dissipation) and when it is positive (energy production). It also assumes that the vortex shedding is synchronized along the whole of the riser.

If we now increase the length of the riser to 1,000 m and keep the other parameters unchanged, we obtain that the modal reduced velocities are given by $U_{Rn} = 66.7/n$. The criterion $5 \leq U_{Rn} \leq 8$ gives n between 8 and 14! Several modes are likely to be involved.

At such depths, currents are not of constant speed (nor direction) in the water column. If one assumes for example, still in the same case, that the current velocity varies from 0.5 m/s near the bottom up to 1.5 m/s at the free surface, one finds that modes 4–7 are concerned near the bottom and modes 12–20 near the free surface. A given mode can be excited only in a given strip of the water column; elsewhere it is damped by the hydrodynamic drag forces, which are much more efficient than the structural damping.

There are numerical models that can handle this kind of situation but they are still far from absolutely reliable. In addition to the problem of managing several modes in competition, the main remaining difficulties are the famous correlation length and the extrapolation to high Reynolds numbers (often transcritical) of experimental results obtained in subcritical regime.

To illustrate the capabilities of these models, a comparison is presented between measurements and calculations carried out with the code *DEEPVIV* of IFPEN. The measurements come from tests carried out at the BGO-FIRST Ocean Basin, as part of the CLAROM "HYDLINES" project. A cable is tensioned obliquely across the basin and covered, in its lower part, with sheaths of varying diameters. The current is uniform but the vortex shedding frequencies vary along the cable, due to the different diameters. The situation is therefore equivalent to that of a cable of constant diameter in a sheared current. The motion of the cable is measured in its aerial part with 20

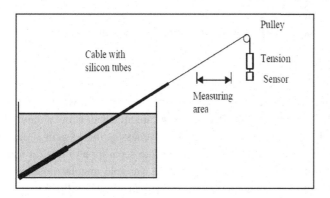

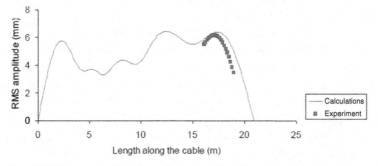

Figure 5.7 Transverse vibrations of a taut cable with varying diameter. Measured and calculated rms transverse displacements.

infrared diodes tracked by an optical system. Figure 5.7 shows the setup and standard deviations of the cross-flow motion along the cable, obtained by *DEEPVIV* and from the measurements, in one of the test cases, with a good agreement.

5.1.4   Mitigation of VIVs

Vortex-induced vibrations are generally detrimental owing to the fatigue they cause to the structures. As such, for risers, the stresses due to the higher-order modes are the most damaging since, at constant vibration amplitude, the curvatures, and therefore the stresses, are greater and the number of cycles increases with the frequency. The increase in drag can also be detrimental, for instance, the mooring loads on a single-column platform (Spar) can be greatly amplified (see Figure 4.32).

The best way to get rid of VIVs is to make sure that the reduced velocities associated with the critical modes of vibration do not take values in the wrong range, between 4 and 10 (or more, in the case of low mass ratios). This can mean adjusting the structural or mooring stiffnesses, or the mass distribution. This remedy is hardly applicable to structures such as risers, which have many modes of vibration.

Another way is to increase energy dissipation, by playing either on the structural damping or the hydrodynamic damping. For example, flexible risers are reputed to be less prone to VIVs than rigid risers, the friction between the constitutive layers dissipating a large amount of energy.

Finally, one can try to act at the very source of the problem, by removing or reducing the excitation. The anti-VIV devices presented below mainly seek to act following this principle.

Anti-VIV Devices

A well-known device, often visible on chimneys, consists in winding plates of small width around the cylinder. Typically 3 **strakes** are attached, about a tenth of diameter wide, at a pitch of 5–10 diameters. The effectiveness of this system is due to the fact that it combines several effects: by imposing the separation points (at sharp angles) it reduces the lift force; by imposing them at variable angular positions along the cylinder it destroys the coherence of the wake and decreases the correlation length; finally, the hydrodynamic damping is increased. Another advantage is that the system is efficient regardless of the direction of the current.

The drawback, in addition to handling and installation problems, is a significant increase in the drag coefficient, which rises to 1.4, even in transcritical regime (but still can be less than under VIV).

Many other devices can be considered as parent to the strakes. For example, the strakes can be replaced by cables or wires, or reproduced discontinuously by small elements parallel to the generator of the cylinder.

One can also vary the pitch, up to the extreme case where it becomes infinite, the strakes being parallel to the generator. The efficiency then depends on the number and location of the strakes, with respect to the direction of the current. The device therefore loses its omnidirectional character. It is then no longer possible to act on the

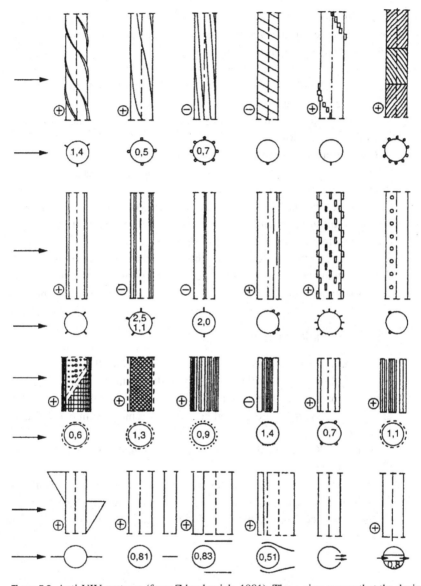

Figure 5.8 Anti-VIV systems (from Zdravkovich, 1981). The + sign means that the device is efficient, the − sign that it is not. The number inside the circle is the drag coefficient.

correlation length. Zdravkovich (1981) presents a compilation of different variants and reports that some are ineffective, even detrimental (Figure 5.8).

Another principle is to surround the cylinder with a second perforated cylinder, a shroud. The flow, by its passage through the openings of the shroud, becomes somehow regularized, the wake is more homogeneous and the appearance of the alternate alley of vortices is shifted further downstream. Typically the diameter of the shroud is 25% larger than that of the cylinder, the open-area ratio is between 30% and 40%. The shape and size of the perforations matter less. Drag coefficients, based on the diameter of the shroud, of the order of 0.6 or 0.7 (therefore 0.75–0.9 based on the diameter of the cylinder), are reported in the literature. The shroud efficiency appears to be as good as that offered by the strakes.

With regard to towing cables, it is common to use ribbons, anchored between the strands, which on the one hand block the communication between the flows on either side of the cable, and on the other hand, increase the hydrodynamic damping. One advantage of the system is that it is compatible with the operations of winding and unwinding the cables.

One can also take advantage of the need to incorporate connectors, or buoyancy elements, into the riser, and play on variations in diameter. This limits the correlation length, and ensures the presence of energy-dissipating zones, insofar as the excitation cannot take place both on the riser parts and on the float parts, when the diameters differ sufficiently. The successful application of this principle to delicate drilling operations is reported by Brooks (1987).

Another solution consists in clothing the riser with fairings, with the practical difficulty that they must be able to rotate freely around the riser. An experimental comparison between different concepts of fairings is given by Baarholm *et al.* (2015).

5.1.5   VIVs in Oscillatory Flow

Vortex-induced vibrations may also occur in oscillatory flows.

At very low K_C values (less than 2 or 4, depending on the Reynolds number), the flow remains attached (for a circular cylinder) and the problem does not arise. For K_C around 4 or 5, vortices are emitted but the flow remains symmetrical and no alternate lift is induced.

At higher K_C values, the flow becomes asymmetrical and vibratory responses are likely to appear.

There are two asymptotic cases, depending on whether the Keulegan–Carpenter number is very high or of order 1. This parameter relates directly to the number of (pairs of) vortices emitted by cycle. In fact, when the velocity of the incident flow is of the form

$$U(t) = A\,\omega\,\sin \omega t = U_m\,\sin \omega t \tag{5.23}$$

and when one assumes that the Strouhal relationship $f\,D/U = S_t$ remains valid, the instantaneous frequency of the vortex shedding is given by

$$f(t) = \frac{S_t\,|U(t)|}{D} = \frac{S_t\,U_m}{D}\,|\sin \omega t| \tag{5.24}$$

The number of shed pairs of vortices over one period is then given by

$$N = \int_0^T f(t)\,\mathrm{d}t = \frac{2}{\pi}\,S_t\,K_C \simeq 0.13\,K_C \tag{5.25}$$

(in the subcritical regime where $S_t \simeq 0.2$).

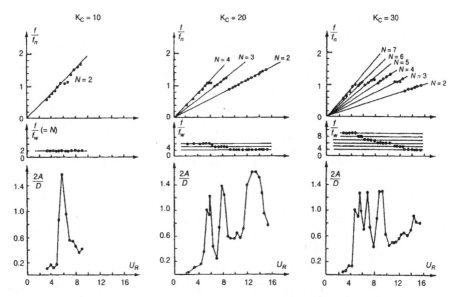

Figure 5.9 VIV response of a circular cylinder in oscillatory flow. Oscillation frequency f (top) and amplitude A (bottom) (from Sumer & Fredsøe, 1988).

If K_C is high enough, the velocity of the flow and therefore the frequency of the vortex shedding change little from one emission of vortices to the next. Vibrations may then occur if the associated reduced velocity falls within the critical interval [5 8]. These vibrations appear in bursts at each flow cycle, when the instantaneous reduced velocity becomes critical. For example, wave gauges used in wave tanks are often seen to vibrate; typically, their diameters are a few millimeters, which leads to K_C of the order of several hundreds.

To estimate the amplitude of the vibratory response, one may follow the approach used in current alone, and write the lift force (on the fixed cylinder) as

$$F_L(t) = \frac{1}{2} \rho \, C_L \, D \, U_m^2 \, \sin^2 \omega t \, \cos \Psi(t) \tag{5.26}$$

where $\Psi(t)$ verifies

$$\frac{\partial \Psi}{\partial t} = 2\pi \, f(t) = S_t \, K_C \, \omega \, |\sin \omega t| \tag{5.27}$$

From where, by integration (Bearman *et al.*, 1984):

$$F_L(t) = \frac{1}{2} \rho \, C_L \, D \, U_m^2 \, \sin^2 \omega t \, \cos[S_t \, K_C \, (1 - \cos \omega t) + \Psi_0] \tag{5.28}$$

the phase Ψ_0 possibly changing by π at the end of each cycle (when $\omega t = n\pi$).

The other case is when the Keulegan–Carpenter number is less than a rather imprecise threshold, of the order of 50 or 100. It is then observed that the vibrations, when they appear, take place at a harmonic Nf of the oscillation frequency of the flow, where N is the nearest integer value of the ratio f_n/f, f_n being the natural frequency in calm water. Even harmonics appear preferentially. Response peaks occur when Nf

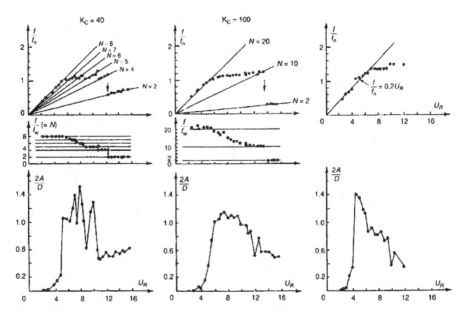

Figure 5.10 Same as Figure 5.9 for higher K_C values and in current only (from Sumer & Fredsøe, 1988).

is close enough to f_n and the reduced velocity $U_m/(D f_n)$ is higher than 4. Figures 5.9 and 5.10, taken from Sumer & Fredsøe (1997), illustrate the regimes obtained vs K_C and the reduced velocity.

5.1.6   In-line VIVs

It has been seen in paragraph 4.4.1 that, for cylinders in sinusoidal motion in-line with the current, negative drag coefficients are obtained, experimentally, for certain (low) values of the relative amplitude A/D and of the parameter $U_0 T/D$, U_0 being the current velocity and T the period of oscillation (Figure 4.26): The cylinder extracts energy from the fluid.

For a cylinder on elastic support, it can be inferred that inline instabilities are likely to appear for corresponding values of the reduced velocity $U_0/(f_n D)$. There are two areas of instability:

- a first zone for reduced velocities between (roughly) 2 and 3, where the vortex shedding remains alternate. The reduced velocities are half those where cross-flow vibrations would appear, which makes sense since the frequency of the in-line load is twice the lift frequency.
- a second zone for lower reduced velocities, between 1 and 2, where the vortex shedding is symmetrical (Figure 5.11).

The amplitude of the inline response is small, typically A/D is less than 0.15. However inline VIVs concern higher elastic modes than the cross-flow VIVs, with the result that the induced bending stresses are comparable.

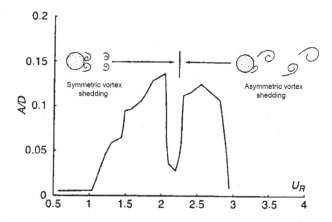

Figure 5.11 In-line VIVs (from Wootton *et al.*, 1972, by permission of CIRIA).

5.2   Galloping and Flutter

Galloping and flutter are other types of instabilities, not due to alternate vortex shedding, that take place at high values of the reduced velocity, beyond the VIV range. Because of the high U_R values, they can efficiently be modeled through quasi-steady approaches.

In galloping, also known as "plunge," only one degree of freedom is involved. Flutter instability is due to mechanical coupling between two degrees of freedom, for instance, heave and pitch in the case of foils.

5.2.1   Galloping

A well-known case of galloping instability is the square prism. On the fixed cylinder, a uniform flow of velocity U_0 exerts loads in the X and Y directions (see Figure 5.12) that can be written as

$$F_X = \frac{1}{2} \rho C_X(\alpha) D U_0^2 \tag{5.29}$$

$$F_Y = \frac{1}{2} \rho C_Y(\alpha) D U_0^2 \tag{5.30}$$

with α the current direction and D the side of the square prism.

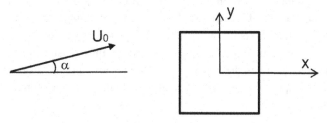

Figure 5.12 Square prism in uniform flow.

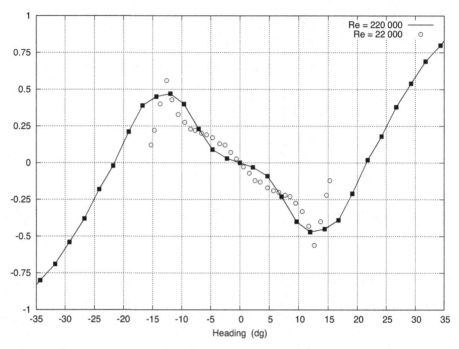

Figure 5.13 Square prism in uniform flow. Transverse force coefficient $C_Y(\alpha)$.

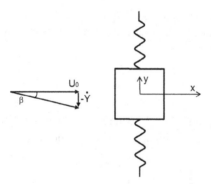

Figure 5.14 Square prism in between two springs.

Figure 5.13 shows the coefficient $C_Y(\alpha)$ obtained from measurements. The lower Reynolds value is from wind tunnel tests reported in Parkinson & Smith (1964). The higher Re values are from Molin *et al.* (2012). It can be seen that C_Y has a peculiar shape in the range $[-15°\ 15°]$ with a sign conflicting with intuition: The projections of the force and current velocity on the Y axis are in opposite directions!

Consider now the square prism in between two springs, as shown in Figure 5.14, and allowed to move only in the Y direction, with the current in the X direction. Under the assumption of quasi-steady hydrodynamic loading, the Y equation of motion can be formulated as

$$(M + M_a)\,\ddot{Y} + 2\,\zeta\,\omega_0\,(M + M_a)\,\dot{Y} + K\,Y = \frac{1}{2}\,\rho\,C_Y(\beta)\,D\,(U_0^2 + \dot{Y}^2) \quad (5.31)$$

Figure 5.15 Experimental setup (from Molin *et al.*, 2012).

with $M + M_a$ the mass plus added mass, ω_0 the natural frequency, ζ the linear damping ratio, K the combined spring stiffness, and β the apparent current heading, as shown in the figure.

When $|\dot{Y}|$ is small compared to U_0, by Taylor expansion, equation (5.31) takes the form

$$(M + M_a)\, \ddot{Y} + \left[2\,\zeta\,\omega_0\,(M + M_a) + \frac{1}{2}\,\rho\,C'_Y(0)\,D\,U_0 \right]\,\dot{Y} + K\,Y = 0 \qquad (5.32)$$

Instability occurs if the total damping is negative, that is

$$C'_Y(0) < -\frac{4\,\zeta\,\omega_0\,(M + M_a)}{\rho\,D\,U_0} \qquad (5.33)$$

$$C'_Y(0) < -4\,\pi\,\frac{K_S}{U_R} \qquad (5.34)$$

where $K_S = 2\,\zeta\,(M + M_a)/(\rho\,D^2)$ is a stability parameter (the Scruton number).

It can be concluded that if $C'_Y(0)$ is negative, instability will inevitably occur beyond some U_R value.[1]

---

[1] It is customary, in aeroelasticity, to use the drag $C_D(\alpha)$ (in-line with the air flow) and lift $C_L(\alpha)$ (perpendicular) coefficients in place of body-fitted coefficients. The criterion for instability is then $C_D + \partial C_L/\partial\alpha < -4\,\pi\,K_S/U_R$.

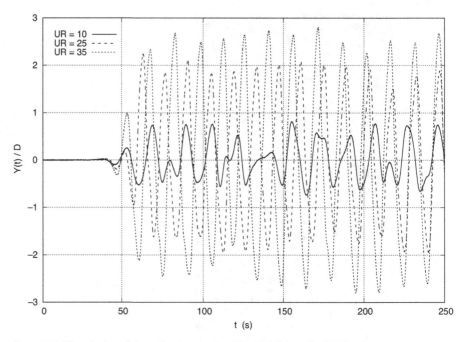

Figure 5.16 Time series of the galloping motion (from Molin *et al.*, 2012).

Figure 5.15 shows the experimental setup used in Molin *et al.* (2012). A 0.3 m × 0.3 m square cylinder, with a draft of 1 m, is suspended from a 5 m long beam. It can oscillate as a pendulum in the transverse direction to the current while the in-line motion is restrained by a cable. Some additional weights, visible in the photograph, are placed at the top of the cylinder in order to achieve a natural period T_0 equal to 10 s in still water. The reduced velocity is varied by gradually increasing the current velocity.

Figure 5.16 shows time series of the sway motion (taken at mid-draft) for reduced velocities $U_R = U_0 T_0/D$ equal to 10, 25 and 35. At $U_R = 10$ the sway motion is slight, and somewhat erratic, but it becomes quite regular at $U_R = 25$ and $U_R = 35$, with oscillation periods respectively equal to 14.3 s and 15 s, notably higher than the still water values. This means that the added mass coefficient has increased about 3 or 4 fold!

Figure 5.17 shows the normalized amplitude of the galloping response A/D vs the reduced velocity $U_0 T/D$, with T either the still water value or the actual oscillation period.

Most polygonal shapes are susceptible to galloping instabilities. In the case of the (smooth) circular cylinder there cannot be galloping since the current load is in-line with the flow, meaning that $C'_Y(0)$ is positive, equal to the drag coefficient C_D. However shapes not so different from circular, such as some cross-sections of riser towers, are prone to galloping.

When the transverse load coefficient $C_Y(\beta)$ has been determined, from model tests or CFD calculations, the galloping response can be calculated from the β value such

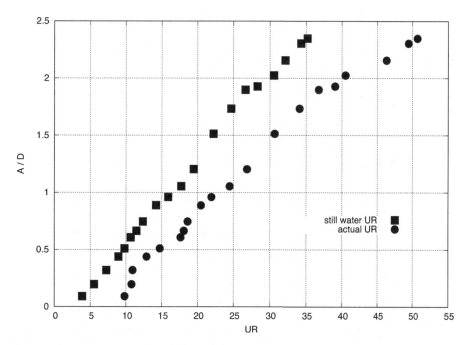

Figure 5.17 Normalized amplitudes of the galloping response (from Molin *et al.*, 2012).

that, over one cycle, the energy balance be zero. It results that the galloping amplitude increases linearly with the current velocity, at variance with VIVs, which are self-bounded.

Galloping may affect rotational modes (Garapin *et al.*, 2016). Galloping instability has also been suggested for multiple riser configurations (Overvik, 1982).

With regard to moored tankers, the shape of the longitudinal current drag coefficient $C_{Xc}(\theta)$, as given by OCIMF (2010), suggests the possibility of galloping in surge, in bow quartering current. This behavior has apparently been observed at least once (Molin & Bureau, 1980).

5.2.2 Flutter Instability of a Flap

Consider a wing section, as shown in Figure 5.18, elastically supported. The wing section has two degrees of freedom, in rotation (α), and in vertical translation (Z). The elastic support consists in a vertical spring, of stiffness K and a rotational spring, of stiffness C, both at a distance l from the center of gravity, as shown in the figure. We assume the wing section to be thin and symmetrical, the lift coefficient is then $2\pi\,\alpha$ for small α angles. The lift force applies at a distance $l+d$ from the center of gravity.

The Z and α equations of motion are

$$M\,\ddot{Z} = \rho\pi L U^2\alpha - K\,(Z+l\,\alpha) \tag{5.35}$$

$$I\,\ddot{\alpha} = \rho\pi L\,(l+d)\,U^2\,\alpha - K\,l\,(Z+l\,\alpha) - C\,\alpha \tag{5.36}$$

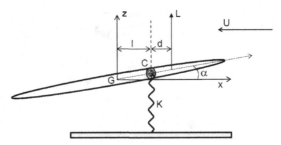

Figure 5.18 Wing section elastically supported in uniform flow.

with M and I the mass and inertia (including added mass and added inertia) and L the length of the wing.

The natural frequencies are the roots of the determinant of the matrix

$$\begin{pmatrix} -M\,\omega^2 + K & K\,l - \rho\,\pi\,L\,U^2 \\ K\,l & -I\,\omega^2 + C + K\,l^2 - \rho\,\pi\,(l+d)\,L\,U^2 \end{pmatrix} \tag{5.37}$$

giving a second-degree equation in ω^2:

$$a\,\omega^4 + b\,\omega^2 + c = 0 \tag{5.38}$$

where a is positive, b is negative and c is given by

$$c = K\,(C - \rho\,\pi\,L\,d\,U^2) \tag{5.39}$$

When c is positive, there are two distinct natural frequencies ω_1 and ω_2. That c be positive is ensured if d is negative, that is the lift force applies behind the hinge point. When d is positive, the coefficient c becomes negative when the current velocity exceeds the critical value $U_c = \sqrt{C/(\rho\,\pi\,L\,d)}$. Then the two frequencies ω_1 and ω_2 become complex, their real parts collapse to the same value, and their imaginary parts are opposite: There are two natural modes, one of them unstable, extracting energy from the fluid.

This instability ("coupled mode flutter") is well-known in aerodynamics. As can be seen, it is a potential threat when two degrees of freedom are coupled through off-diagonal terms in the stiffness matrix.

In marine engineering, this instability should be considered for ship roll stabilization with flaps. It can also be taken advantage of as a way to extract energy from current, or as a propulsive force.

A parent instability is the fishtailing behavior of single-point moored ships, under the action of wind and/or current.

5.3   Wake Instabilities

Different types of instabilities may appear in the case of arrays of cylinders, under the action of current. When a cylinder is in the wake of an upstream cylinder, it may vibrate under the action of the vortices shed by the upstream cylinder, when its natural

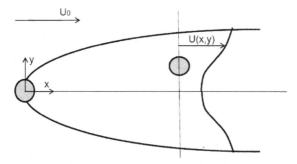

Figure 5.19 Two staggered cylinders in uniform flow.

frequency coincides with the shedding frequency. Unstable dynamic responses may also occur at larger values of the reduced velocity, due to the nonuniformity of the time-averaged wake flow. These behaviors are known as Wake Induced Oscillations (WIOs). They are a well-known problem in the case of electric transmission lines and most of the literature on the topic comes from aeroelasticity.

For the sake of simplicity, we take the case of two cylinders, of identical diameters D, as shown in Figure 5.19. The cylinders are considered to be fixed.

The current load on the upstream cylinder is

$$F_{Du} = \frac{1}{2} \rho C_{D0} D U_0^2 \tag{5.40}$$

with D the diameter and U_0 the current velocity.

It is assumed that the Reynolds number is high enough and the leeward cylinder far enough that the vortices have dissipated and the wake has homogenized. When $u(x,y)$ is the local wake velocity, the load on the downstream cylinder can be written as an in-line drag component and a transverse lift component:

$$F_{Dd} = \frac{1}{2} \rho C_{D0} D u^2(x,y) \tag{5.41}$$

$$F_{Ld} = \frac{1}{2} \rho C_L(x,y) D u^2(x,y) \tag{5.42}$$

Alternatively, one may define local drag and lift coefficients and relate the loads to the outer velocity U_0:

$$F_{Dd} = \frac{1}{2} \rho C_{Dd}(x,y) D U_0^2 \tag{5.43}$$

$$F_{Ld} = \frac{1}{2} \rho C_{Ld}(x,y) D U_0^2 \tag{5.44}$$

Schlichting (1979) has proposed the following velocity profile in the wake of a cylinder, valid at distances such that the wake has homogenized and the pressure can be considered constant, equal to the outer pressure:

$$u(x,y) = U_0 - 0.95 \, U_0 \sqrt{\frac{C_{D0} D}{x}} \, \exp\left\{-11.3 \, \frac{y^2}{C_{D0} D x}\right\} \tag{5.45}$$

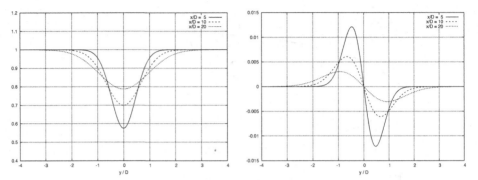

Figure 5.20 Velocity profiles u/U_0 (left) and v/U_0 (right).

The same form has been used by other people, but with coefficients different from 0.95 and 11.3. So we assume a more general expression, in the form

$$u(x,y) = U_0 - a_1 U_0 \sqrt{\frac{C_{D0} D}{x}} \exp\left\{-a_2 \frac{y^2}{C_{D0} D x}\right\} \qquad (5.46)$$

It must be borne in mind that the coefficients a_1 and a_2 are not independent: They are linked by the equality of drag and loss of momentum

$$\frac{1}{2} \rho C_{D0} D U_0^2 \equiv \rho \int_{-\infty}^{\infty} U_0 u(x,y) \, dy \qquad (5.47)$$

giving $a_1 \sqrt{\pi/a_2} \equiv 1/2$.

From the continuity equation the transverse velocity is obtained as

$$v(x,y) = -\int_0^y u_x(x,Y) \, dY = -\frac{1}{2} a_1 U_0 \sqrt{\frac{C_{D0} D}{x}} \frac{y}{x} \exp\left\{-a_2 \frac{y^2}{C_{D0} D x}\right\} \qquad (5.48)$$

directed toward the center line of the wake.

Figure 5.20 shows the velocity profiles $u(x,y)/U_0$ and $v(x,y)/U_0$ from equations (5.45) and (5.48). Here a C_{D0} coefficient equal to 1 has been taken. It can be observed that the magnitude of the transverse velocity is very small, of the order of 1/50 of the inline velocity defect.

The inline drag load is obtained by squaring u. Since the velocity in the wake varies slowly with the horizontal coordinates and is locally close to uniform, it seems legitimate to assume that the drag force on the lee cylinder is actually in the direction (u,v) of the local flow. The lift force is then obtained as $u^2 \times v/u \equiv uv$.

Figure 5.21 therefore shows u^2/U_0^2 and $u v/U_0^2$, which should be directly comparable to experimental drag and lift coefficients as defined in (5.43) and (5.44). Compared to experimental data, the drag coefficient looks reasonable but the lift coefficient is about 20 or 40 times too low!

Blevins (2005), following other investigators (see, e.g., Païdoussis *et al.*, 2011), suggests relating the lift coefficient to the transverse derivative of the drag coefficient:

$$C_{Ld} \simeq a_3 D \, \partial C_D / \partial y \qquad (5.49)$$

with a_3 a negative fitting coefficient.

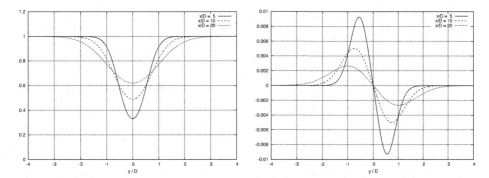

Figure 5.21 Normalized in-line velocity squared u^2/U_0^2 (left) and normalized cross-product $u\,v/U_0^2$ (right).

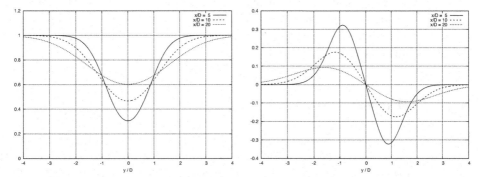

Figure 5.22 Drag (left) and lift (right) coefficients with Blevins' a_i values and $C_{D0} = 1$.

When the local C_{Dd} is taken as $C_{D0}\,u^2/U_0^2$, this gives

$$C_{Ld} \simeq 2a_3\,C_{D0}\,D\,\frac{u\,u_y}{U_0^2} \tag{5.50}$$

It can readily be checked, from equations (5.45) and (5.48), that $u_y = -4\,a_2\,v/(C_{D0}\,D)$ so that

$$C_{Ld} \simeq -8a_3\,a_2\,\frac{u\,v}{U_0^2} \tag{5.51}$$

This is the same as proposed above but with a multiplicative coefficient $-8a_3\,a_2$.

There is a scarcity of experimental data on the matter. Price (1975) reports wind tunnel tests on two staggered cylinders 2.86 cm in diameter at Reynolds numbers around $3 \cdot 10^4$, with low- and high-turbulence intensities (about 1% and 10%). A strong effect of turbulence is observed: the drag and lift coefficients of the lee cylinder are drastically reduced in the high turbulence case.

Blevins (2005) uses the low turbulence data of Price to adjust the coefficients a_1, a_2, and a_3 in order to get a best fit for the drag and lift coefficients. He gets $a_1 = 1$, $a_2 = 4.5$, and $a_3 = -0.59$[2]. The low a_2 value results from the fact that the experimental

---

[2] In Blevins' paper there is a discrepancy between the a_3 coefficient in his equation (7) and the a_3 coefficient in his equation (8): They are not the same. The value -10.6 that he gives must be divided by $4\,a_1\,a_2$.

lift coefficient peaks at a larger y/D value than seen in Figure 5.22. It can be noted that with these a_1 and a_2 values equation (5.47) is not fulfilled, by far.

Figure 5.22 shows the local drag and lift coefficients computed using the a_1, a_2, and a_3 values obtained by Blevins.

Obviously, there is a need for more experimental data. An alternative is to use CFD models, as reported by Wu *et al.* (2002) who obtain numerical drag and lift coefficients in good agreement with Price's experimental values.

When the upstream cylinder is fixed and the downstream cylinder is elastically supported, the static and dynamic stabilities of the downstream cylinder can be investigated, using the analytical expressions of the drag and lift coefficients. This can be done by solving the equations of motion in time domain. It can also be done by assuming small displacements from a reference position and Taylor expanding the drag and lift coefficients. Such an analysis is done by Blevins (2005) who derives regions of static and dynamic instabilities. Blevins ignores the way the drag and lift loads vary with the cylinder's own velocity. More comprehensive analyses can be found in Païdoussis *et al.* (2011).

5.4   References

BAARHOLM R., SKAUGSET K., LIE H., BRAATEN H. 2015. Experimental studies of hydrodynamic properties and screening of riser fairing concepts for deep water applications, in *Proc. ASME 2015 34th International Conference on Ocean, Offshore and Arctic Engineering*, paper OMAE2015-41730.

BLEVINS R.D. 1977. *Flow-Induced Vibrations*. Van Nostrand Reinhold Company.

BLEVINS R.D. 2005. Forces on and stability of a cylinder in a wake, *ASME Journal of Offshore Mechanics and Arctic Engineering*, **127**, 39–45.

BROOKS I.H. 1987. A pragmatic approach to vortex-induced vibrations of a drilling riser, *Proc. 19th Offshore Techn. Conf.*, paper 5522.

CINELLO A., PÉTRIÉ F., RIPPOL T., MOLIN B., LE CUNFF C. 2013. Experimental investigations of VIV at high Reynolds numbers for smooth circular cylinders in single and tandem arrangements, in *Proc. ASME 2013 32nd International Conference on Ocean, Offshore and Arctic Engineering*, paper OMAE2013-10638.

FACCHINETTI M.L., DE LANGRE E., BIOLLEY F. 2004. Coupling of structure and wake oscillators in vortex-induced vibrations, *Journal of Fluids and Structures*, **19**, 123–140.

FENG C.C. 1968. The measurement of vortex-induced effects in flow past a stationary and oscillating circular and D-section cylinders. Masters thesis, University of British Columbia, Vancouver, B.C., Canada.

GARAPIN J., BÉGUIN C., ÉTIENNE S., PELLETIER D., MOLIN B. 2016. Nonlinear model of rotational galloping of square, rectangular and bundle cylinder in cross-flow, in *Actes des 15èmes Journées de l'Hydrodynamique*, Brest (http://website.ec-nantes.fr/actesjh/).

GOVARDHAN R.N., WILLIAMSON C.H.K. 2006. Defining the modified Griffin plot' in vortex-induced vibration: revealing the effect of Reynolds number using controlled damping, *J. Fluid Mech.*, **561**, 147–180.

HARTLEN R.T., CURRIE I.G. 1970. Lift-oscillator model of vortex-induced vibration. *ASCE J. Eng. Mech. Division*, **96**, 577–591.

KHALAK A., WILLIAMSON C. H. K. 1997. Fluid forces and dynamics of a hydroelastic structure with very low mass and damping, *J. Fluids and Structures*, **11**, 973–982.

MOLIN B., BUREAU G. 1980. A simulation model for the dynamic behavior of tankers moored to single point moorings, *Proc. Int. Symp. Ocean Eng. Ship Handling*, SSPA.

MOLIN B., REMY F., RIPPOL T., *et al.* 2012. Experimental investigation on the galloping response of square cylinders at high Reynolds numbers, in *Proc. 6th Intl. Conf. on Hydroelasticity in Marine Technology*, Tokyo.

MOLIN B., REMY F., LE HIR E., RIPPOL T., SCARDIGLI S. 2010. Étude expérimentale des vibrations induites par le détachement tourbillonnaire à grands nombres de Reynolds, in *Actes des 12èmes Journées de l'Hydrodynamique*, Nantes (in French; http://website.ec-nantes.fr/actesjh/).

OIL COMPANIES INTERNATIONAL MARINE FORUM (OCIMF) 2010. Estimating the environmental loads on anchoring systems.

OVERVIK T. 1982. Hydroelastic motion of multiple risers in a steady current, PhD thesis, NTH.

PAÏDOUSSIS M., PRICE S.J., DE LANGRE E. 2011. *Fluid-Structure Interactions. Cross-Flow-Induced Instabilities*. Cambridge University Press.

PARKINSON G.V., SMITH J.D. 1964. The square prism as an aeroelastic nonlinear oscillator, *The Quarterly Journal of Mechanics and Applied Mathematics*, **17**, 225–239.

PRICE S.J. 1975. Wake induced flutter of power transmission conductors, *J. Sound Vib.*, **38**, 125–147.

SCHLICHTING H. 1979. *Boundary Layer Theory*. McGraw-Hill Book Company.

SKOP R.A., GRIFFIN O.M. 1973. A model for the vortex-excited resonant response of bluff cylinders, *J. Sound Vib.*, **27**, 225–233.

SUMER B.M., FREDSØE J. 1988. Transverse vibrations of an elastically mounted cylinder exposed to an oscillating flow, *J. Offshore and Arct. Engineering*, **110**, 387–394.

WOOTTON L., WARNER M., SAINSBURY R. *et al.* 1972. Oscillations of piles in marine structures - a description of the full-scale experiments at Immingham, TN40, CIRIA, London (ISBN: 978-0-901208-41-5) www.ciria.org.

WU W., HUANG S., BARLTROP N. 2002. Current induced instability of two circular cylinders, *Appl. Ocean Res.*, **24**, 287–297.

ZDRAVKOVICH M.M. 1981. Review and classification of various aerodynamic and hydrodynamic means for suppressing vortex shedding, *J. Wind Engineering & Industrial Aerodynamics*, **7**, 145–189.

6 Large Bodies: Linear Theory

6.1 Introduction

In this chapter we consider structures, fixed or floating, for which viscous effects (friction, drag) are negligible in a first approximation. They are therefore bodies with large dimensions compared to the wave amplitude and to the amplitude of their own motions.

We assume that these bodies are subjected to waves only: the current velocity, or the forward speed, is assumed to be nil. When the current velocity is low, its effect on the wave response is relatively negligible. In wave-current interaction, a discriminating parameter is the τ parameter, or Brard number, defined as $\tau = U_C \omega_e / g$ with ω_e the encounter frequency. When τ is much lower than the critical value $\tau_C = 1/4$, the effect of current on the linear response is negligible. For a typical current velocity U_C equal to 1 m/s and a typical wave period T equal to 12 s, the τ parameter is equal to 0.05.

It is possible to include current (or a low forward speed) in the linearized theory of diffraction-radiation, but at the cost of increased complexity of the theoretical model, which is not justified when one is interested only in loads and response taking place at the wave frequency. However current, even low, has a strong effect on wave runup, and on the wave drift forces, which are introduced in Chapter 7. It is therefore in Chapter 7 that the problem of diffraction-radiation with current is addressed.

With regard to this matter, it should be borne in mind that current, or forward speed, are "acceptable" within the potential flow theoretical frame, as long as they do not result in massive separation. At high forward speeds, this means streamlined bodies (ships). For blunt structures, this means a current velocity less than the amplitude of the wave-induced oscillatory velocity (Zhao *et al.*, 1988).

In this chapter potential flow theory is used throughout, not only to derive the kinematics of the incoming waves, but also to tackle their interaction with the body. This means that a no-flow condition is applied at the hull, in the form

$$\nabla \Phi(P,t) \cdot \vec{n} = \vec{U}(P,t) \cdot \vec{n} \qquad (6.1)$$

where $\vec{U}$ is the local body velocity at a point P of the hull. This velocity is usually part of the unknowns, so the equations of motion of the body need to be coupled to the fluid problem.

Linearization means that the free surface and body boundary conditions are simplified, keeping only the leading order terms. A consequence is that all quantities (pressures, loads, responses) will vary in time at the wave frequency, with an amplitude proportional to the wave amplitude.

Linearized free surface conditions have been presented in Chapter 3. As for the no-flow condition (6.1), within a linearized theory, it is applied at the mean position of the wetted part of the hull:

$$\nabla\Phi(P_0,t) \cdot \vec{n_0} = \vec{U}(P_0,t) \cdot \vec{n_0} \tag{6.2}$$

where the subscript $_0$ means "in the position at rest" or "in the time-averaged position."

For the sake of simplicity, we first assume that we are dealing with only one rigid body, so its motion consists in 3 translations and 3 rotations. Due to the small angular motion assumption, these rotations can be taken around fixed axes, so that the motion of a point P, of coordinates (X, Y, Z) in a body-fitted coordinate system $GXYZ$, writes

$$\begin{pmatrix} x(t) \\ y(t) \\ z(t) \end{pmatrix} = \begin{pmatrix} x_G(t) \\ y_G(t) \\ z_G(t) \end{pmatrix} + \begin{pmatrix} \alpha(t) \\ \beta(t) \\ \gamma(t) \end{pmatrix} \wedge \begin{pmatrix} X \\ Y \\ Z \end{pmatrix} \tag{6.3}$$

with (x_G, y_G, z_G) the motion of the center of gravity and (α, β, γ) the angles of rotations around the $G_0 x, G_0 y, G_0 z$ axes, respectively.

Here the origin of the body-fitted coordinate system is taken at the center of gravity G. At rest the axis GZ is vertical in the upward direction; for a ship the axis GX is taken in the longitudinal direction. So $x_G(t)$ is the surge motion, $y_G(t)$ the sway motion, $z_G(t)$ the heave motion, and $\alpha(t), \beta(t), \gamma(t)$ are roll, pitch, and yaw respectively.

The fixed coordinate system $Oxyz$ is translated vertically from the $GXYZ$ coordinate system in its position at rest $(G_0 xyz)$, with Oxy the still free surface (see Figure 6.1).

The no-flow condition (6.1) takes the form:

$$\nabla\Phi \cdot \vec{n_0} = \left[\begin{pmatrix} \dot{x}_G \\ \dot{y}_G \\ \dot{z}_G \end{pmatrix} + \begin{pmatrix} \dot{\alpha} \\ \dot{\beta} \\ \dot{\gamma} \end{pmatrix} \wedge \begin{pmatrix} X \\ Y \\ Z \end{pmatrix} \right] \cdot \begin{pmatrix} n_{0x} \\ n_{0y} \\ n_{0z} \end{pmatrix} \tag{6.4}$$

or $\qquad \nabla\Phi \cdot \vec{n_0} = \begin{pmatrix} \dot{x}_G \\ \dot{y}_G \\ \dot{z}_G \end{pmatrix} \cdot \begin{pmatrix} n_{0x} \\ n_{0y} \\ n_{0z} \end{pmatrix} + \begin{pmatrix} \dot{\alpha} \\ \dot{\beta} \\ \dot{\gamma} \end{pmatrix} \cdot \left[\begin{pmatrix} X \\ Y \\ Z \end{pmatrix} \wedge \begin{pmatrix} n_{0x} \\ n_{0y} \\ n_{0z} \end{pmatrix} \right] \tag{6.5}$

Be M the mass and $\mathbf{I}$ the matrix of inertias (taken at the center of gravity). The equations of translatory and angular motions (still assumed to be small) write

$$M \frac{\mathrm{d}^2}{\mathrm{d}t^2} \begin{pmatrix} x_G \\ y_G \\ z_G \end{pmatrix} = \vec{F} \tag{6.6}$$

$$\mathbf{I} \frac{\mathrm{d}^2}{\mathrm{d}t^2} \begin{pmatrix} \alpha \\ \beta \\ \gamma \end{pmatrix} = \vec{C} \tag{6.7}$$

with $(\vec{F}, \vec{C})$ the torque of the external loads in $Gxyz$.

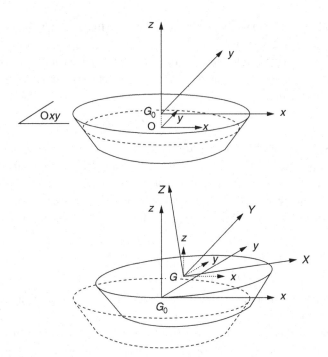

Figure 6.1 Coordinate systems.

The hydrodynamic torque $(\overrightarrow{F_H}, \overrightarrow{C_H})$ is obtained by integrating the fluid pressure over the wetted hull S_B:

$$\overrightarrow{F_H} = \iint_{S_B} p \, \overrightarrow{n} \, \mathrm{d}S \tag{6.8}$$

$$\overrightarrow{C_H} = \iint_{S_B} p \, \overrightarrow{GP} \wedge \overrightarrow{n} \, \mathrm{d}S \tag{6.9}$$

with P the running point on the hull, $\overrightarrow{n}$ the normal vector into the body.

The pressure is obtained from the Bernoulli equation:

$$p = p_0 - \rho g z - \rho \Phi_t - \frac{1}{2} \rho (\nabla \Phi)^2 \tag{6.10}$$

where the vertical coordinate z is taken in $Oxyz$.

Within the frame of a linearized theory, it is consistent to retain only the first-order terms and to write

$$-\rho \iint_{S_B} \left(\Phi_t(P,t) + \frac{1}{2} (\nabla \Phi)^2(P,t) \right) \overrightarrow{n} \, \mathrm{d}S \simeq -\rho \iint_{S_{B_0}} \Phi_t(P_0,t) \, \overrightarrow{n_0} \, \mathrm{d}S \tag{6.11}$$

$$-\rho \iint_{S_B} \left(\Phi_t(P,t) + \frac{1}{2} (\nabla \Phi)^2(P,t) \right) \overrightarrow{GP} \wedge \overrightarrow{n} \, \mathrm{d}S$$

$$\simeq -\rho \iint_{S_{B_0}} \Phi_t(P_0,t) \, \overrightarrow{G_0 P_0} \wedge \overrightarrow{n_0} \, \mathrm{d}S \tag{6.12}$$

where, again, the subscript $_0$ means in the position at rest.

The hydrostatic component $-\rho g z$ of the pressure must be addressed separately. It is of order ε only near the free surface, so the instantaneous position of the hull cannot be assimilated with its position at rest. The hydrostatic loads are obtained from the matrix of hydrostatic stiffnesses $\mathbf{K_H}$[1] in still water (the integration of the hydrostatic pressure, from the mean free surface up to the actual free surface, being of order ε^2, is neglected within a linearized theory).

After introducing the generalized vectors and matrices:

$$\vec{X}(t) = \begin{pmatrix} x_G(t) \\ y_G(t) \\ z_G(t) \\ \alpha(t) \\ \beta(t) \\ \gamma(t) \end{pmatrix} \qquad \vec{N_0} = \begin{pmatrix} n_{0x} \\ n_{0y} \\ n_{0z} \\ Y\,n_{0z} - Z\,n_{0y} \\ Z\,n_{0x} - X\,n_{0z} \\ X\,n_{0y} - Y\,n_{0x} \end{pmatrix} \tag{6.13}$$

$$\mathbf{M} = \begin{pmatrix} M & 0 & 0 & 0 & 0 & 0 \\ 0 & M & 0 & 0 & 0 & 0 \\ 0 & 0 & M & 0 & 0 & 0 \\ 0 & 0 & 0 & I_{44} & I_{45} & I_{46} \\ 0 & 0 & 0 & I_{45} & I_{55} & I_{56} \\ 0 & 0 & 0 & I_{46} & I_{56} & I_{66} \end{pmatrix} \tag{6.14}$$

$$\mathbf{K_H} = \begin{pmatrix} 0 & 0 & 0 & 0 & 0 & 0 \\ 0 & 0 & 0 & 0 & 0 & 0 \\ 0 & 0 & K_{33} & K_{34} & K_{35} & 0 \\ 0 & 0 & K_{34} & K_{44} & K_{45} & 0 \\ 0 & 0 & K_{35} & K_{45} & K_{55} & 0 \\ 0 & 0 & 0 & 0 & 0 & 0 \end{pmatrix} \tag{6.15}$$

and considering that the remaining external loads reduce to the mooring loads, expressed via a stiffness matrix $\mathbf{K_A}$, the equation of motion takes the form

$$\mathbf{M}\frac{d^2\vec{X}(t)}{dt^2} + (\mathbf{K_H} + \mathbf{K_A})\,\vec{X}(t) = -\rho \iint_{S_{B_0}} \Phi_t(P_0,t)\,\vec{N_0}\,dS \tag{6.16}$$

6.2 Linear Response in Regular Waves

We now assume the incoming wave system to be regular, at a frequency ω. Then all quantities vary in time at the frequency ω and we can write:

$$\Phi(x,y,z,t) = \Re\left\{\varphi(x,y,z)\,e^{-i\omega t}\right\} \tag{6.17}$$

$$\vec{X}(t) = \Re\left\{\vec{x}\,e^{-i\omega t}\right\} \tag{6.18}$$

---

[1] See Appendix B.

Equation (6.16) becomes:

$$(-\mathbf{M}\,\omega^2 + \mathbf{K_H} + \mathbf{K_A})\,\vec{x} = \mathrm{i}\,\omega\,\rho\,\iint_{S_{B_0}} \varphi\,\vec{N_0}\,\mathrm{d}S \tag{6.19}$$

The velocity potential is broken down into 3 components:

$$\varphi = \varphi_I + \varphi_D + \varphi_R \tag{6.20}$$

- φ_I the velocity potential of the incoming waves:

$$\varphi_I(x,y,z) = -\mathrm{i}\,\frac{A\,g}{\omega}\,\frac{\cosh k(z+h)}{\cosh kh}\,e^{\mathrm{i}k(x\cos\beta + y\sin\beta)} \tag{6.21}$$

β being the propagating angle with respect to Ox.
- φ_D the velocity potential of the wave system diffracted by the body assumed to be fixed (**diffraction** potential). It satisfies the following no-flow condition at the hull:

$$\nabla\varphi_D \cdot \vec{n_0} = -\nabla\varphi_I \cdot \vec{n_0} \tag{6.22}$$

- φ_R the **radiation** potential, due to the body motion in the absence of incoming waves:

$$\varphi_R = \sum_{j=1}^{6} -\mathrm{i}\,\omega\,x_j\,\varphi_{Rj} \tag{6.23}$$

where x_j stands for the j-th component of the body response $\vec{x}$.

In order that the no-flow condition (6.5) be verified the elementary potentials φ_{Rj} must satisfy

$$\nabla\varphi_{Rj} \cdot \vec{n_0} = N_{0j} \tag{6.24}$$

at the hull, with N_{0j} the successive components of $\vec{N_0}$.

The contribution of the potential φ_R to the hydrodynamic loads depends linearly on the response $\vec{x}$ and is expressed via the matrix of **added masses and added inertias** $\mathbf{M_a}$ and the matrix of **radiation dampings B**:

$$\mathrm{i}\,\omega\,\rho\,\iint_{S_{B_0}} \varphi_R\,\vec{N_0}\,\mathrm{d}S = \left[\mathbf{M_a}(\omega)\,\omega^2 + \mathrm{i}\,\mathbf{B}(\omega)\,\omega\right]\vec{x} \tag{6.25}$$

Finally, the equation of motion takes the form:

$$\left[-\left(\mathbf{M}+\mathbf{M_a}(\omega)\right)\omega^2 - \mathrm{i}\,\mathbf{B}(\omega)\,\omega + \mathbf{K_H} + \mathbf{K_A}\right]\vec{x} = \mathrm{i}\,\omega\,\rho\,\iint_{S_{B_0}} (\varphi_I + \varphi_D)\,\vec{N_0}\,\mathrm{d}S \tag{6.26}$$

Its resolution requires to have determined the matrices $\mathbf{M_a}$ and $\mathbf{B}$, and the right-hand side, denoted as the **excitation loads** or **diffraction loads** (which is ambiguous since for some people, diffraction load is sometimes taken to refer to the contribution of the diffraction potential φ_D alone).

When only the contribution of the incoming velocity potential φ_I is used to derive the excitation loads, this is known as **Froude–Krylov** approximation.

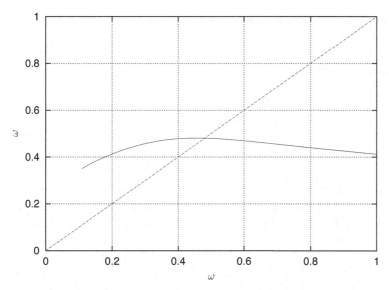

Figure 6.2 Graphical determination of the heave natural frequency for a rectangular barge.

Natural Frequencies

The natural frequencies can be obtained as the roots of the determinant of the matrix

$$- \left(\mathbf{M} + \mathbf{M_a}(\omega) \right) \omega^2 + \mathbf{K_H} + \mathbf{K_A} \tag{6.27}$$

It must be ensured that the obtained natural frequency ω_{j0} for the degree of freedom j coincides with the frequency ω used to compute $\mathbf{M_a}$. This may require an iterative scheme, or a graphical approach, as illustrated in Figure 6.2, where the natural frequency of the heave motion of a rectangular barge is obtained from the intersection of the two curves $y = \sqrt{K_{H33}/(M + M_{a33}(\omega))}$ and $y = \omega$.

The evaluation of the natural frequencies is an important step in the design of offshore systems. In particular, it is important to locate them with regards to the energetic wave frequencies (and, preferably, to avoid that they coincide).

6.3   The Diffraction-Radiation Problem

The diffraction potential φ_D and the 6 elementary radiation potentials φ_{Rj} are solutions of the following Boundary Value Problem:

$$
\begin{aligned}
\Delta \varphi &= 0 & &\text{in the fluid domain} \\
\nabla \varphi \cdot \vec{n}_0 &= f_j & &\text{at the hull } S_{B_0} \\
\varphi_z &= 0 & &\text{at the sea floor } z = -h \qquad (6.28)\\
g\, \varphi_z - \omega^2 \varphi &= 0 & &\text{at the mean free surface } z = 0 \\
\text{Rad}\,(\varphi) & & &x^2 + y^2 \to \infty
\end{aligned}
$$

where, for the diffraction problem

$$f_0 = -\nabla \varphi_I \cdot \vec{n}_0 \tag{6.29}$$

and, for the radiation problems

$$f_j = N_{0j} \tag{6.30}$$

The radiation condition is usually written as the Sommerfeld condition

$$\lim_{R \to \infty} \sqrt{R} \left(\frac{\partial \varphi}{\partial R} - ik\,\varphi \right) = 0 \tag{6.31}$$

where R is the radial distance $\sqrt{x^2 + y^2}$. This condition expresses that, away from the structure, the diffraction and radiation waves propagate in the radial direction with wave number k and amplitudes decaying as $1/\sqrt{R}$ [2]. Asymptotically, the diffraction potential has the form

$$\varphi_D = -i\, \frac{Ag}{\omega} \frac{\cosh k(z+h)}{\cosh kh} \sqrt{\frac{2}{\pi k R}} \, H_D(\theta) \, e^{ikR - i\pi/4} + O(R^{-1}) \tag{6.32}$$

with θ the azimuthal angle.

The angular spreading function $H_D(\theta)$ is known as the Kochin function. Likewise, the radiation potentials have the following far-field expressions:

$$\varphi_{Rj} = \frac{g}{\omega^2} \frac{\cosh k(z+h)}{\cosh kh} \sqrt{\frac{2}{\pi k R}} \, H_{Rj}(\theta) \, e^{ikR - i\pi/4} + O(R^{-1}) \tag{6.33}$$

When the equation of motion (6.26) has been solved the far-field behavior of the total potential can be obtained. An application is the computation of the wave drift forces, considered in Chapter 7.

The Sommerfeld radiation condition is meant to eliminate some non-physical solutions to the Boundary Value Problem (6.28), such as waves that would be standing, not progressive in the radial direction. However there are cases when the solution to the BVP is not unique. In the case of a simply connected body, Fritz John has shown that the solution is unique provided that the hull be contained within the vertical cylinder running through its waterline. A counterexample would be a tapered circular island where so-called **edge waves** can be traveling around, with an exponential decay in the radial direction.

An infinite array of vertical cylinders or, equivalently, one cylinder in between two walls (in a wave flume) is also a case where the solution can be nonunique and where so-called **trapped modes** can be observed (when some relationships between wave number, cylinder diameter, and width of the canal, are fulfilled – see Callan *et al.*, 1991). In the case of a finite number of vertical cylinders, strong resonant amplifications of the free surface elevation can be observed locally at some frequencies (**nearly trapped modes**). This may be a concern for airgap design in the case of semi-submersible platforms or TLPs (Tension Leg Platforms).

---

[2] Here we consider three-dimensional bodies, bounded in the horizontal directions. For two-dimensional bodies, the far-field diffracted and radiated wave systems consist in plane waves, propagating to infinity.

When the elementary radiation potentials φ_{Rj} have been determined, the components of the matrices of added masses-inertias and radiation dampings are obtained from:

$$\left[\mathbf{M_a}(\omega) + i\,\frac{\mathbf{B}(\omega)}{\omega} \right]_{jk} = \rho \iint_{S_{B_0}} \varphi_{Rk}\, N_{0j}\, dS = \rho \iint_{S_{B_0}} \varphi_{Rk}\, \frac{\partial \varphi_{Rj}}{\partial n_0}\, dS \quad (6.34)$$

It results from Green's second identity that the matrices are symmetric:

$$M_{ajk}(\omega) = M_{akj}(\omega) \qquad\qquad B_{jk}(\omega) = B_{kj}(\omega) \qquad (6.35)$$

Physically $\mathfrak{R}\{[M_{ajk}(\omega)\, \omega^2\, x_k + i\, B_{jk}(\omega)\, \omega\, x_k]\, \exp(-i\omega t)\}$ is the hydrodynamic load in the degree of freedom j when the structure oscillates with an amplitude x_k along the degree of freedom k. In the case of a port-starboard symmetry, a forced motion in surge, heave, or pitch induces no load in sway nor roll nor yaw. As a result, the components of the matrices with indices 21, 41, 61, 23, 43, 63, 25, 45, 65 are zero and, by the symmetry of the matrices, also the terms with the inverted indices. In practice, for ship-like bodies, the main coupling is between sway and roll, and, to a lesser extent, between surge and pitch or heave and pitch.

Since they represent an energy dissipation via the radiated wave field, the diagonal terms B_{jj} of the damping matrix are positive. The diagonal terms of the added mass matrix are not always positive, unlike what would happen in an unbounded fluid domain (without free surface). For instance the heave added mass of a catamaran, or of a drillship with moonpool, can be negative when the frequency is close to the natural frequency of the piston mode (vertical motion of the entrapped water). Likewise, the sway added mass can be negative when a sloshing motion resonates between the two hulls; analogous behaviors are found for ships moored alongside quays. Negative added masses also occur in the case of submerged bodies close to the free surface (Newman *et al.*, 1984; Martin & Farina, 1997).

6.3.1 Haskind-Hanaoka Theorem

Consider the j component of the diffraction torque:

$$f_j = i\rho\omega \iint_{S_{B_0}} (\varphi_I + \varphi_D)\, N_{0j}\, dS \qquad (6.36)$$

It can also be written

$$f_j = i\rho\omega \iint_{S_{B_0}} (\varphi_I + \varphi_D)\, \frac{\partial \varphi_{Rj}}{\partial n}\, dS \qquad (6.37)$$

or

$$f_j = i\rho\omega \iint_{S_{B_0}} \left[(\varphi_I + \varphi_D)\, \frac{\partial \varphi_{Rj}}{\partial n} - \varphi_{Rj}\, \frac{\partial(\varphi_I + \varphi_D)}{\partial n} \right] dS \qquad (6.38)$$

Applying Green's second identity, the S_{B_0} integral can be transferred to a vertical cylinder S_∞ surrounding the body (with a change in sign):

$$f_j = -i\rho\omega \iint_{S_\infty} \left[(\varphi_I + \varphi_D)\, \frac{\partial \varphi_{Rj}}{\partial n} - \varphi_{Rj}\, \frac{\partial(\varphi_I + \varphi_D)}{\partial n} \right] dS \qquad (6.39)$$

When the radius of the cylinder goes to infinity, from the asymptotic forms of φ_D (6.32) and φ_{Rj} (6.33), it is easy to show that the term involving their cross-products is nil. There only remains

$$f_j = -\mathrm{i}\rho\omega \iint_{S_\infty} \left[\varphi_I \frac{\partial\varphi_{Rj}}{\partial n} - \varphi_{Rj} \frac{\partial\varphi_I}{\partial n}\right] \mathrm{d}S \tag{6.40}$$

The evaluation of this integral requires the use of the theorem of the stationary phase. The j load component is finally obtained as

$$f_j = -4\mathrm{i}\,\rho\,A\,g\,\frac{C_G}{\omega\,k}\,H_{Rj}(\beta + \pi) \tag{6.41}$$

known as Haskind-Hanaoka's formula.

The j component of the excitation loads is therefore directly proportional to the intensity of the radiated waves in the direction opposite to the wave propagation.

Haskind-Hanaoka's formula is mostly helpful as a check for the accuracy of the numerical results.

6.4   Methods of Resolution

6.4.1   Analytical Methods

In the general case, a purely numerical approach must be used. However there are simple geometries that can be tackled by analytical, or semi-analytical, methods. This is the case for a vertical cylinder, bottom-mounted, and for arrays of vertical cylinders.

For a vertical axisymmetric body a cylindrical coordinate system is appropriate. A first step is to derive elementary solutions to the Laplace equation.

Be Oz the vertical axis, with $z = 0$ the still free surface and R, θ the polar coordinates in the horizontal plane: $x = R\cos\theta$, $y = R\sin\theta$. When the velocity potential φ is expressed as a function of (R, θ, z), the Laplace equation takes the form

$$\Delta\varphi = \varphi_{RR} + \frac{1}{R}\varphi_R + \frac{1}{R^2}\varphi_{\theta\theta} + \varphi_{zz} = 0 \tag{6.42}$$

The potential is now written as a Fourier series with respect to the polar angle θ:

$$\varphi(R,\theta,z) = \sum_{m=0}^{\infty} \left[\varphi_{cm}(R,z)\,\cos m\theta + \varphi_{sm}(R,z)\,\sin m\theta\right] \tag{6.43}$$

For the sake of simplicity, we assume the flow to be symmetric with respect to the plane $y = 0$. It results that the φ_{sm} are nil. For the $\varphi_{cm} = \varphi_m$ terms the Laplace equation is:

$$\varphi_{mRR} + \frac{1}{R}\varphi_{mR} - \frac{m^2}{R^2}\varphi_m + \varphi_{mzz} = 0 \tag{6.44}$$

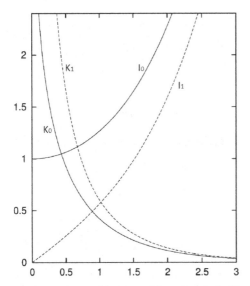

Figure 6.3 The modified Bessel functions I_0, I_1, K_0, K_1.

Elementary solutions in the form $\varphi_m(R,z) = F_m(R)\, G_m(z)$ must verify

$$\frac{F_m''}{F_m} + \frac{1}{R}\frac{F_m'}{F_m} - \frac{m^2}{R^2} = \lambda \tag{6.45}$$

$$\frac{G_m''}{G_m} = -\lambda \tag{6.46}$$

Different kinds of solutions are obtained depending on whether λ is positive, zero, or negative.

Case $\lambda = k^2$

The F_m equation becomes

$$R^2 F_m'' + R F_m' - (k^2 R^2 + m^2) F_m = 0 \tag{6.47}$$

Elementary solutions are the modified Bessel functions $I_m(kR)$ and $K_m(kR)$. They are plotted in Figure 6.3 for $m = 0$ and $m = 1$.

The I_m functions are nil in $R = 0$ except for $m = 0$ and they increase quasi-exponentially with the radial distance. The K_m functions are singular in $R = 0$ and they decrease quasi-exponentially with R. In a fluid domain containing the Oz axis the K_m cannot be retained whereas it is the other way around for a fluid domain extending to infinity. In an annular domain both are acceptable.

The associated equation for $G_m(z)$ is

$$G_m'' + k^2 G_m = 0 \tag{6.48}$$

G_m must also verify the bottom boundary condition $G_{mz} \equiv \partial G_m/\partial z = 0$ in $z = -h$ and the free surface condition in $z = 0$ ($g\, G_{mz} - \omega^2 G_m = 0$).

Elementary solutions take the form

$$G_m = \cos k_n \, (z + h) \tag{6.49}$$

where the k_n wave numbers are the roots of the equation

$$\omega^2 = -g \, k_n \, \tan k_n h \tag{6.50}$$

There is a discrete set of solutions that behave asymptotically as $n\,\pi/h - \omega^2/(n\,\pi\,g)$ when $n \to \infty$ (see Figure A.6 in Appendix A).

Case $\lambda = 0$

The F_m equation is now

$$R^2 \, F_m'' + R \, F_m' - m^2 \, f = 0 \tag{6.51}$$

Solutions are R^m et R^{-m} for $m \geq 1$ and $\ln R$ for $m = 0$. Associated G_m functions are linear functions of z. As a result, they cannot satisfy both the bottom boundary condition and the free surface equation.

Such solutions would be acceptable in a fluid domain bounded by two rigid horizontal planes (for instance, in the fluid domain below a truncated cylinder).

Case $\lambda = -k^2$

The F_m equation is now

$$R^2 \, F_m'' + R \, F_m' + (k^2 \, R^2 - m^2) \, F_m = 0 \tag{6.52}$$

Elementary solutions are the Bessel functions $J_m(kR)$ and $Y_m(kR)$. The J_m functions, like I_m, are nil at the origin except for $m = 0$, whereas the Y_m functions are singular there. Figure 6.4 shows the functions $J_m(kR)$ and $Y_m(kR)$ for $m = 0$ and $m = 1$.

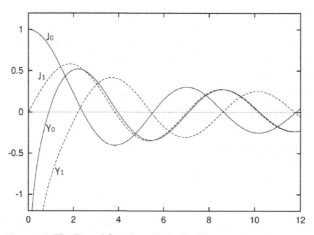

Figure 6.4 The Bessel functions J_0, J_1, Y_0, Y_1.

When kR increases J_m and Y_m asymptotically behave as

$$J_m(kR) \simeq \sqrt{\frac{2}{\pi kR}} \cos\left(kR - \frac{m\pi}{2} - \frac{\pi}{4}\right) \qquad (6.53)$$

$$Y_m(kR) \simeq \sqrt{\frac{2}{\pi kR}} \sin\left(kR - \frac{m\pi}{2} - \frac{\pi}{4}\right) \qquad (6.54)$$

In order for φ to verify the radiation condition (6.31), J_m and Y_m must be associated in the form

$$H_m = J_m + i\, Y_m \qquad (6.55)$$

where H_m is the Hankel function of the first kind.

The associated $G_m(z)$ function is

$$G_m = \cosh k_0(z + h) \qquad (6.56)$$

where k_0 is the unique root of the dispersion equation

$$\omega^2 = g\, k_0 \tanh k_0 h \qquad (6.57)$$

Wave Diffraction by a Vertical Cylinder

Consider a vertical cylinder of radius a, standing on the sea floor. In a cylindrical coordinate system (R, θ, z) centered at the intersection of the cylinder axis with the free surface, the velocity potential of the incoming waves, propagating along the Ox axis, can be expanded as:

$$\varphi_I(R, \theta, z) = -i\, \frac{A g}{\omega} \frac{\cosh k_0(z + h)}{\cosh k_0 h}\, e^{ik_0 R \cos\theta}$$

$$= -i\, \frac{A g}{\omega} \frac{\cosh k_0(z + h)}{\cosh k_0 h} \sum_{m=0}^{\infty} \epsilon_m\, i^m\, J_m(k_0 R) \cos m\theta \qquad (6.58)$$

where J_m is the Bessel function of the first kind and $\epsilon_m = 1$ for $m = 0$, $\epsilon_m = 2$ otherwise.

From the previous analysis, it is straightforward to express the diffraction potential φ_D as

$$\varphi_D(R, \theta, z) = -i\, \frac{A g}{\omega} \frac{\cosh k_0(z + h)}{\cosh k_0 h} \sum_{m=0}^{\infty} \epsilon_m\, i^m\, B_m\, H_m(k_0 R) \cos m\theta \qquad (6.59)$$

The Laplace equation, together with the free surface, bottom, and radiation conditions, is satisfied.

It only remains to verify the no-flow condition on the cylinder, which gives

$$J'_m(k_0 a) + B_m\, H'_m(k_0 a) = 0 \qquad (6.60)$$

This yields the B_m coefficients: $B_m = -J'_m(k_0a)/H'_m(k_0a)$. The total velocity potential, in the fluid domain $R \geq a$, is

$$\varphi = \varphi_I + \varphi_D = \sum_{m=0}^{\infty} \epsilon_m \, i^m - i \, \frac{A \, g}{\omega} \, \frac{\cosh k_0(z+h)}{\cosh k_0 h}$$

$$\left(J_m(k_0 R) - \frac{J'_m(k_0 a)}{H'_m(k_0 a)} H_m(k_0 R) \right) \cos m\theta \qquad (6.61)$$

The horizontal load is obtained by integrating the pressure $i \, \rho\omega\varphi$ on the cylinder. In (6.61) only the $m = 1$ term contributes. One obtains

$$f_x = -2i\pi \, \rho \, A \, g \, a^2 \, \frac{\tanh k_0 h}{k_0 a} \left(J_1(k_0 a) - \frac{J'_1(k_0 a)}{H'_1(k_0 a)} H_1(k_0 a) \right) \qquad (6.62)$$

where $F_x = \Re \{f_x \exp(-i \omega t)\}$. This expression can be reduced by applying the derivation formulas for the Bessel functions

$$\frac{\partial}{\partial z} \begin{pmatrix} J_m(z) \\ Y_m(z) \end{pmatrix} = \begin{pmatrix} J_{m-1}(z) \\ Y_{m-1}(z) \end{pmatrix} - \frac{m}{z} \begin{pmatrix} J_m(z) \\ Y_m(z) \end{pmatrix} \qquad (6.63)$$

and the Wronskian formula:

$$J_m(z) \, Y_{m-1}(z) - J_{m-1} \, Y_m(z) = \frac{2}{\pi z} \qquad (6.64)$$

so that, finally

$$f_x = 4\rho \, g \, A \, a^2 \, \frac{\tanh k_0 h}{(k_0 a)^2} \, \frac{1}{H'_1(k_0 a)} \qquad (6.65)$$

The overturning moment, with respect to a point at the sea floor, is obtained as $c_y = L_A f_x$ where the lever arm L_A is

$$L_A = h \left[1 - \frac{1}{k_0 h} \left(\coth k_0 h - \frac{1}{\sinh k_0 h} \right) \right] \qquad (6.66)$$

In large depth $L_A = h - 1/k_0$: the point of application of the horizontal load is at a distance $1/k_0 = L/(2\pi)$ below the mean sea level.

These results were first obtained by Havelock in infinite depth and extended to finite depth by McCamy & Fuchs. They provide exact values, which are helpful as references for purely numerical methods, or for approximate formulas such as the Morison equation (see Section 4.3.1). Retaining only the inertia term in the latter, one has

$$dF_{\mathrm{Mor}} = 2 \rho \, \pi \, a^2 \, \frac{\partial^2 \Phi_I}{\partial x \, \partial t} \, dz \qquad (6.67)$$

which gives, upon integration from $z = -h$ up to $z = 0$:

$$f_{\mathrm{Mor}} = -2i \, \pi \, \rho \, A \, g \, a^2 \, \tanh k_0 h \qquad (6.68)$$

where $F_{\mathrm{Mor}} = \Re \{f_{\mathrm{Mor}} \exp(-i \omega \, t)\}$

Figure 6.5 shows the modulus, the real and imaginary parts of the nondimensional load $f_x/(\rho \, A \, g \, \pi a^2 \, \tanh k_0 h)$, vs $k_0 a$. As $k_0 a$ decreases to zero, the Morison

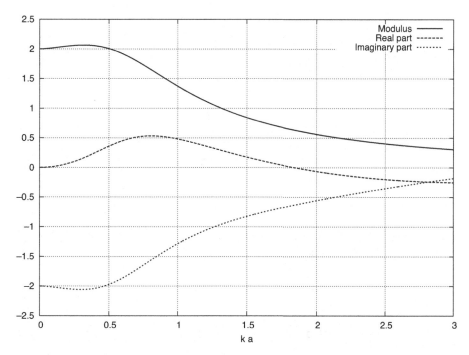

Figure 6.5 Wave diffraction by a vertical cylinder. Modulus, real, and imaginary parts of the nondimensional horizontal load, vs $k_0 a$.

coefficient $(0, -2\,\mathrm{i})$ is recovered. From the figure, this value looks like a reasonable approximation when $k_0 a$ is less than about 0.5, meaning a wave length greater than about 6 diameters.

From equation (6.61) the free surface elevation around the cylinder is obtained as

$$\eta(\theta, t) = A \; \mathfrak{R} \; \left\{ f_\eta(\theta) \, e^{-i \omega t} \right\} \tag{6.69}$$

where

$$f_\eta(\theta) = \frac{2}{\pi k_0 a} \sum_{m=0}^{\infty} \frac{\epsilon_m \; i^{m+1}}{H'_m(k_0 a)} \cos m\theta \tag{6.70}$$

Figure 6.6 shows the modulus of the RAO ("Response Amplitude Operator") of the free surface elevation vs the azimuthal angle counted from upwave, for 3 $k_0 a$ values.

Figure 6.7 shows the modulus of the RAO of the free surface elevation at the upwave point, as a function of $k_0 a$. For $k_0 a \sim 0$, the RAO is equal to 1: the cylinder is completely transparent to the incoming waves. The asymptotic value, for $k_0 a \to \infty$, is equal to 2. It can be seen that the evolution from $k_0 a = 0$ is not monotonous.

Wave Radiation by a Vertical Cylinder

The radiation problem (in surge or pitch) can likewise be solved analytically.

Considering the surge radiation problem, the no-flow condition at the cylinder wall writes

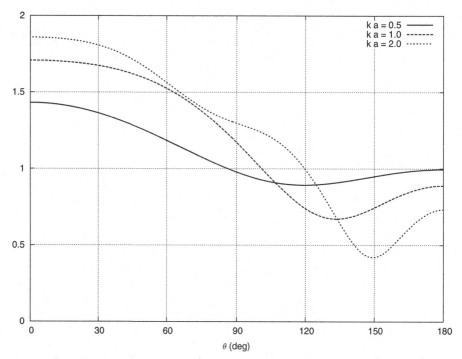

Figure 6.6 Regular wave interaction with a vertical cylinder. RAO of the free surface elevation around the cylinder vs the azimuthal angle (in deg), for $k_0 a = 0.5$, 1, and 2.

$$\left.\frac{\partial \varphi_{R1}}{\partial R}\right|_{R=a} = -n_{0x} = \cos \theta \qquad (6.71)$$

The radiation potential can be searched for under the form

$$\varphi_{R1}(R,\theta,z) = \cos \theta \left\{ A_0 \frac{\cosh k_0(z+h)}{\cosh k_0 h} \frac{H_1(k_0 R)}{k_0 H_1'(k_0 a)} + \sum_{n=1}^{\infty} A_n \cos k_n(z+h) \frac{K_1(k_n R)}{k_n K_1'(k_n a)} \right\} \qquad (6.72)$$

The first term in (6.72) is the progressive mode, radially outgoing and decaying as $R^{-1/2}$ in the far field, whereas the following terms, decaying quasi exponentially, are appreciable only locally, and are denoted as **local**, or **evanescent**, modes (see the piston wavemaker problem in Appendix A).

The no-flow condition gives

$$A_0 \frac{\cosh k_0(z+h)}{\cosh k_0 h} + \sum_{n=1}^{\infty} A_n \cos k_n(z+h) = 1 \qquad (6.73)$$

which must be verified for $-h \leq z \leq 0$.

Multiplying both sides of equation (6.73) with $\cosh k_0(z+h)$ (then $\cos k_n(z+h)$), and integrating from $-h$ to 0, gives the A_0 (then A_n) coefficients as

$$A_0 = \frac{2 \sinh 2k_0 h}{2k_0 h + \sinh 2k_0 h} \qquad A_n = \frac{4 \sin k_n h}{2k_n h + \sin 2k_n h} \qquad (6.74)$$

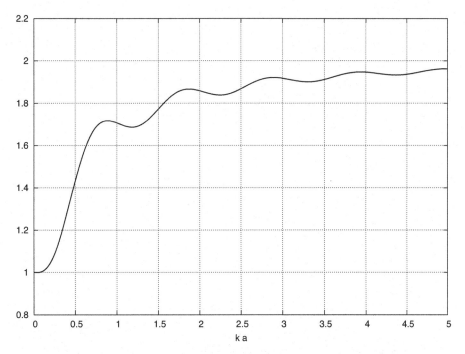

Figure 6.7 Regular wave interaction with a vertical cylinder. RAO of the free surface elevation at the upwave point ($\theta = \pi$), vs $k_0 a$.

The surge added mass M_{a11} and damping B_{11} coefficients are then obtained as

$$M_{a11} + \mathrm{i}\,\frac{B_{11}}{\omega} = -\rho\,\pi\,a\left(A_0\,\frac{\mathrm{H}_1(k_0 a)\,\tanh\,k_0 h}{k_0^2\,\mathrm{H}_1'(k_0 a)} + \sum_{n=1}^{\infty} A_n\,\frac{\mathrm{K}_1(k_n a)\,\sin\,k_n h}{k_n^2\,\mathrm{K}_1'(k_n a)}\right) \quad (6.75)$$

where it can be noted that the evanescent modes only contribute to the added mass, while the progressive mode contributes both to added mass and damping (this is at variance with the two-dimensional wavemaker case where the progressive mode only contributes to damping). Nondimensional added mass and damping coefficients are shown in Figures 6.8 and 6.9, for different values of the a/h ratio. It can be seen that the damping coefficient peaks at a $k_0 a$ value where the slope of the added mass curve is maximum.

The radiation potential in pitch can similarly be obtained. The associated no-flow condition is

$$\left.\frac{\partial \varphi_{R5}}{\partial R}\right|_{R=a} = -(z+h)\,n_{0x} = (z+h)\,\cos\theta \quad (6.76)$$

Then the A_0, A_n coefficients must verify

$$A_0\,\frac{\cosh k_0(z+h)}{\cosh k_0 h} + \sum_{n=1}^{\infty} A_n\,\cos k_n(z+h) = z+h \qquad -h \le z \le 0 \quad (6.77)$$

and can be obtained by the same technique.

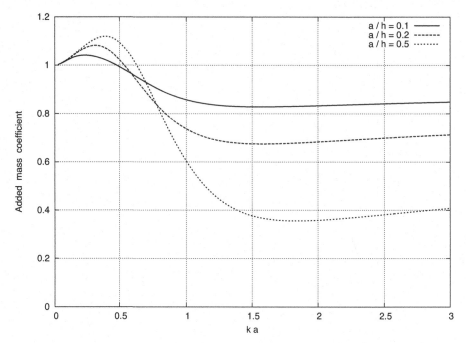

Figure 6.8 Bottom-mounted vertical cylinder. Surge added mass coefficient $M_{a11}/(\rho\pi a^2 h)$ vs $k_0 a$ for different ratios a/h.

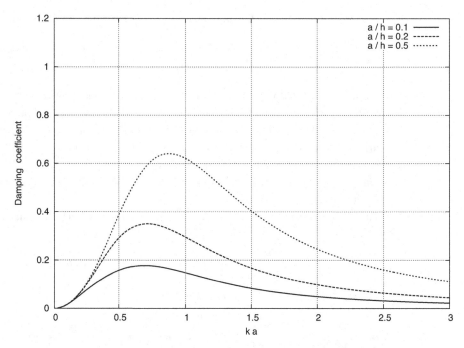

Figure 6.9 Bottom-mounted vertical cylinder. Surge radiation damping coefficient $B_{11}/(\omega\rho\pi a^2 h)$ vs $k_0 a$ for different ratios a/h.

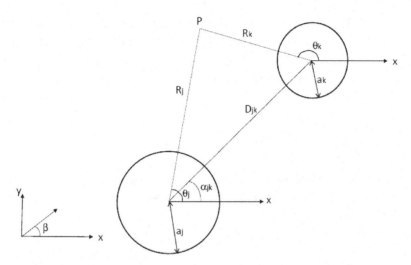

Figure 6.10 Geometry of multiple cylinders.

Wave Diffraction by an Array of Vertical Cylinders

Many offshore structures, such as semis or TLPs, are composed of vertical cylinders, in close proximity. This is also a recurring geometry in coastal engineering, where many structures are supported by piles. Semis also have pontoons but, those being deeply submerged, most of the diffraction phenomena are associated with the columns. Due to their interaction, some nearly resonant phenomena can be observed with the free surface elevation reaching very high values.

When the pontoons are neglected, and the vertical columns are standing on the sea floor, semi-analytical methods can be used to derive the diffracted potential. Here we follow the method of Spring & Monkmeyer (1974), later extended by Linton & Evans (1990).

Consider N_C vertical cylinders, of radii a_j, with their vertical axes centered in x_j, y_j (Figure 6.10).

The incoming wave system propagates at an angle β. Its velocity potential is

$$\varphi_I(x, y, z) = C(z)\, e^{ik_0(x\cos\beta + y\sin\beta)} \tag{6.78}$$

where

$$C(z) = -i\,\frac{A\,g}{\omega}\,\frac{\cosh k_0(z+h)}{\cosh k_0 h} \tag{6.79}$$

The method used in the previous section suggests writing the diffracted potential in the following way:

$$\varphi_D(P) = C(z) \sum_{j=1}^{N_C} \sum_{m=-\infty}^{\infty} A_{jm}\, E_{jm}\, \mathrm{H}_m(k_0 R_j)\, e^{im\theta_j} \tag{6.80}$$

Here E_{jm} is a normalizing coefficient $E_{jm} = \mathrm{J}'_m(k_0 a_j)/\mathrm{H}'_m(k_0 a_j)$ and (R_j, θ_j) are the polar coordinates of a point P in the coordinate system centered in j (see Figure 6.10).

Only the no-flow conditions at the cylinder walls remain to be fulfilled. In the k coordinate system, the incident velocity potential can be expanded as

$$\varphi_I(R_k, \theta_k, z) = C(z) \, I_k \sum_{m=-\infty}^{\infty} J_m(k_0 R_k) \, e^{im(\frac{\pi}{2}-\beta)} \, e^{im\theta_k} \tag{6.81}$$

where

$$I_k = e^{ik_0(x_k \cos\beta + y_k \sin\beta)} \tag{6.82}$$

To express, in the k coordinate system, the diffracted potentials emanating from the $N_C - 1$ other cylinders, the Graf addition theorem (e.g., see Linton & McIver, 2001) is applied:

$$H_m(k_0 R_j) \, e^{im(\theta_j - \alpha_{jk})} = \sum_{n=-\infty}^{\infty} H_{m+n}(k_0 D_{jk}) \, J_n(k_0 R_k) \, e^{in(\pi + \alpha_{jk} - \theta_k)} \tag{6.83}$$

valid under the condition $R_k < D_{jk}$, therefore valid in the neighborhood of the k cylinder.

Changing n into $-n$, and accounting for $J_{-n} = (-1)^n J_n$, one gets:

$$H_m(k_0 R_j) \, e^{im\theta_j} = \sum_{n=-\infty}^{\infty} H_{m-n}(k_0 D_{jk}) \, J_n(k_0 R_k) \, e^{i(m-n)\alpha_{jk}} \, e^{in\theta_k} \tag{6.84}$$

The diffracted potential, from the $N_C - 1$ other cylinders, takes the following form, in the k cylinder coordinate system:

$$\varphi_{D_{j\neq k}} = C(z) \sum_{\substack{j=1 \\ j\neq k}}^{N_C} \sum_{m=-\infty}^{\infty} A_{jm} E_{jm} \sum_{n=-\infty}^{\infty} H_{m-n}(k_0 D_{jk}) \, e^{i(m-n)\alpha_{jk}} \, J_n(k_0 R_k) \, e^{in\theta_k} \tag{6.85}$$

Inverting the summations, and interchanging the indices m and n, gives:

$$\varphi_{D_{j\neq k}} = C(z) \sum_{m=-\infty}^{\infty} \left\{ \sum_{\substack{j=1 \\ j\neq k}}^{N_C} \sum_{n=-\infty}^{\infty} A_{jn} E_{jn} \, H_{n-m}(k_0 D_{jk}) \, e^{i(n-m)\alpha_{jk}} \right\} J_m(k_0 R_k) \, e^{im\theta_k} \tag{6.86}$$

Finally, the no-flow condition at the cylinder k is obtained as

$$A_{km} + \sum_{\substack{j=1 \\ j\neq k}}^{N_C} \sum_{n=-\infty}^{\infty} A_{jn} E_{jn} \, H_{n-m}(k_0 D_{jk}) \, e^{i(n-m)\alpha_{jk}} = -I_k \, e^{im(\frac{\pi}{2}-\beta)} \tag{6.87}$$

for $m = -\infty, \ldots, 0, 1, \ldots, \infty$.

The infinite m series are truncated at an order M such that remaining terms are illusory (typically M is between 5 and 10). A linear system of rank $N_C (2M + 1)$ is thus obtained, and solved numerically, delivering the A_{jm} coefficients.

Figure 6.11 RAO of the free surface elevation around a square 4 cylinder array. Nondimensional radius $k_0 a = 0.63$; nondimensional axis-to-axis distance $k_0 d = 3.25$ (courtesy Y.-M. Scolan).

In the neighborhood of the k cylinder, the total velocity potential (incident + diffracted) can finally be written as

$$\varphi(R_k, \theta_k, z) = C(z) \sum_{m=-M}^{M} A_{km} \left(E_{km} \, \mathrm{H}_m(k_0 R_k) - \mathrm{J}_m(k_0 R_k) \right) \mathrm{e}^{im\theta_k} \qquad (6.88)$$

valid for R_k verifying $R_k < D_{jk} \; \forall j \neq k$

On the wall of the k cylinder, application of the Wronskian formula permits to simplify even further φ:

$$\varphi(R_{k0}, \theta_k, z) = C(z) \frac{-2i}{\pi k_0 R_{k0}} \sum_{m=-M}^{M} \frac{A_{km}}{\mathrm{H}'_m(k_0 R_{k0})} \mathrm{e}^{im\theta_k} \qquad (6.89)$$

from which the hydrodynamic loads are easily obtained.

From (6.78) and (6.80) the velocity potential is obtained in the complete fluid domain. Free surface elevations can then be visualized around the cylinders. An example is provided in Figure 6.11.

6.4.2 Matched Eigenfunction Expansions

The method of matched eigenfunctions expansions applies to simple geometries, such that there exists a coordinate system (cartesian, cylindrical, spherical, etc.) where the fluid domain divides into rectangular subdomains. In two dimensions a simple case is a rectangular barge. This case is taken here as an example.

The fluid domain is decomposed into 3 rectangular subdomains: A, below the barge; B, on the right-hand side; and C, on the left-hand side (see Figure 6.12).

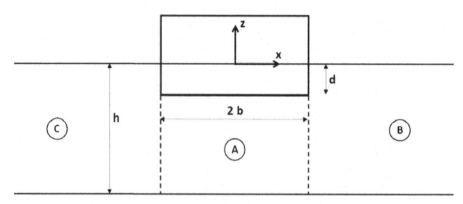

Figure 6.12 Rectangular barge. Different subdomains.

Consider the diffraction problem. Regular incoming waves are propagating along the Ox axis, with the velocity potential

$$\varphi_I = -i\,\frac{Ag}{\omega}\,\frac{\cosh k_0(z+h)}{\cosh k_0 h}\,e^{i k_0 x} \tag{6.90}$$

In domain B a general form of the diffraction potential is

$$\varphi_D = -i\,\frac{Ag}{\omega}\left\{b_0\,\frac{\cosh k_0(z+h)}{\cosh k_0 h}\,e^{i k_0(x-b)} + \sum_{n=1}^{\infty} b_n\,\cos k_n(z+h)\,e^{-k_n(x-b)}\right\} \tag{6.91}$$

In domain C:

$$\varphi_D = -i\,\frac{Ag}{\omega}\left\{c_0\,\frac{\cosh k_0(z+h)}{\cosh k_0 h}\,e^{-i k_0(x+b)} + \sum_{n=1}^{\infty} b_n\,\cos k_n(z+h)\,e^{k_n(x+b)}\right\} \tag{6.92}$$

In (6.91) and (6.92), again, k_0, k_n are the roots of the dispersion equation.

In domain A, the sum $\varphi_I + \varphi_D$ can be expanded as

$$\varphi_I + \varphi_D = -i\,\frac{Ag}{\omega}\left\{a_{01} + a_{02}\,\frac{x}{b} + \sum_{n=1}^{\infty}\left(a_{n1}\,\frac{\cosh \lambda_n x}{\cosh \lambda_n b} + a_{n2}\,\frac{\sinh \lambda_n x}{\sinh \lambda_n b}\right)\cos \lambda_n(z+h)\right\} \tag{6.93}$$

where $\lambda_n = n\pi/(h-d)$ and the constant a_{01} must not be omitted.

All that remains to do is match the velocity potentials and their x derivatives at the boundaries in $x = \pm b$, and ensure $\partial/\partial x(\varphi_I + \varphi_D) = 0$ in $x = \pm b$, $-d \le z \le 0$. Different numerical techniques can be used, such as minimizing a certain norm (Mei & Black, 1969). Advantage can also be taken of the orthogonality of the z functions to eliminate the a_{ij} coefficients and end up with two linear systems for $\vec{B} + \vec{C}$ and $\vec{B} - \vec{C}$ where $\vec{B} = (b_0, b_1, ..., b_N)$ and $\vec{C} = (c_0, c_1, ..., c_N)$ (Cointe et al., 1990).

The surge, heave, and roll radiation problems may similarly be solved. In the case of heave and roll, a particular solution must be added to the expansion in domain A, which is $[(z+h)^2 - x^2]/2(h-d)$ for heave and $[x^3/3 - x(z+h)^2]/2(h-d)$ for roll.

A drawback of the method is that it fails to properly account for the singularity of the velocity potential at the square barge corners. Locally

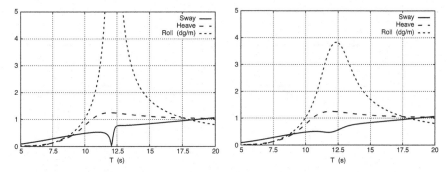

Figure 6.13 Two-dimensional rectangular barge. Calculated RAOs. Left: no roll viscous damping. Right: with viscous damping in roll.

$\varphi \sim \left((x \pm b)^2 + (z + d)^2\right)^{1/3}$, meaning that the potential is finite at the corner but the velocity is singular. Some numerical accuracy problems will result when computing the wave drift force through the pressure integration method, for instance.

An improvement is to use a different set of functions on the vertical cut, which have the proper singular behavior at the corner. The reader is referred to Fernyhough & Evans (1995) or to Faltinsen *et al.*, (2007) where such techniques are implemented.

As an example we take a rectangular barge with the same dimensions as in Section 6.8.2, that is, the beam is 60 m and the draft 12 m, in 100 m water depth. Figure 6.13 shows the calculated RAOs in sway, heave, and roll, without accounting for additional viscous damping (left), and with quadratic additional damping in roll (right). The C_α coefficient is taken equal to 0.15 (see equation (6.127)) and the regular wave amplitude is 1 m. Without viscous damping the peak roll RAO is equal to 17.5 deg/m.

In three dimensions, the eigenfunction expansion method is well-suited to vertical bodies of revolution, using cylindrical coordinates. The wave diffraction problem by a truncated cylinder was tackled by Garrett (1971), while Yeung (1981) addressed the radiation problem. When the body contour is represented as a succession of vertical and horizontal steps, the method can be applied to any arbitrary shape, as Figure 6.14, taken from Kokkinowrachos *et al.* (1986), shows.

6.4.3 Numerical Methods of Resolution

Most numerical software is based on the so-called integral equation method, where the hull is represented as a distribution of elementary singularities, sources, or normal dipoles.

An elementary solution to the Laplace equation, in three dimensions, is $-Q/(4\pi R)$, where $R^2 = x^2 + y^2 + z^2$. The associated radial velocity is $Q/(4\pi R^2)$, so the flux through any closed contour surrounding the origin is Q. This flow is a **source** of intensity Q (or a **sink** if Q is negative). In unbounded fluid the associated velocity decays to zero at infinity. It is then possible to represent the flow around a closed body by distributing a number of sources and sinks inside. Their intensities are obtained by satisfying the no-flow condition on the body surface.

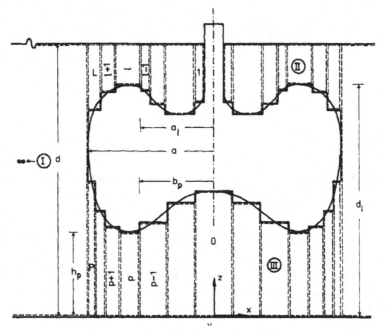

Figure 6.14 Application of the eigenfunction expansion method to general bodies of revolution (Kokkinowrachos *et al.*, 1986).

Another technique consists in distributing, on the body itself, a density of sources $\sigma(Q)$, so that the resulting flow is given by

$$\varphi(P) = \int\!\!\int_{S_B} \sigma(Q)\, G(P,Q)\, dS_Q \tag{6.94}$$

where G, the **Green function**, is simply $-1/(4\pi PQ)$, Q being the running point on the body, P a point in the fluid domain (still assuming unbounded fluid domain). Going to the limit when P is taken on the body itself, the gradient of φ there is given by

$$\nabla\varphi(P) = -\frac{1}{2}\sigma(P)\,\vec{n} + \int\!\!\int_{S_B} \sigma(Q)\, \nabla_Q G(P,Q)\, dS_Q \tag{6.95}$$

with $\vec{n}$ the inward normal vector (into the body).

When the body boundary condition is given in the form $\partial\varphi/\partial n = f(P)$, an integral equation (a Fredholm equation of the second kind) results for the source density.

The wetted hull is then discretized in a number N_P of elementary flat panels where the source densities are usually assumed to be constant (see Figure 6.15). Imposing the body boundary condition at the centroids of the panels gives a linear system of rank N_P. A more refined numerical technique consists in assuming a linear variation of the source densities over the panels. Higher-order representations can also be used (splines and sometimes curved panels).

These numerical methods were first elaborated in aerodynamics. The same technique is used in hydrodynamics, with an adapted Green function in the form $G(P,Q) = -1/(4\pi PQ) + H(P,Q)$ where the additional $H(P,Q)$ term ensures that the free surface

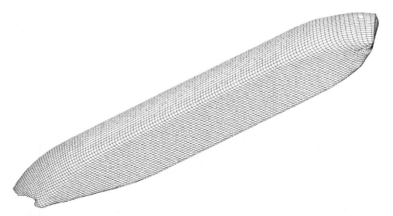

Figure 6.15 Paneling of a tanker hull for diffraction-radiation analysis (courtesy Bureau Veritas).

condition, the no-flow condition at the sea floor, and the radiation condition at infinity, are fulfilled. The additional H is a complicated function, expressed as a Fourier integral in infinite depth. In the 1970s and 1980s, a great deal of effort was devoted to optimizing the computation of the Green functions and developing efficient methods to solve the linear system (which has a full matrix).

Today so-called diffraction-radiation codes have become routine tools for offshore and marine engineers. Most of the commercial codes are based on the integral equation method outlined above, with constant source distributions. (Dipoles may be used to represent elements of zero thickness, for instance strakes or heave plates.)

A variant is to start from Green's second identity applied to φ and G:

$$\int\int_{S_B} \left(\varphi \frac{\partial G}{\partial n} - G \frac{\partial \varphi}{\partial n} \right) \mathrm{d}S \equiv 0 \tag{6.96}$$

The body boundary condition gives an integral equation for φ (a Fredholm equation of the first kind), which is therefore directly obtained. The numerical method of singularities, also known as BIEM for Boundary Integral Equation Method, is further described in Appendix D.

Other numerical models, based on finite elements, or a combination of finite elements and integral equation method, were also developed some decades ago.

6.4.4 Approximate Methods

In naval engineering, it is customary to apply approximate methods of resolution to the diffraction-radiation problem, such as the Froude–Krylov approximation or strip theory. This is partly due to difficulties in computing the Green function with forward speed.

In the case of bodies such as semi-submersible or tension leg platforms, it is possible to take advantage of their relative slenderness to make use of methods inspired by the Morison equation. It has been established in Section 6.4.1 that, in the case of a vertical cylinder, the inertia term in the Morison equation provides a good estimate of

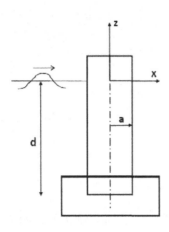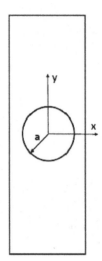

Figure 6.16 Column pontoon assembly.

the wave loading when the diameter is less than about one-sixth of the wave length. For a diameter of 20 m, this means a wave period higher than about 9 s, a value much lower than the usual design wave periods. Moreover, the associated K_C numbers are low, implying that viscous effects can be neglected in a first approximation.

This will be illustrated in the simple, academic case, illustrated in Figure 6.16, of the heave response of an assembly of one submerged rectangular pontoon and one circular column piercing the free surface.

Be a the radius of the column, d its draft. We take d also as the mean immersion of the pontoon, which has a volume V_P. Incoming waves are coming from abeam, along the Ox axis in the figure. We assume infinite depth for simplicity.

The velocity potential of the incoming regular wave system is given by

$$\Phi_I(x,z,t) = \frac{A\,g}{\omega}\,e^{kz}\,\sin(kx - \omega t) \tag{6.97}$$

The wave excitation load in heave on the pontoon is obtained from the inertia term in the Morison equation:

$$dF_{Pz} = (1 + C_a)\,\rho\,S_P\,\Phi_{Izt}\,dL_P \tag{6.98}$$

with C_a the heave added mass coefficient of the rectangular cross-section and S_P its area. The heave load on the pontoon is then

$$F_{Pz} = -(1 + C_a)\,\rho\,V_P\,A\,g\,k\,e^{-kd}\,\cos\omega t \tag{6.99}$$

For the column, the inertia term from the Morison equation cannot be used since the column is not "small" in the vertical direction and it is piercing the free surface. Due to its slenderness, diffraction effects are reduced and a good approximation of the vertical load is achieved by simply integrating the incident pressure over its bottom (a better approximation is achieved by taking the incident pressure at a distance $2a/3$ further below), giving

$$F_{Cz} = \rho \pi a^2 A g e^{-kd} \cos \omega t \tag{6.100}$$

It may be noted that in this approach (known as Hooft method, from Hooft, 1972) interaction effects between column and pontoon are ignored.

The total vertical load is then

$$F_z = \rho \left[V_C - (1 + C_a) k d V_P \right] \frac{A g}{d} e^{-kd} \cos \omega t \tag{6.101}$$

with V_C the volume of the column and V_P the volume of the pontoon.

The added mass of the column is ignored. The pontoon-added mass is

$$M_{aPz} = C_a \rho V_P \tag{6.102}$$

When the column-pontoon assembly is in hydrostatic equilibrium, its mass is

$$M = \rho (V_C + V_P) \tag{6.103}$$

The hydrostatic stiffness is given by

$$K_{33} = \rho g \pi a^2 = \rho g \frac{V_C}{d} \tag{6.104}$$

The equation of heave motion is finally obtained as

$$\rho \left[V_C + (1 + C_a) V_P \right] \frac{d^2 z}{dt^2} + \rho g \frac{V_C}{d} z = \rho \left[V_C - (1 + C_a) k d V_P \right] \frac{A g}{d} e^{-kd} \cos \omega t \tag{6.105}$$

The heave natural frequency is given by

$$\omega_0^2 = \frac{g}{d} \frac{V_C}{V_C + (1 + C_a) V_P} = \frac{g}{d} \frac{1}{1 + \alpha} \tag{6.106}$$

where α is the ratio

$$\alpha = \frac{(1 + C_a) V_P}{V_C} \tag{6.107}$$

Another remarkable frequency is ω_e, given by

$$\omega_e^2 = \frac{g}{d} \frac{1}{\alpha} \tag{6.108}$$

where the right-hand side (the heave excitation load) is nil. This frequency is known as the "equilibrium" (or "cancellation") frequency.

Finally, the heave response is obtained as

$$z = A \frac{\omega_0^2}{\omega_e^2} \frac{\omega^2 - \omega_e^2}{\omega^2 - \omega_0^2} e^{-kd} \cos \omega t \tag{6.109}$$

where $k = \omega^2 / g$

Figure 6.17 shows the RAO of the heave motion, obtained from equation (6.109), for $\omega_0^2 d/g = 0.13$, that would be a heave natural period of 25 s at a draft of 20 m. The RAO shown is absolute value, as a function of ω/ω_0, for different values of ω_e/ω_0 (ω_e/ω_0 = 2, 1.5, 1.25, and 1.125). Note that, in between ω_0 and ω_e, from equation (6.109), the RAO is negative, implying that the heave motion and the free surface

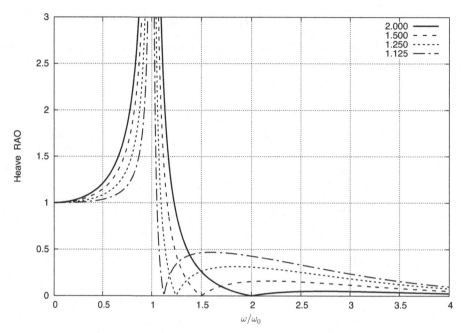

Figure 6.17 Column-pontoon assemble. Heave RAO for different ratios ω_e/ω_0, vs ω/ω_0.

motion are out of phase: the assembly is at its lowest position under the crest and at its highest position over the trough, which is detrimental for airgap and for hydrostatic stability. The safe operating condition is therefore at frequencies higher than the equilibrium frequency.

As a consequence, semi-submersible platforms operating in North Sea conditions aim at an equilibrium period around 20 s, higher than the local design peak wave periods. Taking a draft of 20 m and an equilibrium period of 20 s gives a coefficient α close to 5: the apparent displacement $(1+C_a)\,V_P$ of the pontoon must be about 5 times larger than the displacement of the column. This is the reason why North Sea semis exhibit rather large and flat pontoons, in order to maximize their vertical added mass. For Tension Leg Platforms the strategy is different since the issue is to minimize the loads in the tethers, where both extreme loads and fatigue damage must be taken into account. In North Sea conditions an equilibrium period around 14 s is usually aimed at. Together with a typical draft of 40 m this means a coefficient α equal to 1.2: with a square cross-section ($C_a \simeq 1.2$), the displacement of the pontoons is then less than 30% of the total displacement.

6.5   Loads and Response in Irregular Seas

Within the framework of linear theory, the response of a floating structure, in a regular wave, takes place at the wave frequency and has an amplitude proportional to the wave amplitude.

When the incoming wave elevation, at the point of reference, is given as

$$\eta_I(t) = \Re\left\{A\,e^{-i\omega t}\right\} \tag{6.110}$$

the response of the floating support, in its j degree of freedom, is

$$X_j(t) = \Re\left\{x_j\,e^{-i\omega t}\right\} = \Re\left\{A\,f_j(\omega,\beta)\,e^{-i\omega t}\right\} \tag{6.111}$$

$f_j(\omega,\beta)$ is the transfer function, or **Response Amplitude Operator** (RAO).
In a long-crested wave system of incidence β, of elevation

$$\eta_I(t) = \Re\left\{\sum_i A_i\,e^{i(-\omega_i t+\theta_i)}\right\} \tag{6.112}$$

owing to linearity, the j response of the structure is simply the sum of all elementary sinusoids:

$$X_j(t) = \Re\left\{\sum_i A_i\,f_j(\omega_i,\beta)\,e^{i(-\omega_i t+\theta_i)}\right\} \tag{6.113}$$

The response spectrum is then

$$S_{X_j}(\omega) = S(\omega)\,f_j(\omega,\beta)\,f_j^*(\omega,\beta) \tag{6.114}$$

with f_j^* the complex conjugate of the transfer function.
In a short-crested sea the response spectrum is given by:

$$S_{X_j}(\omega) = \int_0^{2\pi} S(\omega,\beta)\,f_j(\omega,\beta)\,f_j^*(\omega,\beta)\,d\beta \tag{6.115}$$

Moreover, still owing to linearity, the incoming wave elevation being a Gaussian process, the response is a Gaussian process as well. The same statistical analysis as applied to the wave elevation in Chapter 2 can be used. The moments of the response spectrum are obtained as

$$m_{jn} = \int_0^\infty \omega^n\,S_{X_j}(\omega)\,d\omega \tag{6.116}$$

from which the standard deviation of $X_j(t)$ can be derived:

$$\sigma_{X_j} = \sqrt{m_{j0}} \tag{6.117}$$

its mean up-crossing period

$$T_{j2} = 2\pi\,\sqrt{\frac{m_{j0}}{m_{j2}}} \tag{6.118}$$

and the bandwidth parameter

$$\epsilon_j^2 = 1 - \frac{m_{j2}^2}{m_{j0}\,m_{j4}} \tag{6.119}$$

From the ϵ_j value it can be assessed whether the Rayleigh law applies to the distribution of maxima. The mean expected value, during a stationary sea state of duration T, is finally obtained as

$$\overline{X_{j\ max}} = \left[\sqrt{2 \ln \frac{T}{T_{j2}}} + \frac{\gamma}{\sqrt{2 \ln \frac{T}{T_{j2}}}} \right] \sqrt{m_{j0}} \tag{6.120}$$

It should be noted that, compared to equation (2.50) in Chapter 2, the 2 factor is missing. In Chapter 2 the considered parameter was the crest-to-trough value, that is, the double amplitude.

The same approach can be applied to any quantity that linearly relates to the incident wave: excitation forces (for a fixed structure), relative elevation of free surface at the waterline (to study airgap problems), tension in a tendon of a TLP, etc. All that one has to do is derive the RAO, then the response spectrum and its moments, from which the standard deviation and the extreme value during the sea state are obtained.

6.6   Time Domain Resolution of the Equations of Motion

The additivity principle that we have just exploited assumes that all the terms in the equations of motion are linear, both those expressing the hydrodynamic loads and those related to the body's own dynamics. There are situations where one desires, for example, to introduce some viscous damping term, quadratic with the velocity, or a nonlinear restoring force, and to solve the equations of motion in the time domain. The additivity principle still applies to the excitation loads, as long as the motion remains "small". But difficulties arise when one wants to express the radiation loads, because of the frequency dependence of the added masses and radiation dampings.

This sensitivity to frequency of the added masses and dampings reflects the fact that radiation loads do not only depend on the instantaneous acceleration and velocity of the body, but also on their past histories, the memory effect being due to the presence of the free surface. It is therefore necessary to express the radiation loads through the past velocity of the structure, by means of "retardation functions." More precisely, if we consider the equation of motion for the degree of freedom i, put so far in the form:

$$\sum_{j=1}^{6} \left\{ [M_{ij} + M_{aij}(\omega)] \ddot{X}_j + B_{ij}(\omega) \dot{X}_j + K_{ij} X_j \right\} = F_i(t) \tag{6.121}$$

where M_{aij} and B_{ij} depend on the frequency ω, a more general form is

$$\sum_{j=1}^{6} \left\{ [M_{ij} + M_{aij}(\infty)] \ddot{X}_j + \int_{-\infty}^{t} \dot{X}_j(\tau) R_{ij}(t - \tau) \, d\tau + K_{ij} X_j \right\} = F_i(t) \tag{6.122}$$

$M_{aij}(\infty)$ is the infinite frequency added mass, obtained by setting $\varphi_{Rj} = 0$ at the free surface in the Boundary Value Problem (6.28). The usual added masses $M_{aij}(\omega)$ and dampings $B_{ij}(\omega)$ are related to the kernels $R_{ij}(t)$ through the relations

$$M_{aij}(\omega) = M_{aij}(\infty) - \frac{1}{\omega} \int_0^\infty R_{ij}(t) \, \sin \omega t \, dt \qquad (6.123)$$

$$B_{ij}(\omega) = \int_0^\infty R_{ij}(t) \, \cos \omega t \, dt \qquad (6.124)$$

and, conversely:

$$R_{ij}(t) = \frac{2}{\pi} \int_0^\infty B_{ij}(\omega) \, \cos \omega t \, d\omega \qquad (6.125)$$

When one knows the terms of the radiation damping matrix for all values of the frequency ω, from zero to infinity, then, through equation (6.125), the retardation functions $R_{ij}(t)$ can be calculated and the equations of motion can be solved in the time domain. The radiation dampings are usually known at discrete frequencies, within a finite interval; the problem arises of how to interpolate and extrapolate the values. Another issue is how far back in the past should the convolution integrals be calculated. An alternative to the convolutions is the Prony method, where the radiation loads are simulated via coupled oscillators (see, e.g., Babarit *et al.*, 2004).

Finally, it should be mentioned that there are time domain diffraction-radiation codes, based on an unsteady Green function. An advantage of this approach is that the body boundary condition can be delinearized (see, e.g., Magee, 1991). These codes have not entered the commercial stage.

6.7   Multiple or Deformable Bodies

The theoretical model can easily be extended to the case when several bodies coexist (for instance two ships in side-by-side operations). Elementary diffraction and radiation problems must be solved, with all the bodies fixed for diffraction, and all but one for radiation (the concerned body moving successively along its 6 degrees of freedom). With N_B bodies this means $6N_B$ elementary radiation problems, and added mass and damping matrices of ranks $6N_B$. All commercial diffraction-radiation softwares can handle several bodies at a time.

In the case of deformable bodies, the same approach applies. The body motion decomposes into 6 rigid body modes plus some deformation. The first step is to choose a set of modes for the body deformation. In ship elasticity the so-called **dry modes** are usually chosen, that is the modes of deformation "in vacuum," not accounting for hydrostatic restoring forces. Other sets of modes may be used (e.g., see Newman, 1994). Then modal excitation loads, modal added masses, and modal dampings are obtained. Modal stiffnesses are also required, consisting in an elastic part (usually dominant) and an hydrostatic part. The hydrostatics of deformable bodies have been a somewhat controversial topic (see Malenica *et al.*, 2009).

Figures 6.18 and 6.19, taken from Malenica *et al.* (2003), show numerical and experimental results from tests carried out at BGO-FIRST on a flexible barge: the experimental model consists in 12 rectangular modules (except for the bow module, slightly tapered), connected, at deck level, by two steel plates. The 6 dof motion

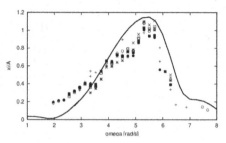

Figure 6.18 Flexible barge in head waves (left) and RAO of the first bending mode (right) (from Malenica *et al.*, 2003).

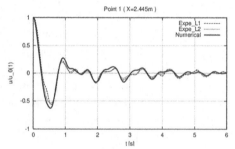

Figure 6.19 Decay test: Initial barge configuration (left) and ensuing vertical motion at the bow (right) (from Malenica *et al.*, 2003).

of every other module is measured with an optical tracking system, enabling the reconstruction of the complete shape of the model.

Figure 6.18 shows a view of the model in head waves and the RAO of the first bending mode, obtained from the numerical code and from tests in irregular waves. Figure 6.19 shows the results of a decay test where the bow module is pulled up with a string, then abruptly released.

6.8   Validity and Limitations Of Linear Theory

Generally speaking, it can be stated that linear theory provides excellent results when its inherent assumptions are respected; or, equivalently, that the possible disagreements between computations and measurements are attributable to the non-respect of one of these assumptions. They are recalled here:

- negligible role of viscous effects;
- low steepness of the incoming waves;
- large dimensions of the considered body, compared to the wave amplitude, and to its response amplitude.

It must also be borne in mind that linear theory only accounts for phenomena (loads, responses) taking place at the wave frequencies. It does not give access to the response taking place at the natural frequencies, when these frequencies are outside the wave frequency range.

In what follows we try to make these considerations more concrete, by illustrating them with particular cases, where experimental and numerical RAOs are compared. This means that the measurements are processed by Fourier analysis (in regular waves) or spectral analysis (in irregular waves) to extract the RAOs.

6.8.1 Wave Loads on Fixed Bodies

This case is relatively uncommon, large offshore structures being most often floating. Practical cases are the concrete gravity platforms in the North Sea, or the Hibernia GBS offshore Newfoundland.

For such structures, the hypothesis of large dimensions compared to the wave amplitude is fairly well respected. Drag forces on the columns of the gravity platforms can nevertheless become appreciable when the flow separates therefrom, that is, when the wave amplitude exceeds the radius. These viscous loads can be added a posteriori, however the choice of the drag coefficient raises some issues. Being 90 degrees out of phase with inertial forces, the viscous loads contribute little to the peak loads.

Another possible problem may be related to a high steepness of the incoming waves, or to an ill-suited wave theory if the water depth is small compared to the wave length. In such a case, higher-order harmonics can be expected to become rapidly appreciable as the steepness increases. It may therefore be appropriate to calculate the second-order diffraction loads appearing at the double frequency. These loads are introduced in Chapter 7.

6.8.2 Wave Response of Floating Bodies

In general, the natural frequencies of the horizontal motions of floating structures are well below the wave frequencies, as a result of their low mooring stiffnesses. Still, high-amplitude resonant responses are usually observed in irregular waves. The prediction of these responses, so-called slow-drift motions, requires the calculation of the second-order low-frequency loads, considered in Chapter 7.

Linear theory only allows prediction of the response taking place at the wave frequencies. This response is superimposed on the slow-drift components, and concerns the six degrees of freedom. For massive structures such as barges and FPSOs linear theory generally applies quite well to the prediction of the wave frequency response. Examples of comparison between calculated and measured responses are given in Figures 6.20 and 6.21, taken from Pinkster (1980). They show the RAOs (moduli and phases) of three floaters: a tanker, a semi-submersible platform, and a barge, in head seas (Figure 6.20) and beam seas (Figure 6.21). Good agreement is observed between measurements and computations except for the roll RAOs of the tanker and the barge, around resonance. For ships and barges roll radiation damping is very weak, and viscous effects play the major role in mitigating resonance.

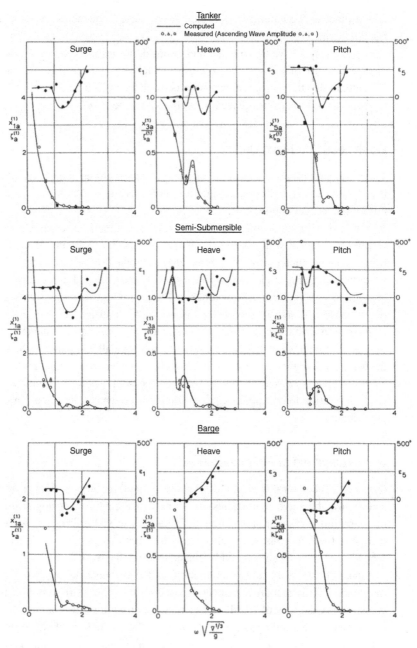

Figure 6.20 Surge, heave, and pitch RAOs of a tanker, a semi, and a barge in head waves. Comparison between regular waves experimental data and calculations (from Pinkster 1980).

Unlike roll, radiation dampings in heave and pitch of barges and FPSOs are appreciable. Even though their natural frequencies usually fall within the wave frequency range, their heave and pitch responses are well predicted by linear theory.

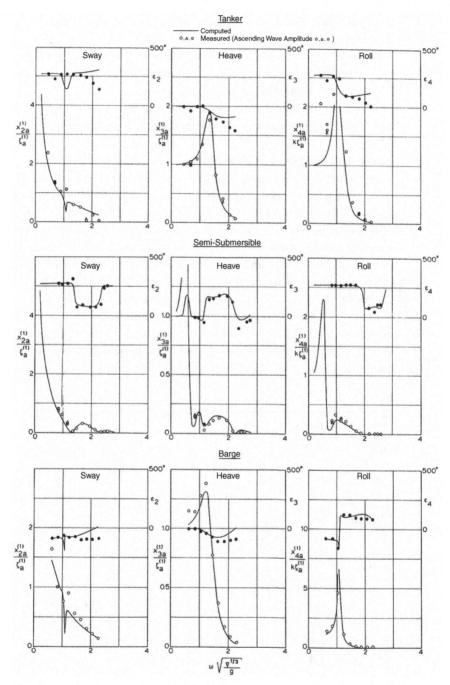

Figure 6.21 Surge, heave, and pitch RAOs of a tanker, a semi, and a barge in beam waves. Comparison between regular waves experimental data and calculations (from Pinkster 1980).

Semi-submersible platforms and Spars are designed in such a way that their natural periods in heave, roll, and pitch be beyond the energetic wave periods. Viscous damping is usually not a concern; however, viscous excitation loads may significantly add to the loads from potential flow theory, for instance, in heave, around the cancellation frequency.

Relatively small bodies are much harder to deal with: viscous effects (flow separation) usually cannot be neglected and nonlinear phenomena associated with large variations of the wetted hull quickly come into play.

We further illustrate these considerations by a few cases where disagreements are found between experimental and numerical results from linear theory.

Barge Roll Response at Resonance

Here we borrow some results from a French JIP that was run under CLAROM, in 2004. More information may be found in Ledoux *et al.* (2004). Extensive experiments were performed in BGO-FIRST, on a barge model 2.5 m long and 0.6 m wide, at a draft of 0.12 m. The barge model had square bilges. The tests were run in irregular waves, at various headings, with significant waveheights ranging from 2 m up to 9 m (assuming a 1:100 scale), and peak periods of 12 s, 16 s, and 20 s, as given in Table 6.1.

The center of gravity is located 13.5 m above the keel line, and the roll/pitch radii of gyration are 19.7 m/63.3 m, giving a roll natural period of 12.5 s.

Figure 6.22 shows the experimental roll and pitch RAOs at a heading of 60 degrees, from all the tests. For the pitch motion little sensitivity to the sea state is observed. As for roll, peak RAO values strongly depend on the sea state, decreasing from about 4 deg/m at $H_S = 2$ m down to about 2 deg/m at $H_S = 9$ m.

Table 6.1

Test name	H_S (m)	T_P (s)	γ
Irr1	2.0	12	1
Irr2	4.0	12	1
Irr3	6.0	12	1
Irr4	2.5	16	1
Irr5	5.0	16	1
Irr6	7.5	16	1
Irr7	3.0	20	1
Irr8	6.0	20	1
Irr9	9.0	20	1

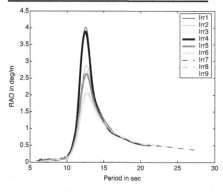 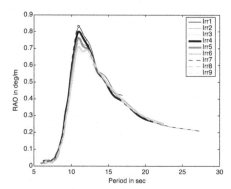

Figure 6.22 Barge experimental roll (left) and pitch (right) RAOs at 60-degree heading.

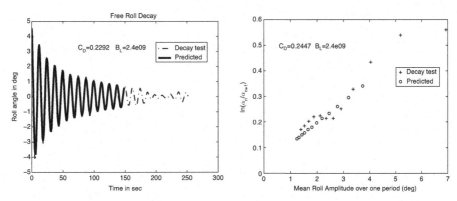

Figure 6.23 Roll decay test. Time series (left) and analysis (right).

The roll damping is usually assumed to consist in a linear radiation damping B_L plus a quadratic viscous damping B_Q:

$$B = B_L + B_Q \, |\dot{\alpha}| \tag{6.126}$$

where the quadratic term is taken as

$$B_Q = \frac{1}{2} \rho \, C_\alpha \, B^4 \, L \tag{6.127}$$

with B the beam of the barge, L its length, and C_α a coefficient to determine.

Experimentally the roll viscous damping may be obtained through decay tests, or forced roll experiments. Time series from tests in irregular waves may also be processed through the Random Decrement Method. Figure 6.23 shows the experimental roll decay test (left) and its analysis (right), leading to a C_α coefficient close to 0.23. This is based on the equivalent linear damping, over a cycle of mean roll amplitude α_n, being $B_{Leq} = B_L + 8/(3\pi) \, \alpha_n \, \omega \, B_Q$. Here the B_L value was enforced as the roll radiation damping from a potential flow code.

The quadratic viscous damping is introduced in the roll equation of motion and stochastic linearization is applied. This means that the RAOs are obtained through iterations over the standard deviation of the roll angular velocity, as shown in Appendix C.

When comparing calculated and measured roll an optimum fit was found by decreasing the C_α value from 0.23 to 0.15. This is likely due to the fact that decay tests lead to overestimated values of the viscous damping: viscous loads have a memory effect. So-called wake models, which take account of previous flow cycles, show that higher viscous loads are attained in decreasing amplitude cycles than in increasing amplitude. This is reflected in Figure 6.23 (right) where it can be seen that the slope of the straight line linking the far two right experimental points (at amplitudes 7 and 5 deg) is much less than the averaged slope.

An alternative to decay or forced roll tests is to run a CFD code, as done in Ledoux *et al.* (2004), or to refer to literature. There is in fact an extensive literature on ship roll damping, much of it based on old small-scale tests in Japan, reported by Ikeda, Tanaka,

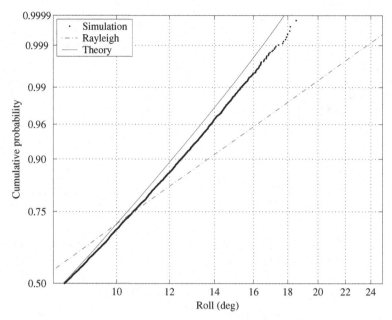

Figure 6.24 Cumulative probability of roll maxima (courtesy M. Prevosto).

and Himeno. Main references can be found in Chakrabarti (2001) who quotes the following formula, proposed by Ikeda (1984), of the quadratic damping for a rectangular cross-section:

$$B_Q = \frac{3}{4}\rho L d^4 \left(H_0^2 + \frac{KG}{d}\right)\left(H_0^2 + \frac{KG^2}{d^2}\right) \tag{6.128}$$

with d the draft, KG the vertical distance from keel line to center of gravity and $H_0 = B/(2d)$.

Applied to our barge case this formula gives a C_α value of 0.13!

As a result of the nonlinear damping, even though the roll response is quite narrow-banded, the distribution of its maxima does not follow the Rayleigh law: it is narrower. An illustration is provided by Figure 6.24, which shows, in Weibull scale, the cumulative probabilities obtained from numerical time domain simulations, from the Rayleigh law, and from a theoretical law proposed by Prevosto (2001). This law is based on the *Linearize & Match (L&M)* technique proposed by Armand & Duthoit (1990). A further improvement, accounting for coupling between roll and sway, can be found in Mendoune Minko *et al.* (2008).

Heave Response of a Truncated Cylinder

An other example is provided in Figure 6.25, which shows the experimental and numerical heave RAOs of a truncated cylinder. The diameter and draft are equal to 8 m. The experimental values are from tests in regular waves. Calculations with the diffraction-radiation code COREV of IFP give a peak heave RAO equal to 12 at resonance (at about 6.5 s). When a quadratic viscous damping in the form $1/2\rho\, C_d\, \pi\, a^2\, |\dot{z}|\, \dot{z}$ is introduced in the heave motion equation, with a drag coefficient

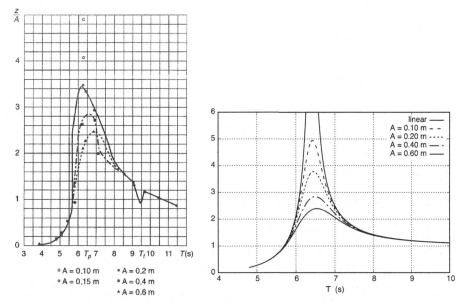

Figure 6.25 Floating truncated cylinder. Experimental (left) and numerical (right) heave RAO.

C_d here equal to 2, a fairly good agreement is obtained with the experimental RAOs. For this case, as in the previous barge case, there is no need to correct the excitation loads given by linearized potential flow theory, it is sufficient to correct the damping.

Pitch Response of a CALM Buoy

There are cases where excitation loads due to viscous effects must be taken into account, simply because the potential flow excitation loads are weak. It is the case, for instance, of the heave response of semi-submersible platforms at the heave cancellation frequency.

Another case, considered in the following, is the roll and pitch response of a CALM buoy, where viscous effects play a role both in damping and in excitation. Here we refer to experiments performed at BGO-FIRST on a large buoy of diameter 23 m and draft 6.25 m (at a scale 1:25). The buoy is fitted with a skirt, 1.25 m wide and 2.5 cm thick, set 1.25 m above the keel line. The center of gravity is on the axis of symmetry, at the free surface level ($KG = 6.25$ m) and the roll-pitch radius of gyration is 4.8 m. The mooring system, consisting in aerial horizontal cables and springs, increases slightly the roll/pitch hydrostatic stiffness, by about 20%. The tests are run in irregular waves, with significant heights ranging from 1 m to 4.5 m and peak periods equal to 7.5 s and 10 s.

Figure 6.26 shows the pitch RAOs obtained from the tests (left) and with the code COREV. Note that, in all figures, the vertical scales are different. Calculations show a strong resonance at a period of 9.5 s and a RAO nil at 7.5 s. This zero value of the RAO is due to a cancellation of the effective moment in pitch, when coupling between surge and pitch is duly accounted for. The experimental RAOs bear no relation to the calculated ones: they vary somewhat with the sea state, and exhibit a peak at around 8 s, where the calculated values are close to zero.

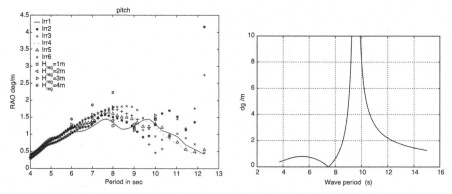

Figure 6.26 CALM buoy with skirt. Experimental (left) and numerical (right) pitch RAO.

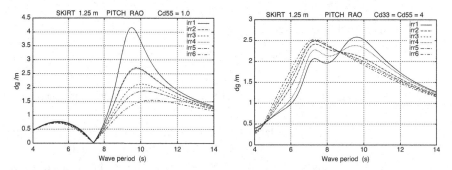

Figure 6.27 CALM buoy with skirt. Calculated pitch RAOs accounting for viscous effects. Left: viscous damping only. Right: viscous damping and excitation.

A peripheral quadratic drag load is then introduced in the numerical model. When the local drag force is derived from the local vertical buoy velocity alone (due to heave and pitch), the predicted RAOs still exhibit a zero value at 7.5 s (Figure 6.27, left). When the local buoy velocity is combined with the local wave-induced velocity (due to incoming waves only), the RAOs shown in the right part of the figure are obtained, closer to the experimental ones, but still a long way from perfect agreement.

6.8.3   Free Surface Elevations

Results such as shown in Figures 6.22 and 6.25 suggest that experimental RAOs decrease when, the wave period and heading being kept the same, the wave amplitude increases. And that numerical RAOs, as given by linear potential flow codes, which omit viscous effects, are some kind of upper bounds. This is generally verified but there are some counterexamples.

One of them is the free surface elevation that takes place on the weather side of a 4-cylinder array when the wave length is close to twice the distance from the front cylinders to the back cylinders. The diffracted waves from the front and back columns are then in phase, resulting in strong reflection (Bragg effect).

Figure 6.28 shows a 4-column TLP model under tests in regular waves. The column diameter is 26.4 m, the axis-to-axis distance is 75 m. The model is held fixed and the free surface elevations are measured at several points below and around the model. At

Figure 6.28 Wave impacts upon the deck of a TLP model.

Figure 6.29 Wave runup along the hull of a barge in regular beam waves.

the critical wave period of 10 s (meaning a wave length equal to 156 m), the experimental RAO of the free surface elevation, in between the two front columns, is equal to 1.8 in small amplitude waves, in good agreement with the numerical model (based on Linton & Evans theory). When the amplitude is 7.5 m (meaning a steepness H/L of 10%), the underside of the deck, at a height of 24 m, is strongly impacted as the figure shows. This suggests a RAO greater than 3.

Similar phenomena may be observed along the hulls of FPSOs or barges in beam waves (see Figure 6.29). In Chapter 8 it is shown that these runups are due to third-order interactions between the incident waves and the waves diffracted and radiated by the body.

6.9   Recovery of Wave Energy

Since the pioneering experiments of Salter (1974) on his "duck" (see Figure 6.30), much effort has been given to the recovery of wave energy. After nearly 50 years it

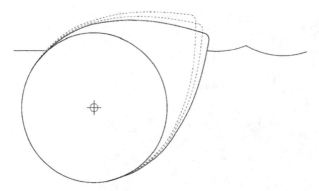

Figure 6.30 Salter's duck (courtesy S. Salter).

is regrettable to see that no WEC (Wave Energy Converter) concept has reached the industrial stage yet, in spite of the wide variety of the systems proposed. An exhaustive historical review is provided by Babarit (2017).

Here we only provide a few basic elements regarding this matter, confining ourselves to cases where wave energy is harnessed via the wave response of a floating body.

For the sake of simplicity, consider a 1 dof system, which obeys the equation (in regular waves):

$$[M + M_a(\omega)] \ddot{X} + B_{\text{rad}}(\omega) \dot{X} + K X = F_0 \cos(\omega t + \theta) = F(t) \tag{6.129}$$

The idea is to introduce extra mass, extra stiffness, and extra damping terms, so that the equation becomes

$$[M + M_a(\omega) + M_{\text{PTO}}] \ddot{X} + [B_{\text{rad}}(\omega) + B_{\text{PTO}}] \dot{X} + [K + K_{\text{PTO}}] X = F_0 \cos(\omega t + \theta) \tag{6.130}$$

where PTO stands for "Power Take Off."

The energy recovered by the PTO system, over one wave period, is given by

$$E_{\text{PTO}} = \int_{t_0}^{t_0+T} B_{\text{PTO}} \dot{X}^2 \, dt = B_{\text{PTO}} \, \omega^2 \, X_0^2 \, \frac{T}{2} \tag{6.131}$$

meaning an averaged extracted power equal to

$$P_{\text{PTO}} = \frac{1}{2} B_{\text{PTO}} \, \omega^2 \, X_0^2 \tag{6.132}$$

where the response amplitude X_0 is given by

$$X_0^2 = \frac{F_0^2}{[-(M + M_a + M_{\text{PTO}}) \omega^2 + K + K_{\text{PTO}}]^2 + \omega^2 (B_{\text{rad}}(\omega) + B_{\text{PTO}})^2]} \tag{6.133}$$

It results that the recovered power is maximized when the system is at resonance, that is, when

$$-[M + M_a(\omega) + M_{\text{PTO}}] \omega^2 + K + K_{\text{PTO}} = 0 \tag{6.134}$$

Under this condition the extracted power is

$$P = \frac{1}{2} \frac{B_{\text{PTO}} F_0^2}{(B_{\text{rad}}(\omega) + B_{\text{PTO}})^2} \tag{6.135}$$

and is maximized when the PTO damping is made equal to the radiation damping:

$$P_{\max} = \frac{1}{8} \frac{F_0^2}{B_{\text{rad}}(\omega)} \tag{6.136}$$

6.9.1 Reversed Piston Wavemaker

As an application, consider the case of a reversed piston wavemaker, that is a mass spring system consisting in a vertical board subjected to regular incoming waves. The board is "dry": there is no water on the other side. The device is under resonating condition, that is, the condition (6.134) above is fulfilled, and the PTO damping is equal to the radiation damping, given by

$$B_{\text{rad}} = \rho \omega \frac{A_0}{k_0^2} \tanh k_0 h \tag{6.137}$$

where the coefficient A_0 is

$$A_0 = \frac{2 \sinh 2k_0 h}{2k_0 h + \sinh 2k_0 h} \tag{6.138}$$

(see Appendix A).

When the board is held in fixed position, the incoming waves are fully reflected, one has a standing wave with the velocity potential

$$\varphi_I + \varphi_D = \frac{-2\mathrm{i}\, A_I\, g}{\omega} \frac{\cosh k_0(z+h)}{\cosh k_0 h} \cos k_0 x \tag{6.139}$$

The excitation load is therefore

$$F(t) = \Re\left\{ f\, \mathrm{e}^{-\mathrm{i}\omega t} \right\} = \Re\left\{ -2\,\rho\, A_I\, g\, \frac{\tanh k_0 h}{k_0}\, \mathrm{e}^{-\mathrm{i}\omega t} \right\} \tag{6.140}$$

From (6.136), (6.137), (6.138), and (6.140), the extracted power is

$$P = \frac{1}{2}\rho g^2 A_I^2 \frac{\tanh k_0 h}{\omega A_0} = \frac{1}{2}\rho g^2 A_I^2 \frac{\tanh k_0 h}{\omega} \frac{2 k_0\, h + \sinh 2\, k_0\, h}{2 \sinh 2\, k_0\, h} = \frac{1}{2}\rho g\, C_G\, A_I^2 \tag{6.141}$$

We have obtained that the extracted power is equal to the energy flux carried by the incoming waves. This means that all the incoming wave energy has been absorbed. In other words the wave system radiated by the board motion just cancels out the wave system reflected by the board held fixed.

It is quite remarkable that total energy absorption has been achieved, without any elaborate control of the board motion: just by tuning it to the wave frequency and equalizing the PTO and radiation dampings.

When there is water on the other side of the board, full absorption can no longer be achieved since waves are generated on that side as well. Only half of the incoming wave energy, at most, can be extracted. This upper bound remains valid for any symmetric body when power absorption is applied to only one degree of freedom. For an asymmetric body of proper shape full absorption is possible, Salter's duck being one example (see Figure 6.30). Full absorption is possible for a symmetric body when the PTO is applied to two resonating degrees of freedom at the same time, heave being one, and surge or pitch the other. An example is a submerged circular cylinder, undergoing orbital motion (Evans *et al.*, 1979). (These comments apply to two-dimensional devices at normal incidences).

6.9.2   Wave-Energy Extraction by Three-Dimensional Axisymmetric Bodies

Many WEC concepts are axisymmetric buoys, the heave response being exploited to extract energy. An upper bound of the recoverable power can then easily be established.

In Section 6.3 the theorem of Haskind–Hanaoka has been introduced, relating the excitation load to the radiation potential:

$$f_i = -4\mathrm{i}\,\rho\,A\,g\,\frac{C_G}{\omega\,k}\,H_{Ri}(\beta + \pi) \tag{6.142}$$

where i is the considered degree of freedom and H_{Ri} is the Kochin function as defined in (6.33).

This relation was obtained by applying Green's second identity to $\varphi_I + \varphi_D$ and φ_{Ri}. By applying Green's second identity to φ_{Rj} and to the complex conjugate of φ_{Ri}, the following expression of the radiation damping B_{ij} is obtained

$$B_{ij} = \frac{2\,\rho\,C_G}{\pi\,k\,\tanh kh}\,\int_0^{2\pi} H_{Ri}^*(\theta)\,H_{Rj}(\theta)\,\mathrm{d}\theta \tag{6.143}$$

For an axisymmetric body the heave Kochin function H_{R3} does not depend on the azimuthal angle. Combining equations (6.142) and (6.143), we get the heave radiation damping as

$$B_{33} = \frac{k}{4\,\rho\,g\,C_G\,A_I^2}\,f_3\,f_3^* \tag{6.144}$$

The maximum recoverable power is then obtained as

$$P_{\max} = \frac{1}{8}\,\frac{f_3\,f_3^*}{B_{33}} = \frac{1}{2\,k}\,\rho\,g\,C_G\,A_I^2 \tag{6.145}$$

that is the energy flux per unit width times $1/k = L/(2\pi)$.

For an axisymmetric body operating in heave the maximum theoretical value of the **capture width** is the wave length divided by 2π, without any assumption regarding the size of the device!

When energy is extracted from the surge or pitch motion, the Kochin function of the radiation potential has the form $H_{Ri}(\theta) = H_{Ri}\cos\theta$. Equation (6.143) then becomes

$$B_{ii} = \frac{k}{8\,\rho\,g\,C_G\,A_I^2}\,f_i\,f_i^* \tag{6.146}$$

with $i = 1$ (surge) or 5 (pitch). The maximum recoverable power is now

$$P_{\max} = \frac{1}{k}\,\rho\,g\,C_G\,A_I^2 \tag{6.147}$$

twice that of the heave case.

When heave is combined with either surge or pitch, the maximum capture width becomes $3L/(2\pi)$, about half the wave length.

These results are remarkable and seem to suggest that small axisymmetric floaters can be very efficient wave absorbers. It must be borne in mind that they have been obtained within the frame of linearized potential flow theory, which has the limitation that the body motion should be small compared to its own dimensions.

Take, for instance, the case of a tethered submerged buoy of spherical shape, like the early CETO designs (e.g., see Babarit *et al.*, 2012). Be d the distance of the center of the buoy to the free surface, and a its radius. When the buoy radius is small compared to the wave length, the vertical excitation load can be obtained from the inertia term in the Morison equation, as

$$F_3 = \rho\,(1 + C_a)\,\forall\,\frac{\partial^2 \Phi_I}{\partial z\,\partial t} \tag{6.148}$$

with $C_a = 1/2$ for a sphere and $\forall = 4\pi\,a^3/3$. This gives

$$f_3 = -2\pi\rho\,g\,k\,a^3\,A_I\,e^{-kd} \tag{6.149}$$

where infinite water depth has been assumed.

Applying equation (6.144), the radiation damping B_{33} can be obtained and then the heave RAO under resonating condition (with the PTO damping equal to the radiation damping):

$$z/A_I = \left|\frac{f_3}{-2\,\mathrm{i}\,B_{33}}\right| = \frac{1}{2\pi}\frac{e^{k\,d}}{k^3\,a^3} \tag{6.150}$$

Taking a wave period of 8 s, a buoy radius of 3 m, and an immersion of 5 m, we get a RAO equal to 32.5! Obviously, the linearity assumptions are going to be violated as soon as the wave amplitude exceeds a few decimeters.

Another difficulty with floating buoys is tuning their heave natural period to the most energetic wave periods, typically in the range [8 s – 12 s]. Take a buoy with a hemispheric hull, of radius a, in hydrostatic equilibrium. Its heave natural frequency follows from

$$\omega_0^2 = \frac{3}{2}\frac{g}{a}\frac{1}{1 + C_a(\omega_0)} \tag{6.151}$$

where C_a is the added mass coefficient $C_a = M_{33}/(2\pi\rho a^3/3)$, shown in Figure 6.31. Assuming deep water ($\omega_0^2 = g\,k_0$), this gives

$$C_a(k_0 a) = \frac{3}{2\,k_0 a} - 1 \tag{6.152}$$

The resonant $k_0 a$ value is obtained from the intersection of the added mass coefficient curve and of the hyperbole $3/(2k_0 a) - 1$ and is found to be 1.05. To achieve a resonant period of 8 s, the sphere radius should be equal to 16.7 m. With a period of 10 s the radius should be 26.1 m. These values are very high, quite impractical!

Some remedies have been devised to overcome this problem. One of them, known as **latching**, consists in artificially increasing the heave natural period, by locking the

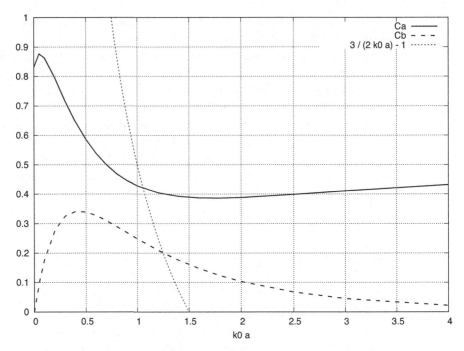

Figure 6.31 Half immersed sphere. Added mass (C_a) and damping (C_b) coefficients in heave.

WEC temporarily in its most extreme positions, when its velocity is nil, and releasing it after some time (e.g., see Babarit *et al.*, 2004).

6.10   Sloshing in Tanks

Many marine and offshore structures contain tanks filled with oil, water, or Liquefied Natural Gas (LNG). When these tanks are completely full, with no free surface, the liquid inside behaves as a solid body under translation, **but it does not behave as a solid body under rotation** (consider a circular or spherical tank: a rotational motion, around its center, does not create any motion of the fluid inside, under the assumptions of potential flow theory).

Figure 6.32 shows the inertia in roll (about its center) of a rectangular tank completely filled (no free surface), of sides h and l. For a square tank the inertia is only about 15% of the solid value.

When there is a free surface, some sloshing response will take place under imposed motion, inducing hydrodynamic loads at the wall. As a result, the wave response of the support may be strongly affected and the coupled response, of the liquid cargos and support, must be considered. This is done in the following section. Here we address the liquid response of a tank under forced motion.

For the sake of simplicity, we consider the two-dimensional case of a rectangular tank in forced horizontal translatory motion. The length of the tank is L, the liquid

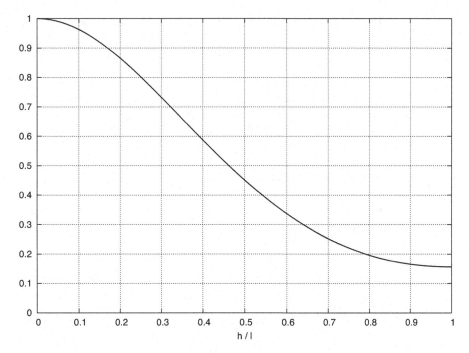

Figure 6.32 Inertia of the liquid contained in a rectangular tank of sides h and l. The inertia is normalized by the solid inertia $\rho\, h\, l\, (h^2 + l^2)/12$.

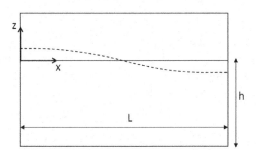

Figure 6.33 Rectangular tank.

height is h, the coordinate system Oxz is centered at the intersection of the mean free surface with the left-hand wall (Figure 6.33).

The tank is first considered fixed. The free surface motion is assumed to be small and linearized potential flow theory is applied. It is easy to see that the velocity potential, defined as

$$\Phi_n(x,z,t) = \frac{A_n\, g}{\omega_n}\, \frac{\cosh \lambda_n\,(z+h)}{\cosh \lambda_n\, h}\, \cos \lambda_n x\, \sin(\omega_n t + \theta) \qquad (6.153)$$

satisfies the Laplace equation and the no-flow conditions in $x = 0$ and $z = -h$. The no-flow condition on the right-hand wall is also verified if λ_n is chosen such that

$$\sin \lambda_n L = 0 \qquad \text{or} \qquad \lambda_n = \frac{n\,\pi}{L} \qquad (6.154)$$

The remaining condition at the free surface yields the associated frequency:

$$\omega_n^2 = g\,\lambda_n\,\tanh \lambda_n h = g\,\frac{n\pi}{L}\,\tanh\frac{n\pi h}{L} \qquad (6.155)$$

The frequencies ω_n of the natural modes in the tank form a discrete set. Associated free surface elevations are sine curves of amplitudes A_n, symmetric with respect to the middle of the tank for n even, antisymmetric for n odd.

This is for a two-dimensional tank. In a three-dimensional rectangular tank, of length L and width B, the natural sloshing modes are of the form

$$\Phi_{mn}(x,y,z,t) = \frac{A_{mn}\,g}{\omega_{mn}}\,\frac{\cosh \nu_{mn}\,(z+h)}{\cosh \nu_{mnn}\,h}\,\cos\lambda_m x\,\cos\mu_n y\,\sin(\omega_{mn}\,t+\theta) \quad (6.156)$$

with $\lambda_m = m\pi/L$, $\mu_n = n\pi/B$, $\nu_{mn}^2 = \lambda_m^2 + \mu_n^2$ and the associated resonant frequencies are given by

$$\omega_{mn}^2 = g\,\nu_{mn}\,\tanh \nu_{mn} h \qquad (6.157)$$

Going back to the two-dimensional case, we now consider that the tank undergoes a forced horizontal motion, of unit velocity, at a frequency ω. The boundary value problem to solve is

$$\Delta\varphi = 0 \qquad\qquad 0 \le x \le L \qquad -h \le z \le 0 \quad (6.158)$$
$$\varphi_x = 1 \qquad\qquad x = 0 \quad \text{and} \quad x = L \qquad (6.159)$$
$$\varphi_z = 0 \qquad\qquad z = -h \qquad\qquad (6.160)$$
$$g\,\varphi_z - \omega^2\,\varphi = 0 \qquad\qquad z = 0 \qquad\qquad (6.161)$$

The solution may be found as a particular solution φ_P, verifying the inhomogeneous Neumann condition at the vertical walls, plus a superposition of natural modes. Here the particular solution is chosen as the solid response: $\varphi_P = x - L/2$. Writing $\varphi = \varphi_P + \psi$, the following ψ problem is obtained:

$$\Delta\psi = 0 \qquad\qquad 0 \le x \le L \qquad -h \le z \le 0 \quad (6.162)$$
$$\psi_x = 0 \qquad\qquad x = 0 \quad \text{and} \quad x = L \qquad (6.163)$$
$$\psi_z = 0 \qquad\qquad z = -h \qquad\qquad (6.164)$$
$$g\,\psi_z - \omega^2\left(\psi + x - \frac{L}{2}\right) = 0 \qquad\qquad z = 0 \qquad\qquad (6.165)$$

The solution is immediate when one projects $x - L/2$ on the set $[\cos\lambda_n x]$ (which is complete and orthogonal over $[0\ L]$):

$$x - \frac{L}{2} = \sum_{n=1}^{\infty} x_n\,\cos\lambda_n x \qquad (6.166)$$

Multiplying both sides with $\cos\lambda_m x$, then integrating in x from 0 to L, gives the x_n coefficients:

$$x_n = \frac{2}{\lambda_n^2\,L}\,[(-1)^n - 1] \qquad (6.167)$$

which are nonzero when n is odd.

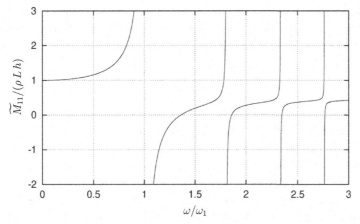

Figure 6.34 Rectangular tank in forced horizontal oscillations. Added mass $\widetilde{M}_{11}/(\rho\, L\, h)$ vs ω/ω_1

When ψ is written as

$$\psi = \sum_{n=1}^{\infty} \alpha_n \, \frac{\cosh \lambda_n \,(z + h)}{\cosh \lambda_n \, h} \, \cos \lambda_n \, x \qquad (6.168)$$

the free surface condition yields the α_n coefficients:

$$\alpha_n = \frac{2}{\lambda_n^2 \, L} \, [(-1)^n - 1] \, \frac{\omega^2}{\omega_n^2 - \omega^2} \qquad (6.169)$$

The hydrodynamic load at the wall is obtained, by integrating the pressure, as $F = -\widetilde{M}_{11}\, \ddot{X}$, X being the forced motion at frequency ω and $\widetilde{M}_{11}$ the internal added mass given by

$$\widetilde{M}_{11} = -\rho \int_{-h}^{0} \left(-\frac{L}{2} + \psi(0, z) \right) \, dz + \rho \int_{-h}^{0} \left(\frac{L}{2} + \psi(l, z) \right) \, dz \qquad (6.170)$$

$$\widetilde{M}_{11} = \rho\, L\, h + 4\rho \sum_{n=1}^{\infty} \frac{1 - (-1)^n}{\lambda_n^3 \, L} \, \tanh \lambda_n \, h \, \frac{\omega^2}{\omega_n^2 - \omega^2} \qquad (6.171)$$

Figure 6.34 shows, in the case $L = 2h$, the nondimensional added mass $\widetilde{M}_{11}/(\rho\, L\, h)$ of the tank vs the ratio ω/ω_1 of the oscillation frequency to the natural frequency of the first sloshing mode. At zero frequency no dynamic effect takes place: The free surface condition reduces to $\partial\varphi/\varphi z = 0$, the free surface is equivalent to a rigid lid, and the fluid moves as a solid body. When the oscillation frequency crosses the natural frequency of any odd mode, the added mass becomes singular, and jumps from $+\infty$ to $-\infty$. This is a result of linear theory; in practice, the free surface would break and the added mass would remain finite. Due to breaking, some energy gets dissipated; as a result, part of the hydrodynamic loads opposes the tank velocity, meaning some damping effect on the tank motion.

The solution to the boundary value problem (6.158) through (6.161) may be obtained through alternative ways, for instance by using another particular solution.

One may also use a different set of eigenfunctions, satisfying the free surface and bottom boundary conditions, as in the piston wavemaker problem considered in Appendix A. Then the velocity potential, satisfying (6.158) through (6.161), is obtained as

$$\varphi(x,z) = \frac{A_0}{k_0} \frac{\cosh k_0(z+h)}{\cosh k_0 h} \frac{\sin k_0(x-L/2)}{\cos k_0 L/2} + \sum_{n=1}^{\infty} \frac{A_n}{k_n} \cos k_n(z+h) \frac{\sinh k_n(x-L/2)}{\cosh k_n L/2}$$

(6.172)

with the same A_n coefficients as in the wavemaker case.

The liquid motion due to a forced angular motion (in roll, pitch or yaw) may be similarly derived. In these cases the particular solution satisfying the no-flow conditions at the walls cannot be taken as the solid motion, which is rotational. It is customary to use the so-called Stokes–Joukowski potentials, which represent the internal flow when the mean free surface is made solid (Dodge, 1966; Faltinsen & Timokha, 2009). There are other possible choices, for instance the "infinite frequency" potentials satisfying $\varphi_P = 0$ at the free surface, used in the following section.

One usually seeks to reduce the liquid motion in the tanks. One way is to keep the tanks close to fully loaded or empty conditions. LNG-carriers usually sail at filling heights larger than 70% or less than 10%. The tanks may also be reduced in size, or compartmentalized, in order that their sloshing frequencies be larger than the wave frequencies. One may also reduce the resonant sloshing response by fitting the tanks with solid or perforated baffles.

A contrario advantage may be taken of the internal fluid motion in tanks to reduce the wave response of the floating support. The natural frequency of the fluid motion in the tank is then made to coincide with the natural frequency of the considered degree of freedom. So-called anti-roll tanks (usually U-tubes) are widely used in some classes of ships, and also in road transportation (in sitern trucks).

6.11   Coupling between Sloshing and Sea-Keeping

The wave response of floating supports induces sloshing motion in the tanks. Conversely, in the case of FSRUs or FLNGs, for instance, LNG sloshing in the tanks affects the dynamic behavior of the supports.

One way to account for the coupling effects is to add, to the support matrix of added masses and inertias (due to the outer fluid domain), added mass/inertia matrices associated with the internal tanks. Modifications of the hydrostatic stiffnesses in roll and pitch must also be duly accounted for (see Appendix B). In this way it is difficult to account for energy dissipation in the tanks, due to free surface breaking or friction at the walls (the latter being usually quite negligible). One possible way is to modify the free surface condition in the tank, or the wall boundary condition, by adding dissipative terms (see Appendix D).

Another method is to introduce the effects of the sloshing modes in the tanks as coupled mass spring oscillators. This technique has been pioneered in the aerospace industry (Dodge 1966). The free surface elevation in each tank is written as a superposition of modal shapes:

$$\eta_i(x,y,t) = \sum_n A_{in}(t)\, \zeta_{in}(x,y) \tag{6.173}$$

where i refers to the tank number and n to the mode number. The modal shapes ζ_{in} are normalized so that they take a maximum value of one over the free surface. They are related to the modal potentials φ_{in} through

$$\zeta_{in}(x,y) = \frac{\omega_{in}}{g}\, \varphi_{in}(x,y,0) \tag{6.174}$$

Here we follow the theoretical model proposed by Molin *et al.* (2002, 2008). Mass spring equations for the modal amplitudes A_{ni} are therein derived in the form

$$\ddot{A}_{in} + \omega_{in}^2\, A_{in} = \sum_{j=1}^{6} D_{inj}\, \ddot{X}_{ij} \tag{6.175}$$

where the ω_{in} are the natural sloshing frequencies in tank i, X_{ij} $(j = 1,6)$ are the 6 dof motion of tank i (in its own coordinate system), and the D_{inj} coefficients are given by

$$D_{inj} = -\frac{g}{\omega_{in}} \frac{\iint_{S_{T_i}} \varphi_{in}\, N_{ij}\, \mathrm{d}S_i}{\iint_{S_{F_i}} \varphi_{in}^2\, \mathrm{d}S_i} \tag{6.176}$$

with S_{T_i} the tank wall and S_{F_i} the tank free surface.

The D_{inj} coefficients are obtained by using auxiliary potentials different from those used by Dodge (1966) or Faltinsen & Timokha (2009), that is the **infinite frequency** internal potentials (with homogeneous Dirichlet condition at the free surface).

The hydrodynamic loads transferred from tank i to the floating support are given by

$$\overrightarrow{F_i} = \sum_n A_{in}(t)\, \overrightarrow{f_{in}} - \mathbf{M_{ai}}(\infty)\, \overset{..}{\overrightarrow{X_i}} \tag{6.177}$$

where $\mathbf{M_{ai}}(\infty)$ is the infinite frequency 6×6 added mass matrix and

$$f_{inj} = \rho\omega_{in} \iint_{S_{T_i}} \varphi_{in}\, N_{ij}\, \mathrm{d}S_i \tag{6.178}$$

Empirical linear and quadratic damping terms may then be introduced in the mass spring equations, which take the form

$$\ddot{A}_{in} + B_{1in}\, \dot{A}_{in} + B_{2in}\, \dot{A}_{in}\, |\dot{A}_{in}| + \omega_{in}^2\, A_{in} = \sum_{j=1}^{6} D_{inj}\, \ddot{X}_{ij} \tag{6.179}$$

In Molin *et al.* (2002) some experiments are reported on a rectangular barge model, fitted with two rectangular tanks (see Figure 6.35). The tests are done in irregular beam waves. The tanks are 80 cm long (in the transverse direction) and 25 cm wide. In the results shown here the filling heights are 19 cm for tank 1 and 39 cm for tank 2. The natural frequencies of the first sloshing modes are respectively 4.95 rad/s and 5.92 rad/s. With the liquid cargo being frozen the roll natural frequency would be 4.8 rad/s.

Figure 6.36 shows the RAO of the relative free surface elevation by the wall of the 39 cm tank, from the tests and from the numerical model. Figure 6.37 shows the

Figure 6.35 Barge model in the wave basin.

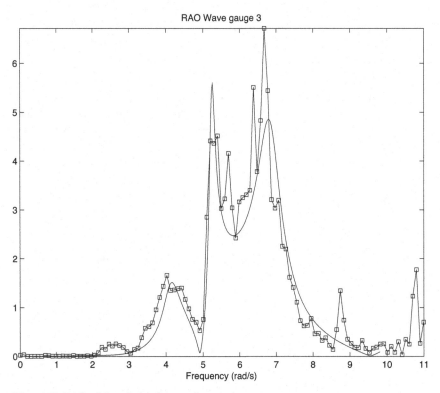

Figure 6.36 RAO of the relative free surface motion.

RAO of the roll motion. The measured values are from the same test, with a significant wave height of 6.6 cm and a peak period of 1.6 s. It can be observed that the roll RAO exhibits 3 peaks, and that it goes to nearly zero at the natural frequencies of the two sloshing modes. These nonzero values are due to dissipation inside the tanks.

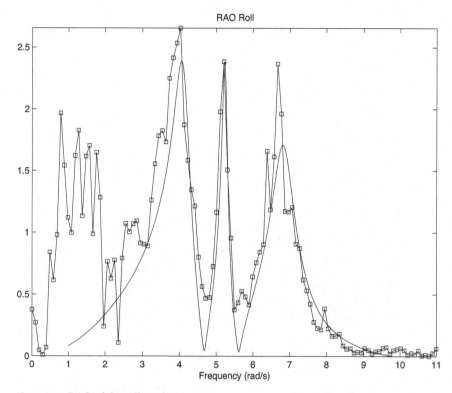

Figure 6.37 RAO of the roll motion (rad/m).

6.12 Moonpool and Gap Resonances

"Moonpools" are vertical openings, from deck to keel, through the hulls of some ships and offshore structures, such as drillships or Offshore Construction Vessels (OCVs). The wind turbine floater developed by Ideol features a large moonpool, which they claim to be beneficial to the seakeeping performance (Figure 6.38). "Gaps" are the narrow spaces in between two ships moored side by side, or between a ship and a quay. Moonpools and gaps are prone to resonance problems, under outer wave action and/or wave-induced motion of the floating supports. Moonpool resonance may also occur under the single effect of forward speed.

Resonant modes in moonpools consist in sloshing modes, as in tanks, and in an additional mode known as the **piston**, or pumping, or Helmholtz, mode, where the entrapped water heaves up and down more or less like a solid body. When resonating, these modes may hinder marine operations such as drilling and installation of subsea equipment. Hence it is useful to have access to their natural frequencies at the design stage.

6.12.1 Piston Mode in Moonpools with Vertical Walls

A theoretical model to predict the resonant frequencies of the piston and sloshing modes in rectangular moonpools, with vertical walls from deck to keel, was proposed

Figure 6.38 The wind turbine prototype "Floatgen" on the SEM-REV site offshore of le Croisic in south Brittany (credits Ideol V. Joncheray).

by Molin (2001). In this model the water depth is considered infinite, the support is held fixed, and its length and beam are assumed to be infinite. It is shown that a good approximation to the natural frequency of the piston mode is obtained when assuming the water entrapped in the moonpool to be "frozen," idealized as a solid body. The natural frequency is then obtained as

$$\omega_0 \simeq \sqrt{\frac{g}{d + b\, f_0(b/l)}} \qquad (6.180)$$

where

$$f_0(r) = \frac{1}{\pi} \left\{ \sinh^{-1}\left(r^{-1}\right) + r^{-1}\, \sinh^{-1}(r) + \frac{1}{3}\left(r + r^{-2}\right) - \frac{1}{3}\left(1 + r^{-2}\right)\sqrt{r^2 + 1} \right\} \qquad (6.181)$$

Here d is the draft, and b and l are the width and length of the moonpool. The quantity $\rho b^2\, l\, f_0(b/l)$ is nothing but the zero frequency limit of the heave added mass of a flat plate of length l and breadth b at the free surface.

Figure 6.39 shows the function f_0 as a function of the ratio b/l.

For a circular moonpool, the natural frequency is obtained as

$$\omega_0 \simeq \sqrt{\frac{g}{d + 8\,a/(3\pi)}} \qquad (6.182)$$

with a the radius.

Due to the assumption of infinite horizontal extension of the support, equations (6.180) and (6.182) slightly underestimate the natural frequencies. In the circular

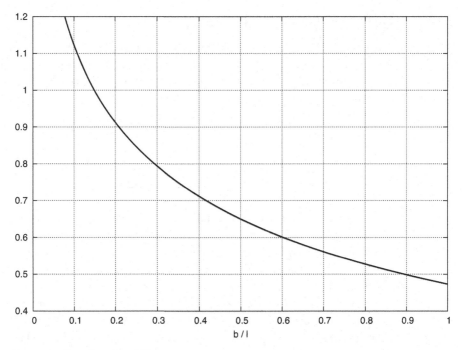

Figure 6.39 Function f_0 vs b/l.

moonpool case, an upper bound, corresponding to the case of the piston mode inside a vertical column of zero thickness, is

$$\omega_0 \simeq \sqrt{\frac{g}{d + 2\,a/\pi}} \qquad (6.183)$$

When the draft d is of the same order as the diameter ($d \sim 2a$), the relative difference between (6.182) and (6.183) is only about 4%.

6.12.2   Piston Mode in Moonpools with Slowly-Varying Cross-Section

When the cross-section is slowly varying over the height, the vertical component of the flow velocity inside the moonpool is dominant over the horizontal components, and the natural frequency of the piston mode can be estimated as

$$\omega_0 = \sqrt{\frac{g}{\int_{-d'}^{0} \frac{S(0)}{S(z)}\, dz}} \qquad (6.184)$$

where $S(z)$ is the cross-section and d' an increased draft to account for added mass effects from the lower fluid domain. So $d' = d + b\,f_3(b/l)$ for a rectangular shape, and $d' = d + 8\,a/(3\pi)$ for a circular one, where b, l, or a refer to the lower end of the moonpool.

6.12.3   Sloshing Modes in Moonpools

When the draft is deep enough that no flow takes place below the keel line, the natural frequencies and modal shapes of the sloshing modes are the same as in a deep tank (see Section 6.10). In a rectangular moonpool of length l and width b, the sloshing mode frequencies are given by

$$\omega_{mn}^2 = g\, \nu_{mn} \tag{6.185}$$

with

$$\nu_{mn}^2 = \frac{m^2\, \pi^2}{l^2} + \frac{n^2\, \pi^2}{b^2} \tag{6.186}$$

In a circular moonpool, the natural frequencies are obtained from the zeros of the derivative of the Bessel functions J_m. The same formula (6.185) applies, where ν_{mn} verifies $J_m'(\nu_{mn} a) = 0$ with a the radius.

These approximations are applicable provided that the draft be greater than the length l or the radius a of the moonpool.

When the draft decreases, it is found that the natural frequencies increase from their infinite draft values. This is opposite to what happens in a tank where the natural frequencies decrease with decreasing water height. It is due to the flow and the associated kinetic energy taking place below the hull.

In Molin (2001), a theoretical model is proposed that provides the natural modes in a rectangular moonpool. The flow in the moonpool is decomposed as a superposition of elementary sloshing modes, similar to (6.156), plus a uniform vertical flow:

$$\Phi(x,y,z,t) = \mathcal{R} \left\{ \sum_{m=0}^{\infty} \sum_{n=0}^{\infty} \cos \lambda_m x \cos \mu_n y (A_{mn} \cosh \nu_{mn} z + B_{mn} \sinh \nu_{mn} z)\, e^{-i\omega t} \right\} \tag{6.187}$$

with $\lambda_m = m\pi/l$, $\mu_n = n\pi/b$, $\nu_{mn}^2 = \lambda_m^2 + \mu_n^2$, and, when $m = n = 0$, the hyperbolic functions are replaced with $A_{00}\, z/d + B_{00}$.

A matching condition with the flow below the moonpool is then derived as

$$\Phi(x,y,-d,t) = \frac{1}{2\pi} \int_0^l dx' \int_0^b dy' \, \frac{\Phi_z(x',y',-d,t)}{\sqrt{(x-x')^2 + (y-y')^2}} \tag{6.188}$$

This condition assumes that the water depth, and also the width and length of the support, is infinite. In Molin et al. (2018) a different matching method is proposed that relaxes, to some extent, these restrictions.

From equations (6.187) and (6.188), an eigen value problem can be formulated from which the natural frequencies and associated modal shapes are derived.

When only one mode is retained in (6.187), so-called single-mode approximations are derived that yield approximate values of the resonant frequencies. For the piston mode equation (6.180) is then obtained. For the longitudinal sloshing modes the resonant frequencies are given by

$$\omega_{n0}^2 \simeq g\, \lambda_n \, \frac{1 + J_{Nn}(\lambda_n b)\, \tanh \lambda_n d}{J_{Nn}(\lambda_n b) + \tanh \lambda_n d} \tag{6.189}$$

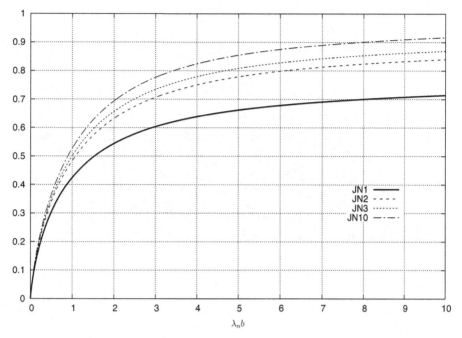

Figure 6.40 Functions J_{Nn} vs $\lambda_n b$.

where $\lambda_n = n\pi/l$ and J_{Nn} is given by

$$J_{Nn} = \frac{2}{n\pi^2 r}\left\{\int_0^1 \frac{r^2}{u^2\sqrt{u^2+r^2}}\left[1+(u-1)\cos(n\pi u) - \frac{\sin(n\pi u)}{n\pi}\right]du + \frac{1}{\sin\theta_0} - 1\right\}$$

(6.190)

with $r = b/l$ and $\tan\theta_0 = r^{-1}$.

Figure 6.40 shows the J_{Nn} functions vs $\lambda_n b$ for $n = 1, 2, 3$, and 10.

In the circular moonpool case, analogous single-mode approximations can be found in Molin *et al.* (2018).

6.12.4 Sloshing Modes in Gaps

Gaps are the narrow spaces in between two ships side-by-side (for instance during LNG transfer, as shown in Figure 6.41), or between a ship and a quay. They are like open-ended moonpools. Single-mode approximations for the natural frequencies of the sloshing modes can be derived by replacing the Neumann condition ($\partial\varphi/\partial n = 0$) at the moonpool ends with a Dirichlet condition ($\varphi = 0$). Then, following the same method as in the rectangular moonpool case, the resonant frequencies are obtained as

$$\omega_{n0}^2 \simeq g\,\lambda_n\,\frac{1 + J_{Dn}(\lambda_n b)\,\tanh\lambda_n d}{J_{Dn}(\lambda_n b) + \tanh\lambda_n d}$$

(6.191)

Figure 6.41 LNG-carrier moored alongside the Prelude FLNG (courtesy Shell).

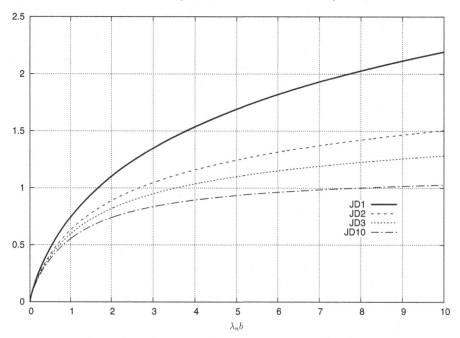

Figure 6.42 Functions J_{Dn} vs $\lambda_n b$.

where $\lambda_n = n\pi/l$ and J_{Dn} is given by

$$
J_{Dn} = \frac{2}{n\pi^2 r} \left\{ \int_0^1 \frac{r^2}{u^2\sqrt{u^2+r^2}} \left[1 + 2u + (u-1)\cos(n\pi u) - \frac{3}{n\pi}\sin(n\pi u) \right] du \right.
$$
$$
\left. - \frac{1}{\sin\theta_0} + 1 + 2r \ln\frac{1+\cos\theta_0}{1-\cos\theta_0} \right\} \tag{6.192}
$$

where $r = b/l$ and $\tan\theta_0 = r^{-1}$.

The functions J_{Dn} are shown in Figure 6.42 for $n = 1$, 2, 3, and 10. In a particular case, taken from Molin *et al.* (2009), Figure 6.43 shows, as vertical dashed bars, the

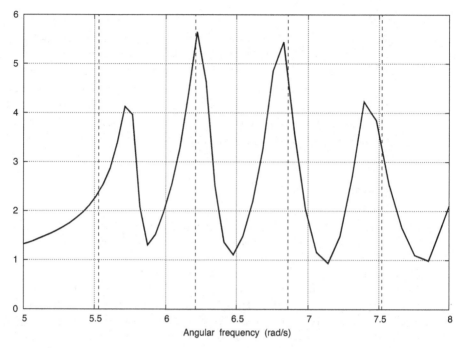

Figure 6.43 Gap. RAOs of the free surface elevation.

resonant frequencies given by (6.191), together with the calculated RAOs of the free surface elevation at some point in the gap, with a linearized potential flow solver. It can be seen that the calculated RAOs reach very high values (up to 6 here). Equation (6.191) appears to underestimate somewhat the location of the first resonant frequency. This is due to the infinite width assumption of the two rectangular barges (better prediction can be achieved with the alternate formulas given in Molin *et al.*, 2018). From the second mode the agreement is better.

6.12.5 Resonant Modes in Finite Depth

The effect of finite depth on the resonant frequencies for moonpools and gaps is considered in Molin *et al.* (2018). Alternate single-mode approximations are proposed therein that account for finite depth and also for the finite horizontal extensions of the supports.

6.12.6 Damping of Moonpool and Gap Resonance

Calculations based on linearized potential flow theory usually give unrealistically large RAOs of the free surface motion. Both for moonpools and gaps, damping associated with wave radiation plays a minor role, and most of the energy dissipation comes from viscous effects at the hull, friction, and pressure forces due to flow separation. This is quite similar with the damping processes of ship roll resonance.

When the bilges are sharp or fitted with bilge keels, damping due to flow separation is dominant over friction, and the damping force is a priori quadratic with the local relative velocity of the flow with respect to the hull. Nevertheless the RAO over-prediction problem is often tackled by introducing linear damping terms into the free surface equation inside the gap or moonpool, or within the no-flow condition at the hull. Another technique consists in covering the free surface with a finite number of fictitious, massless plates (treating then an $N + 1$ body diffraction-radiation problem), and assigning a combination of linear and quadratic damping terms to their heave, roll, and pitch equations of motion (e.g. see Molin et al., 2009). There remains the problem of determining the drag coefficients to use.

6.13   References

ARMAND J.-L., DUTHOIT C. 1990. Distribution of maxima of non-linear ship rolling, in *2nd Intl. Symposium on Dynamics of Marine Vehicles and Structures in Waves*, Elsevier, 305–316.

BABARIT A. 2017. *Ocean Wave Energy Conversion*, ISTE Press, Elsevier.

BABARIT A., DUCLOS G., CLÉMENT A.H. 2004. Comparison of latching control strategies for a heaving wave energy device in random sea, *Appl. Ocean Res.*, **26**, 227–236.

BABARIT A., HALS J., MULIAWAN M.J., KURNIAWAN A., MOAN T., KROKSTAD J. 2012. Numerical benchmarking study of a selection of wave energy converters, *Renewable Energ.*, **41**, 44–63.

CALLAN M., LINTON C.M., & EVANS D.V. 1991. Trapped modes in two-dimensional waveguides, *J. Fluid Mech.*, **229**, 51–64.

CHAKRABARTI S. 2001. Empirical calculation of roll damping for ships and barges, *Ocean Engineering*, **28**, 915–932.

COINTE R., GEYER P., KING B., MOLIN B., TRAMONI M.-P. 1990. Nonlinear and linear motions of a rectangular barge in a perfect fluid, in *Proc. 18th ONR Symposium on Naval Hydrodynamics*, Ann Arbor, MI.

DODGE F.T. 1966. Analytical representation of lateral sloshing by equivalent mechanical systems, in *The Dynamic Behavior of Liquids in Moving Containers*, NASA SP 106, ch. 6, H.N. Abramson editor.

EVANS, D.V., JEFFERY, D.C., SALTER, S.H., TAYLOR, J.R.M., 1979. Submerged cylinder wave energy device: theory and experiment, *Appl. Ocean Res.*, **1**, 3–12.

FALTINSEN O.M., ROGNEBAKKE O.F., TIMOKHA A.N. 2007. Two-dimensional resonant piston-like sloshing in a moonpool, *J. Fluid Mech.*, **575**, 359–397. doi:10.1017/S002211200600440X.

FALTINSEN O.M., TIMOKHA A.N. 2009. *Sloshing*, Cambridge University Press.

FERNYHOUGH M., EVANS D.V. 1995. Scattering by a periodic array of rectangular blocks, *J. Fluid Mech.*, **305**, 263–279.

GARRETT C.J.R. 1971. Wave forces on a circular dock, *J. Fluid Mech.*, **46**, 129–139.

HOOFT J.P. 1972. Hydrodynamic aspects of semisubmersible platforms. Ph.D. Thesis, Delft Technical University.

IKEDA, Y. 1984. Roll damping of ships. In: *Proceedings of Ship Motions, Wave Loads and Propulsive Performance in a Seaway*, First Marine Dynamics Symposium, The Society of Naval Architecture in Japan, 241–250 (in Japanese).

KOKKINOWRACHOS K., MAVRAKOS S., ASORAKOS S. 1986. Behavior of vertical bodies of revolution in waves, *Ocean Engineering*, **13**, 505–538.

LEDOUX A., MOLIN B., DE JOUETTE C., COUDRAY T. 2004. FPSO roll damping prediction from CFD and 2D and 3D model tests investigations, ISOPE-I-04-118, The Fourteenth International Offshore and Polar Engineering Conference, Toulon.

LINTON C.M., EVANS D.V. 1990. The interaction of waves with arrays of vertical circular cylinders, *J. Fluid Mech.*, **215**, 549–569.

LINTON C.M., MCIVER P. 2001. Handbook of Mathematical Techniques for Wave/Structure Interactions, Chapman & Hall/CRC.

MAGEE A.R. 1991. Large amplitude ship motions in the time domain, PhD thesis, U. Michigan.

MALENICA Š., MOLIN B., REMY F., SENJANOVIC I. 2003. Hydroelastic response of a barge to impulsive and non-impulsive wave loads, *Proc. 3rd Int. Conf. on Hydroelasticity*, Oxford, UK.

MALENICA Š., MOLIN B., TUITMAN J.T., BIGOT F., SENJANOVIC I. 2009. Some aspects of hydrostatic restoring forces for elastic bodies, in *Proc. 24th International Workshop in Water Waves and Floating Bodies*, Zelenogorsk, Russia (www.iwwwfb.org).

MALENICA Š., ZALAR M, CHEN X.B. 2003. Dynamic coupling of seakeeping and sloshing. In *Proc.13th int. Offshore and Polar Eng. Conf.*, Honolulu, Hawaii.

MARTIN P.A., FARINA L. 1997. Radiation of water waves by a heaving submerged horizontal disc, *J. Fluid Mech.*, **337**, 365–379.

MEI C.C., BLACK J.L. 1969. Scattering of surface waves by rectangular obstacles in waters of finite depth, *J. Fluid Mech.*, **38**, 499–511.

MENDOUNE MINKO I.D., PREVOSTO M., LE BOULLUEC M. 2008. Distribution of maxima of non-linear rolling in case of coupled sway and roll motions of a floating body in irregular waves, in *Proc. of the ASME 27th International Conference on Offshore and Arctic Engineering*, OMAE2008-57935, Estoril.

MOLIN B. 2001. On the piston and sloshing modes in moonpools, *J. Fluid Mech.*, **430**, 27–50.

MOLIN B., REMY F., RIGAUD S., DE JOUETTE C. 2002. LNG-FPSO's: frequency domain, coupled analysis of support and liquid cargo motions, *Proc. IMAM Conference*, Rethymnon.

MOLIN B., REMY F., KIMMOUN O., STASSEN Y. 2002. Experimental study of the wave propagation and decay in a channel through a rigid ice-sheet, *Applied Ocean Res.*, **24**, 247–260.

MOLIN B., REMY F., LEDOUX A., RUIZ N. 2008. Effect of roof impacts on coupling between wave response and sloshing in tanks of LNG-carriers, in *Proc. of the ASME 27th International Conference on Offshore and Arctic Engineering*, OMAE2008-57039, Estoril.

MOLIN B., REMY F., CAMHI A., LEDOUX A. 2009. Experimental and numerical study of the gap resonances in between two rectangular barges, in *Proc. 13th Congress of Intl. Maritime Assoc. of Mediterranean (IMAM)*, Istanbul.

MOLIN B., ZHANG X., HUANG H., REMY F. 2018. On natural modes in moonpools and gaps in finite depth, *J. Fluid Mech.*, **840**, 530–554.

NEWMAN J.N. 1994. Wave effects on deformable bodies, *Applied Ocean Res.*, **16**, 47–59.

NEWMAN J.N., SORTLAND B., VINJE T. 1984. Added mass and damping of rectangular bodies close to the free surface, *J. Ship Research*, **28**.

PINKSTER J.A. 1980. Low frequency second order wave exciting forces on floating structures, Ph.D. Dissertation, Delft.

PREVOSTO M. 2001. Distribution of maxima of non-linear barge rolling with medium damping, in *Proc. 11th Int. Offshore & Polar Eng. Conf.*, **III**, 307–316.

SALTER H. 1974. Wave power, *Nature*, **249**, 720–724.

SPRING B.H., & MONKMEYER P.L. 1974. Interaction of plane waves with vertical cylinders, *Proc. 14th Conf. on Coastal Engineering*, ASCE, 1828–1847.

YEUNG R.W. 1981. Added mass and damping of a vertical cylinder in finite-depths water, *Applied Ocean Res.*, **3**, 119–133.

ZHAO R., FALTINSEN O.M., KROKSTAD J.R., AANESLAND V. 1988. Wave-current interaction effects on large-volume structures, *Proc. 5th Int. Conf. Behaviour of Offshore Structures*, BOSS'88, **2**, 623–638.

7   Large Bodies: Second-Order Effects

7.1   Introduction

In Chapter 6 the wave body interaction problem was considered within the scope of a linearized theory: The wave slope and amplitude are assumed to be small, the phenomena considered (loads, response, etc.) take place at the same frequencies as the waves.

In situ measurements and model scale experiments show that loads and responses do not occur only at the wave frequencies. Incoming waves induce nonzero time-averaged loads, known as drift forces, that the mooring systems must withstand. Resonant responses take place at the natural frequencies of the mechanical system composed by the floating body and its mooring. Among them, low-frequency components, known as slow-drift motion, are particularly noticeable. In the case of stiff systems such as Tension Leg Platforms (TLPs), resonant responses in heave, roll, and pitch, known as springing, take place in sea states with peak periods about twice the heave, roll, or pitch natural periods.

The excitation loads responsible for these responses are "second-order." This follows from writing all quantities coming into play in the form

$$F = F^{(0)} + \varepsilon \, F^{(1)} + \varepsilon^2 \, F^{(2)} + \varepsilon^3 \, F^{(3)} + \dots \tag{7.1}$$

where ε is a small parameter linearly related to the wave amplitude (in Stokes theory ε is the steepness kA). The first term $F^{(0)}$ corresponds to the state of rest, or to current alone (or forward speed for ships). The following term $\varepsilon \, F^{(1)}$ is the first order of approximation, that is linearized theory. In this chapter we are concerned with the next term $\varepsilon^2 \, F^{(2)}$. As will be seen subsequently, second-order quantities relate quadratically to first-order quantities. This means that, if the incoming wave is regular, with amplitude A and frequency ω, the second-order loads appear at the double frequency $2\,\omega$ and also at the zero frequency (i.e., nonzero time-averaged loads), and all are proportional to the wave amplitude squared. In irregular seas the second-order loads take place at the sums $\omega_i + \omega_j$ and at the differences $|\omega_i - \omega_j|$ of the carrier frequencies of the wave system. A wide frequency domain is thus covered, as Figure 7.1 shows.

In this chapter, we start with the regular wave case, and we successively address the drift forces and the double frequency loads. Then we move on to bichromatic sea (two frequencies ω_1 and ω_2), which is the base case to consider to be able to generalize to irregular seas. Finally, we address the practical problem of calculating the low-frequency responses of moored structures, where the difficulty lies as much

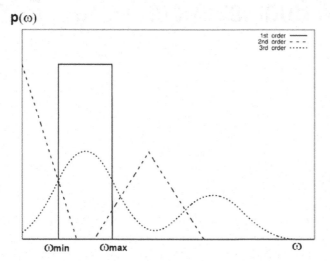

Figure 7.1 The carrier frequencies of the incoming wave system are taken to be uniformly distributed in the interval $[\omega_{min}\,\omega_{max}]$. The second-order frequencies $|\omega_i \pm \omega_j|$ then appear in two triangles, from 0 to $\omega_{max} - \omega_{min}$ (difference frequencies) and from $2\,\omega_{min}$ up to $2\,\omega_{max}$ (sum frequencies). The distribution of the third-order component frequencies $|\omega_i \pm \omega_j \pm \omega_k|$ is also shown.

in properly identifying and evaluating the damping components as in evaluating the excitation loads.

7.2   Drift Forces in Regular Waves

7.2.1   Mean Wave Loads Upon a Vertical Wall

Consider a regular wave fully reflected by a vertical wall, at normal incidence (see Figure 7.2). The first-order wave elevation is a standing wave:

$$\eta^{(1)}(x,t) = A\,\cos(kx - \omega t) + A\,\cos(-kx - \omega t) = 2A\,\cos kx\,\cos \omega t \qquad (7.2)$$

and the first-order velocity potential is

$$\Phi^{(1)}(x,z,t) = -\frac{2A\,g}{\omega}\,\frac{\cosh k(z + h)}{\cosh kh}\,\cos kx\,\sin \omega t \qquad (7.3)$$

The horizontal load upon the wall is obtained by integrating the pressure over the wetted part of the wall:

$$F(t) = \int_{-h}^{\eta(t)} -\rho\left(\Phi_t + g\,z + \frac{1}{2}(\nabla\Phi)^2\right)\,\mathrm{d}z \qquad (7.4)$$

with

$$\eta = \eta^{(1)} + \eta^{(2)} + \dots \qquad\qquad \Phi = \Phi^{(1)} + \Phi^{(2)} + \dots \qquad (7.5)$$

where $\eta^{(1)}$ is of order ε, $\eta^{(2)}$ of order ε^2, etc.[1]

---

[1] From here on we no longer write the ε, which are implicit.

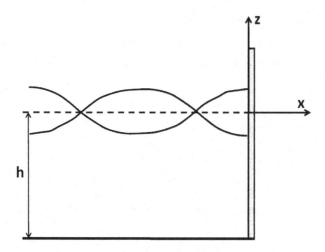

Figure 7.2 Standing wave in front of a vertical wall.

Similarly, the load decomposes into $F = F^{(0)} + F^{(1)} + F^{(2)} + \ldots$, where $F^{(0)}$ is the still water hydrostatic load. Retaining only the second-order component we get

$$F^{(2)} = -\rho \left(\Phi_t^{(1)} z + \tfrac{1}{2} g \, z^2 \right) \Big]_0^{\eta^{(1)}} \quad -\rho \int_{-h}^0 \left(\Phi_t^{(2)} + \tfrac{1}{2} (\nabla \Phi^{(1)})^2 \right) dz \quad (7.6)$$

$$= \tfrac{1}{2} \rho g \, \eta^{(1)2} \quad -\rho \int_{-h}^0 \left(\Phi_t^{(2)} + \tfrac{1}{2} \Phi_z^{(1)2} \right) dz \quad (7.7)$$

In Chapter 3 it has been established that the second-order potential $\Phi^{(2)}$ takes place at the double frequency 2ω (and, in this case of a standing wave, is independent of the space variables). It does not contribute to the time-averaged value of the second-order load, which is obtained as

$$\overline{F^{(2)}} = 2 \rho g \, A^2 \overline{\cos^2 \omega t} - \int_{-h}^0 2 \rho g \, A^2 k \, \frac{\sinh^2 k(z+h)}{\sinh kh \, \cosh kh} \, \overline{\sin^2 \omega t} \, dz \quad (7.8)$$

$$\overline{F^{(2)}} = F_d = \rho g \, A^2 - \tfrac{1}{2} \rho g \, A^2 \left[1 - \frac{2kh}{\sinh 2kh} \right] = \tfrac{1}{2} \rho g \, A^2 \left[1 + \frac{2kh}{\sinh 2kh} \right] = \rho g \, A^2 \frac{C_G(\omega)}{C_P(\omega)} \quad (7.9)$$

where C_G and C_P are, respectively, the group and phase velocities.

At second-order of approximation a nonzero time-averaged load, the **drift force**, appears, proportional to the wave amplitude squared. The main contribution to the drift force comes from the integration of the first-order pressure $-\rho g \, z - \rho \, \Phi_t^{(1)}$ on the varying wetted surface at the waterline. This load component is directed into the wall, in the direction of wave propagation, while the contribution from the quadratic term in the Bernoulli equation is in the opposite direction, toward the fluid.

When the incoming waves propagate at an angle β with respect to the normal vector into the wall, the drift force, per unit length, is obtained as

$$F_d = \rho g \, A^2 \frac{C_G(\omega)}{C_P(\omega)} \cos^2 \beta \quad (7.10)$$

7.2.2　Bottom-Mounted Vertical Cylinder

The first-order problem was solved in Chapter 6 where an analytical expression of the velocity potential was given (equation (6.61)). The drift force can then also be derived analytically, starting from

$$F_d = \frac{1}{T} \int_0^T dt \left\{ -\frac{1}{2} \rho g \int_0^{2\pi} \eta^{(1)2} a \cos\theta \, d\theta + \rho \int_{-h}^0 dz \int_0^{2\pi} \frac{1}{2} \nabla \Phi^{(1)2} a \cos\theta \, d\theta \right\} \quad (7.11)$$

$$F_d = -\frac{1}{4} \rho \frac{\omega^2}{g} \int_0^{2\pi} \varphi \, \varphi^* a \cos\theta \, d\theta + \frac{1}{4} \rho \int_{-h}^0 dz \int_0^{2\pi} \nabla\varphi \cdot \nabla\varphi^* a \cos\theta \, d\theta \quad (7.12)$$

where $\Phi^{(1)}(R, z, \theta, t) = \Re \left\{ \varphi(R, z, \theta) \, e^{-i\omega t} \right\}$, φ^* being the complex conjugate.
The drift force is finally obtained as

$$F_d = \rho g \, A^2 \, a \, \frac{4}{\pi^2 \, (ka)^3} \left(1 + \frac{2kh}{\sinh 2kh} \right) \sum_{m=0}^{\infty} \frac{\left[1 - m(m+1)/(k^2 a^2)\right]^2}{(J'^2_m + Y'^2_m)(J'^2_{m+1} + Y'^2_{m+1})} \quad (7.13)$$

where the derivatives of the Bessel functions are to be evaluated in ka, a the radius of the cylinder.

Figure 7.3 shows the nondimensional drift force $F_d/(\rho g A^2 a)$ vs ka, in infinite depth and for $h/a = 1$ and $h/a = 2$. In the high-frequency range, the curves have the same asymptotic value 2/3: when the wave length becomes very small compared to the radius, locally the cylinder wall reflects the incoming wave as a straight wall

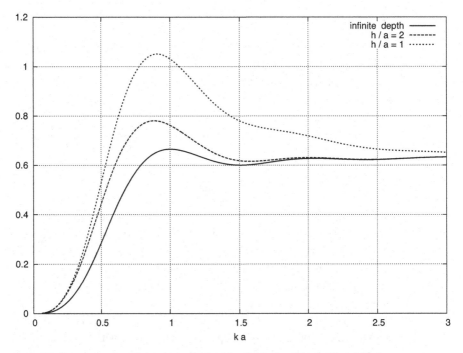

Figure 7.3 Bottom-mounted circular cylinder. Nondimensional horizontal drift force $F_d/(\rho g A^2 a)$ vs ka for different h/a ratios.

would do, and (7.10) can be applied locally, giving for the integrated load

$$F_{das} = \frac{1}{2}\rho g A^2 a \int_{-\pi/2}^{\pi/2} \cos^3 \theta \, d\theta = \frac{2}{3}\rho g A^2 a \qquad (7.14)$$

where the integration has been carried out on the weather side (the lit side) of the cylinder.

From Figure 7.3, it can be observed that this asymptotic value actually provides a reasonable estimate of the drift force from $ka \sim 1.5$ (in deep water), that is a wave length less than 2 diameters. In the general case, it is quite easy and useful to compute this asymptotic value (which only depends on the shape of the waterline) as it provides a good estimate of the drift force in the high-frequency range (when numerical errors due to insufficiently refined paneling may occur).

Conversely, in the low-frequency range, the drift force is approximated by

$$\frac{F_d}{\rho g a \, A^2} \simeq \frac{5\pi^2}{16} \left(1 + \frac{2kh}{\sinh 2kh}\right) (k a)^3 \qquad \text{for } ka \to 0 \qquad (7.15)$$

The deep water case is shown again in Figure 7.4, together with its low- and high-frequency asymptotes.

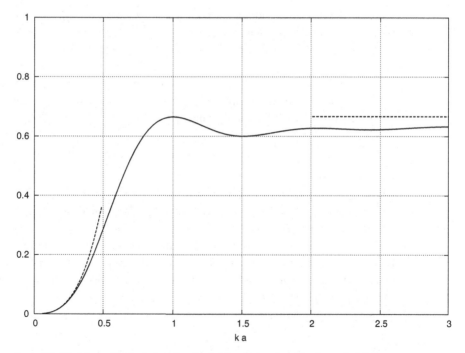

Figure 7.4 Circular cylinder in infinite depth. Nondimensional horizontal drift force $F_d/(\rho g A^2 a)$ vs ka in deep water, together with its asymptotes for $ka \to 0$ and $ka \to \infty$.

7.2.3　General Case

In the general case, different numerical methods may be used to compute the wave drift force in regular waves. They all assume that the first-order diffraction-radiation problem has been solved, and that the total potential has been derived, as

$$\Phi^{(1)} = \Re \left\{ \left[\varphi_I + \varphi_D - \sum_{j=1}^{N_{\text{dof}}} i\,\omega\,x_j\,\varphi_{Rj} \right] e^{-i\omega t} \right\} \tag{7.16}$$

A first method, known as **near-field** method, consists in directly integrating the pressure upon the hull, retaining all second-order terms (Pinkster & van Oortmerssen, 1977). This gives:

$$\overrightarrow{F_d} = \frac{1}{T} \int_0^T dt \left\{ \int_{\Gamma_0} \frac{1}{2} \rho g \left(\eta^{(1)2} - 2\eta^{(1)}\,\zeta^{(1)} \right) \frac{\overrightarrow{n_0}}{\cos\alpha} d\Gamma + \overrightarrow{A}^{(1)} \wedge \iint_{S_{B_0}} -\rho\,\Phi_t^{(1)}\,\overrightarrow{n_0}\,dS \right.$$
$$\left. + \iint_{S_{B_0}} -\rho \left[\frac{1}{2} \left(\nabla\Phi^{(1)} \right)^2 + \overrightarrow{P_0 P}^{(1)} \cdot \nabla\Phi_t^{(1)} \right] \overrightarrow{n_0}\,dS + \overrightarrow{F}_{HS}^{(2)} \right\} \tag{7.17}$$

In this expression, Γ_0 is the waterline, α is the angle between the normal vector $\overrightarrow{n_0}$ (inside the body) and the still water plane, $\zeta^{(1)}$ is the first-order vertical motion (so $\eta^{(1)} - \zeta^{(1)}$ is the relative free surface elevation), $\overrightarrow{A}^{(1)}$ is the first-order angular motion, $\overrightarrow{P_0 P}^{(1)}$ is the first-order motion at a point P_0 on the hull, and $\overrightarrow{F}_{HS}^{(2)}$ is the second-order component of the hydrostatic restoring force in still water (hence purely vertical).

An analogous expression can be obtained for the drift moments in yaw, roll, and pitch.

Another formulation results from considerations regarding the change of momentum of the fluid contained between the moving body and a surrounding fixed control surface Σ. The drift force is then obtained as

$$\overrightarrow{F_d} = \frac{1}{T} \int_0^T dt \left\{ -\int_{\Gamma_\Sigma} \frac{1}{2} \rho g\,\eta^{(1)2}\,\overrightarrow{n}\,d\Gamma + \iint_\Sigma \rho \left[\frac{1}{2} \left(\nabla\Phi^{(1)} \right)^2 \overrightarrow{n} - \nabla\Phi^{(1)} \frac{\partial\Phi^{(1)}}{\partial n} \right] dS \right\} \tag{7.18}$$

where Γ_Σ is the intersection between Σ and the free surface.

This expression only applies to the horizontal components in surge and sway. An analogous expression can be obtained for the drift moment in yaw.

The control surface can be taken as a vertical cylinder with its radius going to infinity. The z integrations can then be performed analytically from the far-field expression of the velocity potential, and only azimuthal integrations remain to be carried out numerically. One obtains:

• for the horizontal components

$$\begin{pmatrix} F_{dx} \\ F_{dy} \end{pmatrix} = -\frac{1}{2\pi} \rho g A^2 \frac{1}{k} \left(1 + \frac{2kh}{\sinh 2kh} \right) \left\{ \int_0^{2\pi} H(\theta)H^*(\theta) \begin{pmatrix} \cos\theta \\ \sin\theta \end{pmatrix} d\theta + 2\pi\,\Re\,(H(\beta)) \begin{pmatrix} \cos\beta \\ \sin\beta \end{pmatrix} \right\} \tag{7.19}$$

- for the moment in yaw

$$C_{dz} = \frac{1}{2} \rho g \, A^2 \frac{1}{k^2} \left(1 + \frac{2kh}{\sinh 2kh}\right) \Im \left\{2 H'^*(\beta) + \frac{1}{\pi} \int_0^{2\pi} H(\theta) \, H'^*(\theta) \, d\theta\right\}$$

(7.20)

with H the Kochin function as defined previously in Chapter 6 (summing up diffraction and radiation components).

This is known as the **far-field** method, or Maruo-Newman method.[2]

The far-field momentum method may also be applied in two dimensions (in a vertical plane). One obtains that the drift force is simply related to the reflection and transmission coefficients R and T:

$$F_d = \frac{1}{2} \rho g \frac{C_G}{C_P} A^2 \, (1 + R R^* - T T^*)$$

(7.21)

If energy is conserved, then $R R^* + T T^* = 1$, giving

$$F_d = \rho g \frac{C_G}{C_P} A^2 \, R R^*$$

(7.22)

The control surface Σ may be coincident with S_{B_0}. Then an expression different from (7.17) is obtained:

$$\vec{F}^{(2)} = \frac{1}{T} \int_0^T dt \left\{ \frac{1}{2} \rho g \int_{\Gamma_0} \eta^{(1)2} \frac{\vec{n}_0}{\cos \alpha} \, d\Gamma - \rho \iint_{S_{B_0}} \right.$$
$$\left[\frac{1}{2} \left(\nabla \Phi^{(1)}\right)^2 \vec{n}_0 - \left(\nabla \Phi^{(1)} \cdot \vec{n}_0\right) \nabla \Phi^{(1)} \right] dS + \vec{F}_{HS}^{(2)}$$
$$\left. - \rho g \int_{\Gamma_0} \frac{\vec{X}^{(1)} \cdot \vec{n}_0}{\cos \alpha} \, \eta^{(1)} \, \vec{k} \, d\Gamma \right\}$$

(7.23)

valid also for the vertical component (Molin & Hairault, 1983).

The control surface may also be taken at an intermediate position. This is known as the **middle-field** formulation (Chen, 2007).

Finally, another formulation, based on the Lagally theorem, can be used when the diffraction-radiation problem is solved via the integral equation method, with a source distribution. It can be established that the horizontal drift force is simply obtained as

$$\vec{F}_d = -\rho \frac{1}{T} \int_0^T dt \left\{ \iint_{S_{B_0}} \sigma \vec{V} \, dS \right\}$$

(7.24)

where σ is the source density of the perturbed part of the flow (diffraction + radiation) and where the velocity $\vec{V}$ is

$$\vec{V} = \nabla \Phi_I + \iint_{S_{B_0}} \sigma(Q,t) \, \nabla_P H(P,Q) \, dS_Q$$

(7.25)

with H the regular part of the Green function (Ledoux et al., 2006).

---

[2] Maruo (1960) introduced this method for F_{dx} and F_{dy} and Newman (1967) extended it to the drift moment in yaw.

The drift moment in yaw may also be obtained and, for a completely submerged body, the vertical components of the drift torque. When there is more than one body, the diffraction-radiation flow components from the other bodies must be included in the incoming velocity $\nabla \Phi_I$.

There are pros and cons to the different formulations of the drift force: the far-field method is considered to be the most accurate numerically but it yields only the horizontal components and it cannot discriminate when there are several bodies. The near-field method provides the 6 components of the drift force but its numerical accuracy is sometimes poor, in particular, when the hull has sharp corners where the potential flow is singular. The middle-field method offers an alternative. The Lagally formulation is very accurate, except in the vicinity of the irregular frequencies.

Some codes compute the drift force following both the near-field and far-field methods, or near-field and Lagally. When the obtained values agree it is a good signal (but not a proof!) of numerical convergence.

Figures 7.5 and 7.6 show numerical and experimental drift forces on a large tanker, 310 m long, in 82.5 m water depth. The experimental points are taken from the PhD theses by Pinkster (1980) and Wichers (1988). The numerical results are from 3 different methods, near-field (NF), middle-field (MF), and far-field (FF), with a fair agreement. The asymptotic values of the drift forces, as the frequency goes to infinity, from equation (7.26), are also shown. The paneling of the hull can be seen in Figure 6.15.

General characteristics of the drift force

As is made clear by the far-field formulation, the horizontal components of the drift force are related to the amount of waves diffracted and radiated by the structure, particularly in the direction opposite to the wave direction. Therefore the drift force drops to zero at low frequencies where the structure is transparent to the waves.

Conversely, when the wave length becomes very small, the structure acts like a curved wall totally reflecting the incoming waves, and the wave drift force tends to an asymptotic value, given by

$$\overrightarrow{F}_{d\,as} = \frac{1}{2} \rho g A^2 \int_{\Gamma_{\text{lit}}} \left[\overrightarrow{n_0} \cdot \begin{pmatrix} \cos \beta \\ \sin \beta \end{pmatrix} \right]^2 \overrightarrow{n_0} \, d\Gamma \qquad (7.26)$$

where Γ_{lit} is the part of the waterline which is "lit," exposed to the waves (see Figure 7.7). Here it has been assumed that the hull is wall-sided at the waterline. It has been seen, in the case of the circular cylinder, that this asymptotic value yields a good approximation at wave lengths shorter than two diameters. In the case of shiplike bodies, a good approximation of the sway drift force, at high wave frequencies, is $1/2 \rho g A^2 L \sin^2 \beta$ where L is the length and β the wave direction, as Figure 7.6 shows.

In the intermediate wave frequency range, peaks of the drift force are usually associated with resonances (where the radiated wave fields are important). In the case of semis and TLPs a peak usually occurs at a wave length equal to twice the column

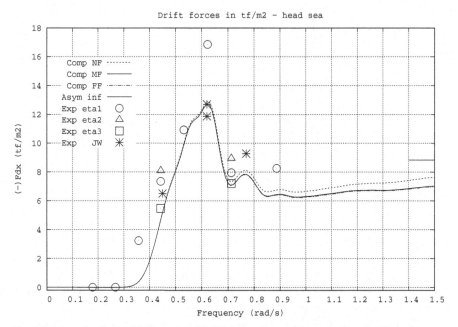

Figure 7.5 Longitudinal drift force on a VLCC in head waves (courtesy Bureau Veritas).

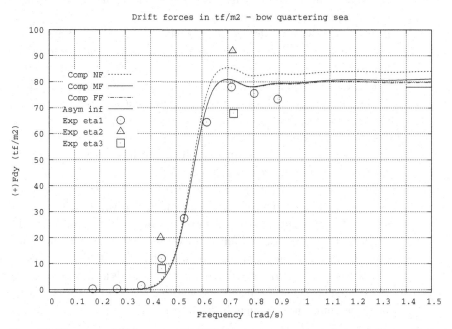

Figure 7.6 Transverse drift force on a VLCC in bow quartering waves (courtesy Bureau Veritas).

spacing, where waves diffracted by the back and front columns are in phase (Bragg effect).

As the water depth becomes more shallow the drift forces increase. From the far-field expression of the drift force it can also be deduced that its projection in the

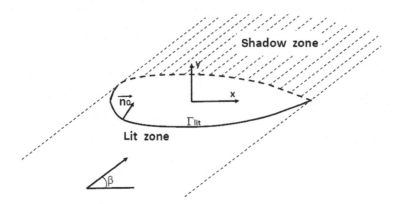

Figure 7.7 Lit side of the waterline.

direction of wave propagation is positive. This is usually confirmed experimentally, although there have been some observations of opposite behavior, attributed to viscous effects (e.g., see Huse, 1977).

Vertical drift forces are generally of minor interest, except for a few specific cases. An example, in another field, is the stability of submarines at low submergence which, under waves, are attracted toward the free surface. These vertical drift forces have nothing to do with the diffracted or radiated wave fields and are simply due to the exponential decay of the wave kinematics with the vertical coordinate. It follows that the average pressure, associated with the term $-1/2\,\rho\,(\nabla\Phi)^2$ of the Bernoulli equation, is lower (in algebraic value) above the submarine than below, hence an overall upward force. In the case of submarines, the vertical drift force can be calculated with more rudimentary numerical methods, ignoring free surface effects, such as the Rainey equations (see Section 7.5.3).

The case of the drift moments in roll and pitch is more complex since they result from local loads both horizontal and vertical.

In the case of massive structures such as barges or FPSOs, excellent agreement is usually found between experimental and numerical values of the drift force and the quadratic relationship, with respect to wave amplitude, is well fulfilled (see Figures 7.5 and 7.6). In the case of semi-submersible platforms some contribution from viscous effects may come into play. For fully submerged structural elements such as the pontoons the time-averaged values of the viscous loads are zero or nearly zero. For surface piercing components, that is, vertical columns (see Figure 7.8), the time-averaged value (assuming the column to be fixed) is given by

$$\overline{F_{\text{drag}}} = \frac{2}{3\pi}\,\rho\,C_D\,D\,A^3\,\omega^2 \tag{7.27}$$

(with D the diameter) apparently cubic in the wave amplitude, but actually even more nonlinear since the drag coefficient C_D depends on the Keulegan–Carpenter number K_C: at full-scale Reynolds numbers, for a circular column, separation occurs from $K_C \sim 4$ or 5, and the drag coefficient is quasi nil at lower K_C values. When separation

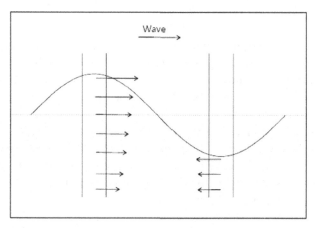

Figure 7.8 Drag forces on a vertical cylinder, under crest and under trough.

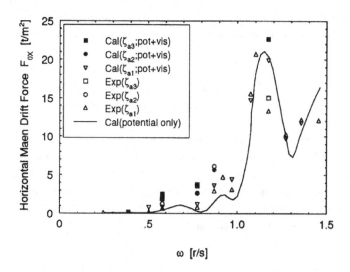

Figure 7.9 Drift force on a semi-submersible platform in head waves (from Dev, 1996).

occurs, the drag coefficient jumps from nearly zero to $O(1)$ values. In the North Sea there have been some observations of sudden offsets of semis in high-wave groups, apparently induced by viscous drift forces.

Figure 7.9, taken from Dev (1996), shows experimental and calculated values of the drift force upon a semi-submersible platform in head waves. When a viscous correction, in a form similar to equation (7.27), is added to the potential flow calculation, reasonable agreement is obtained with the measured values.

7.2.4   Effect of Current

We have just seen that, in regular wave only, within the scope of a theory that assumes perfect fluid and irrotational flow, there appear nonzero mean loads. In current only there also appear nonzero mean loads, due to viscous effects and flow separation. What happens in wave plus current, and how can the two representations of the fluid and flow be reconciled?

In the general case, empirical methods must be followed, with CFD and experiments valuable alternatives. However, when the current velocity is low enough that it is everywhere lower than the wave-induced oscillatory velocity, the periodic reversal of the flow annihilates separation and potential flow theory remains applicable. This is the case that we consider in the following.

There are now three parameters that are assumed to be low:

- $\varepsilon = k\,A$, the wave steepness
- $F_n = U_C/\sqrt{g\,L}$, the Froude number, U_C the current velocity, and L a characteristic length
- $\tau = U_C\,\omega_e/g$, usually named the Brard number, ω_e being the **encounter frequency**: in a body-fitted coordinate system and in a system traveling with the current, the apparent frequencies of the waves are different. They are related via

$$\omega_e = \omega_0 + k\,U_C\,\cos\beta \qquad (7.28)$$

with $\omega_0^2 = g\,k\,\tanh\,kh$. The relative difference between ω_0 and ω_e is of order τ.

Here we assume the current to propagate in the x direction and β is the wave direction with respect to x.

The velocity potential is first split into a steady part $U_C\,\overline{\phi^{(0)}}$ due to the current alone, and an unsteady part Φ. The Froude number being assumed to be low, no wave making occurs and a good approximation of $U_C\,\overline{\phi^{(0)}} = U_C\,(x + \overline{\phi})$ is the "double body flow" where the free surface is assimilated to a horizontal wall. It can then be established that the unsteady component $\Phi(x,y,z,t) = \Re\{\varphi(x,y,z)\,\exp(-i\,\omega_e\,t)\}$ of the velocity potential verifies the following free surface condition

$$-\frac{\omega_e^2}{g}\,\varphi + \frac{\partial\varphi}{\partial z} = i\,\tau\left[2\,\nabla_0\overline{\phi^{(0)}}\cdot\nabla_0\varphi - \frac{\partial^2\overline{\phi^{(0)}}}{\partial z^2}\,\varphi\right] \qquad (7.29)$$

where ∇_0 stands for the horizontal gradient $(\partial/\partial x,\ \partial/\partial y)$, and where terms of orders higher than ε and $\varepsilon\,\tau$ have been discarded. As $\overline{\phi}$ decays rapidly away from the body, beyond some distance the free surface condition reduces to

$$-\frac{\omega_e^2}{g}\,\varphi + \frac{\partial\varphi}{\partial z} = 2\,i\,\tau\,\frac{\partial\varphi}{\partial x} \qquad (7.30)$$

When the body is not fixed, that is, free to respond to the waves, there also arise complications in the body boundary condition (so-called "m-terms").

Some diffraction-radiation software has been extended to the numerical resolution of this problem. In some cases, the elementary diffraction and radiation potentials are decomposed as

$$\varphi_D = \varphi_{D0} + \tau\,\varphi_{D1} \qquad\qquad \varphi_{Rj} = \varphi_{R0j} + \tau\,\varphi_{R1j} \qquad (7.31)$$

with the parameter τ being formally kept. In other software the problem is solved for a given finite value of τ. Then the drift forces and their $O(\tau)$ corrections can be obtained in various ways, as in the zero current case, via the near-field and far-field methods. For simple geometries, consisting in one or several vertical cylinders, the formal decomposition to orders 0 and τ allows an analytical solution (Matsui *et al.*, 1991; Emmerhoff & Sclavounos, 1992; Malenica *et al.*, 1995).

An effect of current is a modification of the wave pattern. When the incoming waves and current propagate in the same direction, higher free surface elevations are observed at the weather side of the body while they decrease along its flanks. One interpretation is that, ahead of the body, the incoming waves get refracted by the local modification $U_C\,\overline{\phi}$ of the current, opposing them; as a result, their amplitudes increase. Along the sides the local flow $U_C\,\overline{\phi}$ adds to the current velocity lengthening the waves and decreasing their amplitudes. An illustrative example is provided in Figure 7.10, in the case of the generic FPSO[3] shown in Figure 7.11.

It makes sense then that drift forces are sensitive to current velocity since we have seen that, according to the near-field formulation, the main component of the drift force comes from the relative wave elevation along the waterline (see equation (7.17)).

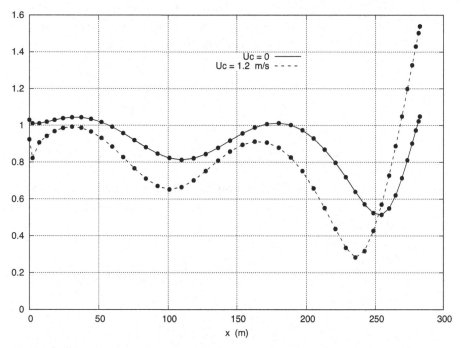

Figure 7.10 RAO of the free surface elevation along the hull of a FPSO, without current, and with a head current of 1.2 m/s (from Malenica & Chen, 1997).

---

[3] The FPSO consists in a rectangular box with a semicircular bow. Its length is 257.5 m, its beam 51 m, and its draft 20 m.

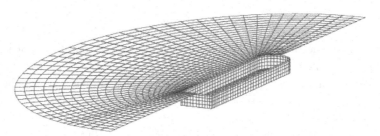

Figure 7.11 Discretization of the hull and surrounding free surface for diffraction-radiation analysis in waves and current (from Malenica & Chen, 1997).

Aranha's Formula

Even though there is software accounting for wave-current interaction, it requires more effort to run than the regular diffraction-radiation codes. For instance the free surface needs to be paneled over a large domain around the body (see Figure 7.11 and Appendix D).

It so happens that an empirical formulation has been proposed that provides, in most cases, a rather good approximation of the way drift forces are modified by current. It is known as Aranha's formula, from Aranha (1991, 1994, 1996). See also Clark *et al.* (1993).

Aranha's formula is as follows: be $\overrightarrow{F_d}(\omega,\beta,0,0) = A^2 \overrightarrow{f_d}(\omega,\beta,0,0)$ the wave drift force without current, in a regular wave of frequency ω and incidence β. Then with a current of velocity U_C and direction α superimposed, the normalized drift force $\overrightarrow{f_d}$ becomes

$$\overrightarrow{f_d}(\omega,\beta,U_C,\alpha)=[1+4\,\tau\cos(\alpha-\beta)]\overrightarrow{f_d}\Big(\omega[1+\tau\cos(\alpha-\beta)],\beta$$
$$+2\tau\sin(\alpha-\beta),0,0\Big)+O(\tau^2) \tag{7.32}$$

where $\tau = U_C\,\omega/g$.

Note that here α (resp. β) is the direction where the current (resp. the waves) is (are) going.

An interpretation of (7.32) would be that the current has the effect of increasing or decreasing the wave amplitude, and modifying its frequency and direction, as though through refraction.

Equivalently to equation (7.32) one can write

$$\overrightarrow{f_d}(\omega,\beta,U_C,\alpha) = \overrightarrow{f_d}(\omega,\beta,0,0) + \begin{pmatrix} b_{d11} & b_{d12} \\ b_{d21} & b_{d22} \end{pmatrix} \begin{pmatrix} U_C\,\cos\alpha \\ U_C\,\sin\alpha \end{pmatrix} + O(\tau^2) \tag{7.33}$$

where the components of the matrix $\mathbf{b_d}$ are given by:

$$b_{d11} = \left(\frac{4\omega}{g}\,f_{dx} + \frac{\omega^2}{g}\,\frac{\partial f_{dx}}{\partial\omega}\right)\cos\beta - \frac{2\omega}{g}\,\frac{\partial f_{dx}}{\partial\beta}\,\sin\beta \tag{7.34}$$

$$b_{d12} = \left(\frac{4\omega}{g}\,f_{dx} + \frac{\omega^2}{g}\,\frac{\partial f_{dx}}{\partial\omega}\right)\sin\beta + \frac{2\omega}{g}\,\frac{\partial f_{dx}}{\partial\beta}\,\cos\beta \tag{7.35}$$

$$b_{d21} = \left(\frac{4\omega}{g}\,f_{dy} + \frac{\omega^2}{g}\,\frac{\partial f_{dy}}{\partial\omega}\right)\cos\beta - \frac{2\omega}{g}\,\frac{\partial f_{dy}}{\partial\beta}\,\sin\beta \tag{7.36}$$

$$b_{d22} = \left(\frac{4\omega}{g} f_{dy} + \frac{\omega^2}{g} \frac{\partial f_{dy}}{\partial \omega} \right) \sin \beta + \frac{2\omega}{g} \frac{\partial f_{dy}}{\partial \beta} \cos \beta \qquad (7.37)$$

These are for infinite depth. Extensions to finite depth can be found in Malenica *et al.* (1995). For instance b_{d11} becomes

$$b_{d11} = \left\{ \left[\left(\frac{\partial f_{dx}}{\partial \omega} + \frac{1}{\gamma} \frac{\partial \gamma}{\partial \omega} f_{dx} \right) \omega + \frac{2}{\gamma} f_{dx} \right] \cos \beta - \frac{1}{\gamma} \frac{\partial f_{dx}}{\partial \beta} \sin \beta \right\} \frac{k}{\omega} \qquad (7.38)$$

where γ is the group-to-phase velocity ratio

$$\gamma = \frac{C_G}{C_P} = \frac{1}{2} \left(1 + \frac{2kh}{\sinh 2kh} \right) \qquad (7.39)$$

These expressions also apply to the case of a structure with some steady velocity $(\dot{X}, \dot{Y})$, which is kinematically equivalent to a current $(U_C \cos \alpha, U_C \sin \alpha) = -(\dot{X}, \dot{Y})$. The b_{Dij} coefficients (multiplied by the amplitude squared) can then be interpreted as damping terms, known as **wave drift damping**.

Aranha's formulas provide no information on the way the wave drift moment in yaw is affected by current.

So far nobody has succeeded in proving the well-foundedness of Aranha's formulas. It would appear that they provide exact results when the velocity potential only consists in wave propagating components (no evanescent modes), that is for fixed wall-sided bodies. In the case of an array of 4 fixed vertical cylinders in infinite depth, it has been found that the agreement with quasi-analytical results (from Emmerhoff & Sclavounos, 1992) is perfect (Clark *et al.*, 1993). For one cylinder in finite depth the formula is also exact (using equation (7.38)) when the cylinder is fixed, but it is no longer so when the cylinder is free to surge and sway (Malenica *et al.*, 1995).

Figures 7.12 and 7.13 refer to the FPSO geometry of Figure 7.11. They show b_{d11} in head waves, as obtained from Aranha's formula and from Hydrostar. When the FPSO is held fixed, the agreement is quite good. This can be related to the fact that the FPSO hull is wall-sided and has a deep draft. When the FPSO is free to respond

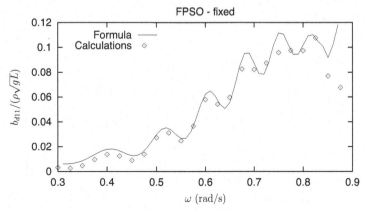

Figure 7.12 Wave drift damping in surge $b_{d11}/(\rho \sqrt{gL})$ for a FPSO fixed in waves, from Aranha's formula and from Hydrostar (courtesy Bureau Veritas).

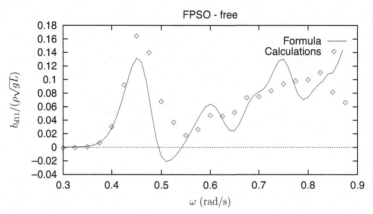

Figure 7.13 Wave drift damping in surge $b_{d11}/(\rho\sqrt{gL})$ for a FPSO freely floating in waves, from Aranha's formula and from Hydrostar (courtesy Bureau Veritas).

to the waves, the agreement deteriorates, although Aranha's formula still provides an acceptable first estimate.

7.3   Double Frequency Loads in Regular Waves

In regular waves, the second-order loads consist in a nonzero time-averaged component, the drift force, and a component fluctuating in time at the double frequency 2ω. The computation of this double frequency component is mostly of academic interest, but it does illustrate some important features that are also present in the second-order sum frequency loads in irregular seas.

Compared to the drift force case considered above, the additional complexity is that the second-order potential $\Phi^{(2)}$, taking place at the double frequency 2ω, must be included in the computation, and that its contribution usually cannot be neglected. This second-order potential consists in incident, diffracted, and radiated components, as at first-order:

$$\Phi^{(2)}(x,y,z,t) = \Re\left\{\left(\varphi_I^{(2)}(x,y,z) + \varphi_D^{(2)}(x,y,z) - \sum_{j=1}^{6} 2\mathrm{i}\,\omega\, x_j^{(2)}\,\varphi_{Rj}(2\omega)\right) e^{-2\mathrm{i}\omega t}\right\}$$

(7.40)

The second-order incident potential $\varphi_I^{(2)}$ has been given in Chapter 3

$$\varphi_I^{(2)} = -\frac{3\,\mathrm{i}\,A^2\,\omega}{8}\,\frac{\cosh 2k_0(z+h)}{\sinh^4 k_0 h}\,e^{2\mathrm{i}(k_0\,x\,\cos\beta + k_0\,y\,\sin\beta)}$$

(7.41)

The elementary radiation potentials φ_{Rj} are the same as in the first-order problem, except that they are at the double frequency 2ω. The second-order diffraction potential

$\varphi_D^{(2)}$ verifies the following boundary value problem

$$\Delta\varphi_D^{(2)} = 0 \qquad \text{in the fluid domain}$$
$$g\,\varphi_{Dz}^{(2)} - 4\omega^2\,\varphi_D^{(2)} = \alpha_{FS}^{(2)} \qquad z = 0 \qquad (7.42)$$
$$\nabla\varphi_D^{(2)} \cdot \vec{n_0} = -\nabla\varphi_I^{(2)} \cdot \vec{n_0} + \alpha_B^{(2)} \qquad \text{on the body surface} \quad (7.43)$$
$$\varphi_{Dz}^{(2)} = 0 \qquad z = -h$$

$$\text{radiation condition}$$

The right-hand side of the free surface equation has the following expression

$$\alpha_{FS}^{(2)} = -\frac{i\omega}{2g}\varphi^{(1)}\left(g\varphi_{zz}^{(1)} - \omega^2\varphi_z^{(1)}\right) + i\omega\nabla\varphi^{(1)2} + \frac{i\omega}{2g}\varphi_I^{(1)}\left(g\varphi_{Izz}^{(1)} - \omega^2\varphi_{Iz}^{(1)}\right) - i\omega\nabla\varphi_I^{(1)2}$$

$$(7.44)$$

where $\varphi^{(1)} = \varphi_I^{(1)} + \varphi_D^{(1)} - \sum_{j=1}^6 i\omega\,x_j^{(1)}\,\varphi_{Rj}(\omega)$.
Here use has been made of the identity

$$\mathcal{R}\left\{A\,e^{-i\omega t}\right\} \times \mathcal{R}\left\{B\,e^{-i\omega t}\right\} = \frac{1}{2}\mathcal{R}\left\{A\,B\,e^{-2i\omega t}\right\} + \frac{1}{2}\mathcal{R}\left\{A\,B^*\right\} \qquad (7.45)$$

The right-hand side $\alpha_{FS}^{(2)}$ can be interpreted as a dynamic pressure applied all over the free surface, generating a wave system that interacts with the body. It decays slowly with the radial distance (as $R^{-1/2}$), the leading order term being of the form $\exp\{i\,k_0\,R\,[1 + \cos(\theta - \beta)]\}/\sqrt{k_0 R}$. As a result, it can be established (Molin, 1979) that the far field of the second-order diffraction potential $\varphi_D^{(2)}$ consists of two kinds of waves, both decaying as $R^{-1/2}$:

1. "Free waves" traveling in the radial direction, with a wave number k_2 related to 2ω through the usual dispersion equation

$$4\,\omega^2 = g\,k_2\,\tanh k_2 h \qquad (7.46)$$

2. "Locked waves" (or "bound waves"), traveling locally in the intermediate direction between the incoming waves and the radial vector.

This double wave system is clearly visible in Figure 7.14, which shows the second-order wave pattern around an array of four vertical cylinders. The wave number and dimensions of the array are the same as in Figure 6.11: $kd = 3.25$; $ka = 0.63$.

The term $\alpha_B^{(2)}$ that appears in the body boundary condition (7.43) results from developing to second-order the no-flow condition, giving

$$A_B^{(2)} = \left(\vec{V}^{(1)} - \nabla\Phi^{(1)}\right) \cdot \left(\vec{A}^{(1)} \wedge \vec{n_0}\right) - \left[\left(\overrightarrow{P_0 P}^{(1)} \cdot \nabla\right)\nabla\Phi^{(1)}\right] \cdot \vec{n_0} = \mathcal{R}\left\{\alpha_B^{(2)}\,e^{-2i\,\omega t}\right\}$$

$$(7.47)$$

Numerical methods have been established that can solve the second-order diffraction problem and yield second-order pressures on the body surface, or second-order free surface elevations as in Figure 7.14. When only global loads are needed, Haskind's theorem may be applied, as follows.

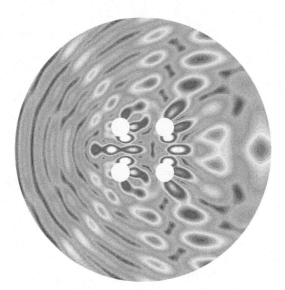

Figure 7.14 Second-order diffraction of a regular wave by an array of four vertical cylinders. Second-order wave pattern (from Scolan, 1989).

Be $\psi = \varphi_{Rj}(2\omega)$ the radiation potential for the considered degree of freedom j at frequency 2ω. It verifies the following boundary value problem:

$$
\begin{aligned}
\Delta\psi &= 0 & &\text{in the fluid domain} \\
g\,\psi_z - 4\,\omega^2\,\psi &= 0 & &z = 0 \\
\nabla\psi \cdot \vec{n_0} &= N_{0j} & &\text{on the body surface} \quad (7.48) \\
\psi_z &= 0 & &z = -h \\
\sqrt{R}\,(\psi_R - \mathrm{i}k_2\,\psi) &= 0 & &R^2 = x^2 + y^2 \to \infty
\end{aligned}
$$

The following integral needs to be evaluated

$$
I_{Dj} = \iint_{S_{B_0}} \varphi_D^{(2)} N_{0j}\,\mathrm{d}S = \iint_{S_{B_0}} \varphi_D^{(2)} \frac{\partial\psi}{\partial n}\,\mathrm{d}S \tag{7.49}
$$

From Green's second identity:

$$
I_{Dj} = \iint_{S_{B0}} \psi\,\frac{\partial\varphi_D^{(2)}}{\partial n}\,\mathrm{d}S - \iint_{z=0 \cup S_\infty \cup z=-h} \left(\varphi_D^{(2)}\,\frac{\partial\psi}{\partial n} - \psi\,\frac{\partial\varphi_D^{(2)}}{\partial n} \right)\mathrm{d}S \tag{7.50}
$$

where S_∞ is a vertical cylinder surrounding the body.

The integral over the sea floor is nil. Taking S_∞ to infinity it is possible to show, from the asymptotic behaviors of ψ and $\varphi_D^{(2)}$, that its contribution is also zero. Then it remains

$$
I_{Dj} = -\iint_{S_{B0}} \psi\left(\frac{\partial\varphi_I^{(2)}}{\partial n} + \alpha_B^{(2)} \right)\mathrm{d}S + \frac{1}{g}\iint_{z=0} \alpha_{FS}^{(2)}\,\psi\,\mathrm{d}S \tag{7.51}
$$

The integrand of the free surface integral in (7.51) contains known quantities when the first-order problem has been solved. Still its convergence with radial distance is

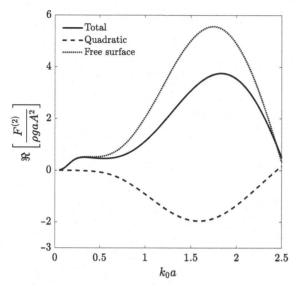

Figure 7.15 Second-order double frequency loads on a vertical cylinder. Deep water ($h = 80\,a$). Real parts (credits Y.M. Choi).

slow and asymptotic expansions of the integrand must be sought for in its evaluation (e.g., see Eatock Taylor & Hung, 1987).

The contribution from $\alpha_B^{(2)}$ also creates problems as $\alpha_B^{(2)}$ involves double-space derivatives of the first-order potential. These can be reduced through applying some integral transforms (see Chen, 1988). Still it is not legitimate to do a Taylor expansion of the velocity potential in the vicinity of sharp corners, and it is not clear what errors can be made there. An alternative, proposed by Shao & Faltinsen (2010), consists in writing the boundary value problem in a body-fitted coordinate system.

When the body is held fixed $\alpha_B^{(2)}$ is nil. The case of a bottom-mounted cylinder is easy to deal with since all first-order quantities are known analytically and the free surface integral reduces to a radial integration. Figures 7.15 and 7.16 show the real and imaginary parts of the double-frequency horizontal load, in deep water ($h = 80\,a$). In this deep water case the contribution from the incident second-order potential is quasi nil. From the figures, it can be seen that the contribution from the second-order diffraction potential (the free surface integral) cannot be neglected when $k_0 a$ is greater than about 0.10 or 0.15.

As for the contribution from the first-order quantities (the waterline component plus the contribution from $-1/2\,\rho\,\left(\nabla\Phi^{(1)}\right)^2$), when $k_0 a$ goes down to zero, it is asymptotically given by

$$f^{(2)} \simeq -\frac{5}{4}\,\mathrm{i}\,\pi\,\rho\,g\,k_0\,a^2\,A^2 \tag{7.52}$$

where $F^{(2)} = \Re\,\left\{ f^{(2)} \exp(-2\,\mathrm{i}\,\omega\,t) \right\}$.

This equation provides a good approximation of the total second-order load for $k_0 a < 0.10$ and $k_0 h > 3$. The application of the Morison equation, integrated up to the instantaneous free surface, gives a different expression, which is therefore wrong. The

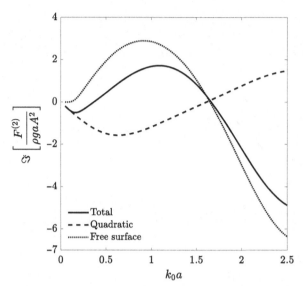

Figure 7.16 Second-order double frequency loads on a vertical cylinder. Deep water ($h = 80\,a$). Imaginary parts (credits Y.M. Choi).

reason is that the Morison equation assumes uniform incident flow in each horizontal plane (the same velocity in all x and y). To be consistent to second-order, one must abandon Morison and use the **Rainey equations**, described in Section 7.5.

A characteristic of the second-order pressures is that they penetrate deeply the water column, on the upwave side, due to the partially standing wave system that develops there. It has been seen in Chapter 3 that, in the case of two regular waves of same frequency and opposite directions, the second-order potential becomes independent of the space variables. The same phenomenon occurs here. Newman (1990) has shown that the decrease of the second-order diffraction potential, with the vertical coordinate, is in $1/z$. Figure 7.17 shows, for $k_0 a = 2$, and as a function of $k_0 h$, the modulus of the first-order horizontal load, that of the second-order component resulting from the first-order terms, and that of the second-order component due to the potential $\varphi_D^{(2)}$, via the free surface integral. The first two components stabilize for $k_0 h > 3$, while the last one keeps on increasing logarithmically.

Another illustration is provided by Figure 7.18, which shows the second-order vertical load acting on one column of the Snorre TLP, as obtained from experiments and from computations. It is a truncated cylinder, with a diameter of 25 m and a draft of 37.5 m, tested at a scale of 1:40 at SINTEF Ocean (formerly Marintek). As the wave frequency increases, the second-order vertical force (divided by the wave amplitude squared) does not decrease, unlike the first-order one, but tends to increase while oscillating.

It is understandable that these vertical forces, appearing even in the case of large drafts, have somewhat worried the TLP designers, faced with the risks of resonance

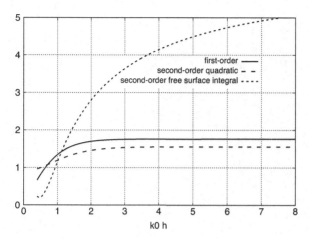

Figure 7.17 First-order $F^{(1)}/\rho g a^2 A$ and second-order $F^{(2)}/\rho g a A^2$ loads on a vertical cylinder. Sensitivity to water depth for $k_0 a = 2$ (credits Y.M. Choi).

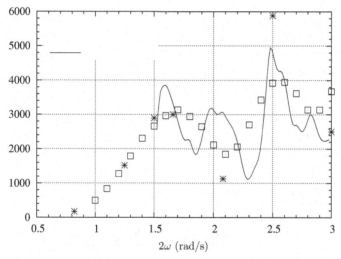

Figure 7.18 Second-order load in heave (in N/m^2) vs twice the wave frequency on one column of the Snorre TLP. Calculated values (square symbols), measured values in regular waves (star symbols) and in irregular waves (line, from bispectral analysis) (taken from Moe, 1993).

in heave, roll, or pitch. The challenge is to calculate the second-order loads in waves with periods that are twice the natural period considered, for example approximately 4.5 s for the Snorre TLP. For these calculations very fine meshes are required, to allow the auxiliary radiation potential ψ at the half period (2.2 s!) to be obtained with a good accuracy, on the hull (Figure 7.19) and also on the free surface. These are therefore extremely demanding calculations, difficult to carry out in preliminary designs.

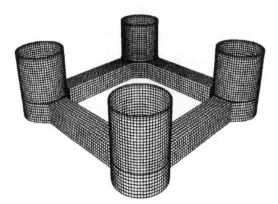

Figure 7.19 Paneling of the Snorre TLP for second-order computations (taken from Moe, 1993).

7.4   Second-Order Loads in Bichromatic Seas

In the same way that a regular wave is the base case to consider in a linear theory, a bichromatic sea (two regular waves superimposed) is the base case at second order.

For the sake of simplicity, we assume that the two wave components are collinear and propagate along the Ox axis. At first-order the free surface elevation is given by

$$\eta^{(1)}(x,t) = A_1 \cos(k_1 x - \omega_1 t) + A_2 \cos(k_2 x - \omega_2 t) \qquad (7.53)$$

The first step is to solve, for each component, the first-order diffraction-radiation problem, and to compute the linear response. Then the total first-order velocity potential is obtained:

$$\Phi^{(1)}(x,y,z,t) = \Re \left\{ \left(\varphi_{I1}^{(1)} + \varphi_{D1}^{(1)} - \sum_{j=1}^{6} i\omega_1 x_{1j}^{(1)} \varphi_{R1j} \right) e^{-i\omega_1 t} \right.$$
$$\left. + \left(\varphi_{I2}^{(1)} + \varphi_{D2}^{(1)} - \sum_{j=1}^{6} i\omega_2 x_{2j}^{(1)} \varphi_{R2j} \right) e^{-i\omega_2 t} \right\} \qquad (7.54)$$

with the body motion being

$$\vec{X}^{(1)}(t) = \Re \left\{ \vec{x}_1^{(1)} e^{-i\omega_1 t} + \vec{x}_2^{(1)} e^{-i\omega_2 t} \right\} \qquad (7.55)$$

The second-order loads are usually obtained by the pressure integration method, as

$$\vec{F}^{(2)} = \int_{\Gamma_0} \frac{1}{2} \rho g \left(\eta^{(1)2} - 2\eta^{(1)} \zeta^{(1)} \right) \frac{\vec{n_0}}{\cos \alpha} \, d\Gamma + \vec{A}^{(1)} \wedge \iint_{S_{B_0}} -\rho \, \Phi_t^{(1)} \, \vec{n_0} \, dS$$
$$+ \iint_{S_{B_0}} -\rho \left[\Phi_t^{(2)} + \frac{1}{2} \left(\nabla \Phi^{(1)} \right)^2 + \overrightarrow{P_0 P}^{(1)} \cdot \nabla \Phi_t^{(1)} \right] \vec{n_0} \, dS + \vec{F}_{HS}^{(2)} \quad (7.56)$$

(see equation (7.17) for the wave drift force).

All quantities that appear here are quadratic expressions of the first-order solution, including $\Phi^{(2)}$ (due to its body and free surface boundary conditions, similar to equations (7.42) and (7.43)).

It may then be concluded that $\overrightarrow{F}^{(2)}$ has the general form

$$\overrightarrow{F}^{(2)}(t) = A_1^2 \overrightarrow{f_d}(\omega_1) + A_2^2 \overrightarrow{f_d}(\omega_2) + \Re \left\{ A_1^2 \overrightarrow{f_+}^{(2)}(\omega_1, \omega_1) \, e^{-2i\omega_1 t} + A_2^2 \overrightarrow{f_+}^{(2)}(\omega_2, \omega_2) \, e^{-2i\omega_2 t} \right.$$

$$\left. + 2 A_1 A_2 \overrightarrow{f_-}^{(2)}(\omega_1, \omega_2) \, e^{-i(\omega_1-\omega_2)t} + 2 A_1 A_2 \overrightarrow{f_+}^{(2)}(\omega_1, \omega_2) \, e^{-i(\omega_1+\omega_2)t} \right\} \tag{7.57}$$

where $\overrightarrow{f_d}$ stands for the normalized drift force (in unit wave amplitude).

The final quantities of interest are the so-called **Quadratic Transfer Functions** (or QTFs), $\overrightarrow{f_+}^{(2)}(\omega_1,\omega_2)$ and $\overrightarrow{f_-}^{(2)}(\omega_1,\omega_2)$, from which other quantities in (7.57) can be derived since, for instance

$$\overrightarrow{f_d}(\omega) = \overrightarrow{f_-}^{(2)}(\omega,\omega) \tag{7.58}$$

Here the angular dependence of the QTFs has been omitted. Strictly they must be written as $\overrightarrow{f_+}^{(2)}(\omega_1,\omega_2,\beta_1,\beta_2)$ and $\overrightarrow{f_-}^{(2)}(\omega_1,\omega_2,\beta_1,\beta_2)$ as the two basic wave components may travel in different directions.

To compute the QTFs, the theoretical and numerical issues are about the same as to obtain the double frequency loads in regular waves. There seems to be no alternative to the pressure integration method, with possibly numerical convergence problems at sharp corners. However, for difference frequency QTFs, an extension of the method based on the Lagally theorem has been proposed by Rouault et al. (2007). Chen et al. (2007), likewise, have extended their middle-field method to this computation.

7.4.1 Sum Frequency Second-Order Loads

The sum frequency second-order loads are of interest to study the response of stiff systems: vibrations of ship hulls, vertical resonance of tension leg platforms (springing), and bending of gravity base structures. The numerical task for the calculation of these loads is tremendous since, on the one hand, the contribution from the second-order diffraction potential cannot be neglected and, on the other hand, the paneling of the hull (and of the free surface) must be refined enough to capture phenomena taking place at small spatial scales (wave lengths of the order of a few meters). The QTFs $\overrightarrow{f_+}^{(2)}$ must then be evaluated for all couples ω_i, ω_j such that $\omega_i + \omega_j$ be close to a natural frequency. Some software offers this option.

7.4.2 Difference Frequency Second-Order Loads

These loads are responsible for the low-frequency (or slow-drift) response of moored structures in the horizontal plane. Due to the usually low stiffnesses of their moorings, their natural frequencies in surge, sway, and yaw are typically of the order of 0.01 Hz, well below the wave frequencies. Second-order response in heave, roll, and pitch may also be observed with Spars and large semis when their natural frequencies are below the wave frequencies. Designs of moorings and riser systems must account for these low-frequency motions, which usually dominate the wave frequency response (see Figures 1.15 and 1.17).

The difference frequency QTFs may be obtained in a similar way as the sum frequency QTFs. Numerical computations show that, as long as the difference frequency $|\omega_1 - \omega_2|$ is sufficiently low, the contribution to the QTF from the free surface integral can be discarded. This is the main difference with sum frequency QTFs for which the free surface integral cannot be neglected.

Most diffraction-radiation software nowadays provides "exact" calculations of the difference frequency QTFs, usually without the contribution from the free surface integral. However, these computations still require extensive CPU effort and approximations are frequently used.

Newman's Approximations

An approximation commonly used, known as **Newman's approximation** (from Newman, 1974), consists in taking advantage of the fact that the knowledge of $\overrightarrow{f}_-^{(2)}(\omega_1, \omega_2)$ is requested only at $|\omega_1 - \omega_2|$ values close to the natural frequencies, usually very low, and that, as $\omega_2 \to \omega_1$, then $\overrightarrow{f}_-^{(2)}(\omega_1, \omega_2) \to \overrightarrow{f_d}(\omega_1)$ (under the condition that the two wave components be collinear: $\beta_1 \equiv \beta_2$).

Different forms of Newman's approximations have been proposed:

$$\overrightarrow{f}_-^{(2)}(\omega_1, \omega_2) \simeq \overrightarrow{f_d}(\omega_1) \qquad \text{(Newman, 1974)} \tag{7.59}$$

$$\overrightarrow{f}_-^{(2)}(\omega_1, \omega_2) \simeq \overrightarrow{f_d}(\frac{\omega_1 + \omega_2}{2}) \qquad \text{(Pinkster, 1975)} \tag{7.60}$$

$$f_{-j}^{(2)}(\omega_1, \omega_2) \simeq \sqrt{|f_{dj}(\omega_1) f_{dj}(\omega_2)|} \ \text{sign}(f_{dj}) \qquad \text{(Molin \& Bureau, 1980)} \tag{7.61}$$

Under these approximations the QTF is real: the imaginary part (of order $|\omega_1 - \omega_2|$) is considered negligible.

Newman's approximation applies well in deep water, on condition that the resonant frequencies be very low and that the wave drift forces be appreciable, that is, for massive structures such as FPSOs. Improvements can be made by including the contribution from the incident second-order potential following a Morison-type approach, assuming the body to be small with respect to the accompanying long waves, as outlined in the following paragraph.

Contribution from the Second-Order Incident Potential

In Chapter 3 the expression of the second-order difference frequency potential associated to a bichromatic sea has been given. When the two wave components travel in the same direction $\beta = 0$ it reduces to

$$\Phi_{I-}^{(2)} = \frac{q_-}{-(\omega_1 - \omega_2)^2 + g(k_1 - k_2)\tanh(k_1 - k_2)h} \frac{\cosh(k_1 - k_2)(z + h)}{\cosh(k_1 - k_2)h}$$
$$\sin\left[(k_1 - k_2)x - (\omega_1 - \omega_2)t\right] \tag{7.62}$$

where

$$q_- = -\frac{1}{2}A_1 A_2 \left(\frac{\omega_1^3}{\sinh^2 k_1 h} - \frac{\omega_2^3}{\sinh^2 k_2 h}\right) - A_1 A_2 \,\omega_1\, \omega_2\, (\omega_1 - \omega_2)\left(\frac{1}{\tanh k_1 h \,\tanh k_2 h} + 1\right)$$

$$(7.63)$$

When the considered body is "small" with respect to the long-wave scale (i.e., $|k_1 - k_2| L < 1/4$),[4] with L a typical body dimension (the width in beam waves or the length in head waves), a Morison approach can be used, relating the associated hydrodynamic load to the horizontal acceleration:

$$F_{-x}^{(2)} = \rho\,(1 + C_{axx})\,\forall\,\frac{\partial^2 \Phi_{I-}^{(2)}}{\partial x\,\partial t}(0,0,z_B,t) \tag{7.64}$$

with C_{axx} the zero frequency added mass coefficient, $\forall$ the displacement and z_B the vertical coordinate of the center of buoyancy.

This approximation is justified in many cases and it is easy to implement. It provides a purely imaginary component to the QTF, which adds up to the real component from Newman's approximation.

In deep water the contribution from $\Phi_{I-}^{(2)}$ is usually negligible. As the water depth decreases, it behaves asymptotically as the water depth h to the power $-5/2$. Moreover, shallow water moorings are usually stiffer than deep water moorings, meaning that the slow-drift resonant frequencies are increased. In the end, for instance in the case of LNG or FSRU terminals, the second-order incident potential provides the dominant contribution to the second-order loading.

Practical experience suggests that applying the equations above leads to mooring designs that are much too conservative. There are several possible reasons for this: one is that (7.62) assumes constant depth from infinity. The shoaling of the long wave accompanying the bichromatic group must be accounted for; this is addressed in the following paragraph. Another factor is directional spreading: when the primary waves are not collinear, separated by an angle β, the difference frequency wave number is $\sqrt{(k_1^2 + k_2^2 - 2k_1 k_2 \cos\beta}\ = \sqrt{(k_1 - k_2)^2 + 4k_1 k_2 \sin^2\beta/2}$ (see Chapter 3). As a result, the second-order acceleration $\left(\partial^2\Phi_{I-}^{(2)}/\partial x\,\partial t,\ \partial^2\Phi_{I-}^{(2)}/\partial y\,\partial t\right)$ is much reduced and points in a direction different from the carrier waves; this effect is discussed and illustrated in Molin & Fauveau (1984).

Shoaling of the accompanying long waves

There is extensive literature on the transformation of so-called subharmonic, or infra-gravity, waves in the coastal zone (e.g., see Battjes *et al.*, 2004, for an extensive review).

In Liu *et al.* (2011) a simple model is proposed that applies to a two-dimensional varying bathymetry, in between two semi-infinite constant depth zones, where bichromatic wave systems are propagated. The bottom slope is assumed to be small, so

---

[4] The value 1/4 is taken here with reference to the vertical cylinder case. See Figure 6.5.

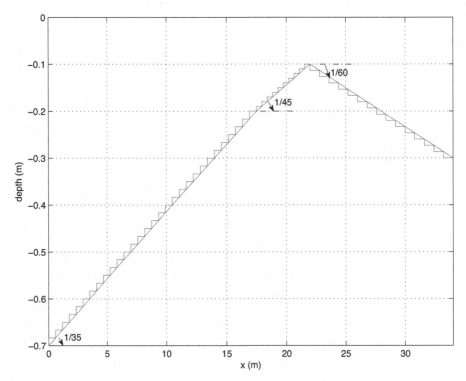

Figure 7.20 Extended bathymetry inspired by Van Noorloos' experiments (from Liu *et al.*, 2011).

the first-order wave components shoal following conservation of energy: $A_i(h) = A_{i0}\sqrt{C_{Gi0}/C_{Gi}(h)}$ with C_{Gi} the group velocity and A_{0i}, C_{Gi0} the deep water values. The varying bathymetry domain is idealized as a succession of rectangular subdomains where the second-order potential is obtained semi-analytically. Finally, the solution is derived through matching the potentials and their x-derivatives at the successive boundaries. Good agreement is reported with experiments by Van Noorloos (2003) – see also Van Dongeren *et al.* (2004).

As an illustration, consider the bathymetry shown in Figure 7.20, inspired by the experiments of Van Noorloos (2003). A bichromatic wave propagates from a semi-infinite flat domain of depth 0.70 m, over the bump, and then into a second semi-infinite domain of depth 0.30 m. The two wave periods are 1.55 s and 2.0 s, and the amplitudes are 0.06 m and 0.008 m.

Figure 7.21 shows the amplitude of the incoming long wave along the varying bathymetry, as obtained from the step model (results from a parent model, inspired by Schäffer (1993), are also shown), compared to the value from the flat bottom expression (7.62). It can be seen that at the top of the bump the flat bottom value is about 4 times greater than the value from the other two models. After the bump the ratio reverses, the step and Schäffer models provide values higher than the flat bottom expression: this is due to the release of second-order **free waves** at the depth transition (e.g., see Mei & Benmoussa, 1984).

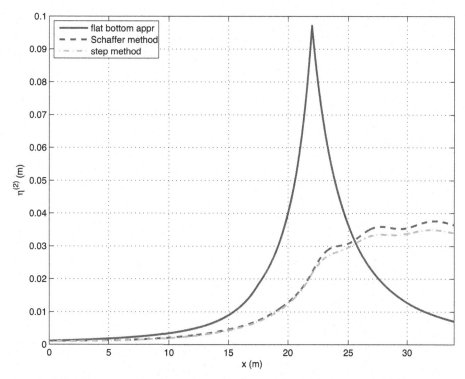

Figure 7.21 Second-order free surface elevation amplitude, from flat bottom expression and from the step model (from Liu *et al.*, 2011).

Figure 7.22 shows the phase shift between the subharmonic long wave and the envelope of the first-order wave profile. The phase shift increases along the slope from zero until it reaches about 70 degrees at the top of the bump. This means that the load contribution from the subharmonic long wave is not only reduced in magnitude, but is also shifted in time, as referenced to the wave group elevation.

As a result, when, in constant depth, the complex difference frequency QTF is separated into its real and imaginary parts, the real part P from Newman's approximation and the imaginary part Q due to the accompanying long wave:

$$f_-^{(2)}(\omega_1, \omega_2) = P(\omega_1, \omega_2) + i\, Q(\omega_1, \omega_2) \tag{7.65}$$

then, in varying bathymetry, the imaginary part must be corrected by a factor $R(\omega_1, \omega_2)\ \exp\{i\ \alpha(\omega_1, \omega_2)\}$ so that the QTF becomes

$$f_-^{(2)}(\omega_1, \omega_2) = P(\omega_1, \omega_2) + i\, Q(\omega_1, \omega_2) \times R(\omega_1, \omega_2)\ e^{i\,\alpha(\omega_1, \omega_2)}$$
$$= P - Q\, R\, \sin\alpha + i\, Q\, R\, \cos\alpha \tag{7.66}$$

When $R < 1$ the modulus of the QTF is decreased compared to its flat bottom reference value. It can even become lower than its Newman approximation P alone when $0 \le \alpha \le \pi/2$. As a matter of fact, in Liu *et al.* (2011), some experiments are reported, on a barge model moored at a beach, where measured slow-drift motions are lower in some cases than those calculated with the Newman approximation alone.

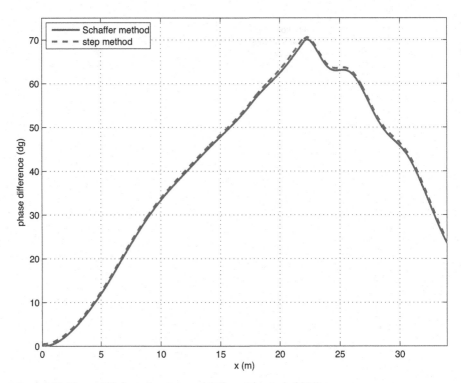

Figure 7.22 Phase shift from the step model (from Liu *et al.*, 2011).

An important teaching from Figure 7.21 is that the flat bottom expression is not always conservative, for instance when a ship is moored in a deeper water region on the lee side of a shoal where free subharmonic waves may be emitted.

7.5   Rainey's Equations

So-called Rainey's equations are an extension of the Morison equation, more precisely of the inertia component in the Morison equation. They apply to slender bodies, in unbounded fluid: Free surface effects are ignored, and there are no radiated nor diffracted waves.

"Slender" means that the transverse dimensions (the radius for a circular cylinder) are small with respect to the axial dimension and with respect to the scale of the incoming flow (the wave length): the parameter ka is assumed to be small ($ka \ll 1$ with k the wave number and a the radius).

These are the same assumptions as in the Morison formula, where the incoming flow is locally assumed to be uniform. The difference is that a higher-order approximation of the incident flow field is considered. More precisely, when $\Phi_I(x,y,z,t)$ is the velocity potential of the incoming flow, in the Morison approach Φ_I is locally taken, in the neighborhood of a point (x_0,y_0,z_0), as:

$$\Phi_I(x,y,z,t) \simeq \Phi_I(x_0,y_0,z_0,t) + (x-x_0)\, U + (y-y_0)\, V + (z-z_0)\, W \qquad (7.67)$$

where U, V, W are taken in (x_0, y_0, z_0).

In the Rainey case, higher-order gradients are included:

$$\begin{aligned}
\Phi_I(x,y,z,t) \simeq\; & \Phi_I(x_0,y_0,z_0,t) + (x-x_0)\, U + (y-y_0)\, V + (z-z_0)\, W \\
& + \frac{1}{2}\,(x-x_0)^2\, U_x + \frac{1}{2}\,(y-y_0)^2\, V_y + \frac{1}{2}\,(z-z_0)^2\, W_z \\
& + (x-x_0)(y-y_0)\, U_y + (x-x_0)(z-z_0)\, U_z + (y-y_0)(z-z_0)\, V_z
\end{aligned}$$

$$(7.68)$$

Rainey (1989, 1995) considers the general case of cylinders with arbitrary cross-section. He derives the hydrodynamic loads through conservation of the kinetic energy. The problem may be tackled more classically by locally deriving the perturbation potential and integrating the pressure (including the quadratic term in the Bernoulli equation). The latter approach was earlier taken by Madsen (1986) in the particular case of circular cylinders.

In this theory, it appears that there are two small parameters, ka the slenderness and kA the wave steepness. Compared to the inertia term in the Morison equation, higher-order components in ka and kA are expected. It turns out that, in the case of **circular** cylinders, there only appear $k^2 A^2$ terms, that is, load components second-order in the wave steepness.

Rainey equations also provide second-order end loads, for truncated cylinders.

Here we restrict ourselves to the case of circular cylinders. We use two coordinate systems: a fixed coordinate system (x,y,z) and a local system (X,Y,Z) with Z along the cylinder axis. Be (U,V,W) the incoming flow velocity components in (X,Y,Z), and $(\dot{U},\dot{V})$ the X and Y components of the Eulerian acceleration. To obtain $(U,V,W,\dot{U},\dot{V})$ one therefore starts from $\Phi_I(x,y,z,t)$ in the fixed coordinate system, derives its gradient $\nabla\Phi_I$ and Eulerian acceleration $\partial/\partial t(\nabla\Phi_I)$, and projects these vectors in the body-fitted coordinate system (X,Y,Z).

The distributed loads (along the cylinder) are obtained as:

$$\mathrm{d}F_X = \rho\,\pi\,a^2 \left\{ 2\dot{U} + 2UU_X + 2VU_Y + 2WU_Z - \ddot{X} + (U-\dot{X})\,W_Z - 2\,\dot{\beta}\,(W-\dot{Z}) \right\}\mathrm{d}Z$$

$$(7.69)$$

$$\mathrm{d}F_Y = \rho\,\pi\,a^2 \left\{ 2\dot{V} + 2UV_X + 2V\,V_Y + 2W\,V_Z - \ddot{Y} + (V-\dot{Y})\,W_Z + 2\,\dot{\alpha}\,(W-\dot{Z}) \right\}\mathrm{d}Z$$

$$(7.70)$$

where a is the radius and $\dot{\alpha}$, $\dot{\beta}$ are the angular velocity components around X and Y.

At the cylinder ends local loads must be added:

$$F_{Xe} = \rho\,\pi\,a^2\,(U-\dot{X})\,(W-\dot{Z}) \qquad (7.71)$$

$$F_{Ye} = \rho\,\pi\,a^2\,(V-\dot{Y})\,(W-\dot{Z}) \qquad (7.72)$$

$$F_{Ze} = \pi\,a^2\,p_I - \frac{1}{2}\,\rho\,\pi\,a^2\left((U-\dot{X})^2 + (V-\dot{Y})^2\right) \qquad (7.73)$$

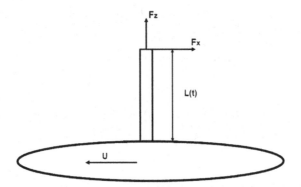

Figure 7.23 Missile launching from a submarine.

where p_I is the pressure in the incoming field

$$p_I = -\rho\,\Phi_{It} - \frac{1}{2}\rho\,(U^2 + V^2 + W^2) \tag{7.74}$$

and the Z axis points into the cylinder.

The end loads (7.71) and (7.73) may easily be understood from energy arguments. Consider the problem illustrated in Figure 7.23: A submarine sailing at a velocity U launches a missile, idealized here as a circular cylinder. The kinetic energy in the fluid, due to the missile, is $E_C = 1/2\,\rho\pi\,a^2\,L\,U^2$. The variation in kinetic energy is equal to the work produced by the hydrodynamic loads: $dE_C/dt = F_X\,U - F_Z\,W$. It is easy to check that this is verified by equations (7.71) and (7.73); since the geometry by the submarine wall does not change, the horizontal load must be applied at the tip of the missile. When the missile is completely ejected, an opposite load applies at its lower end, so that d'Alembert's paradox is verified, but a torque results equal to $\rho\pi\,a^2\,U\,W\,L$: the Munk moment (see equation (4.11)).

In the case of a fixed vertical cylinder in two-dimensional flow ($V \equiv 0$) the distributed x force reduces to

$$\begin{aligned}
dF_x &= \rho\pi\,a^2\left(2\dot{U} + 2U\,U_x + 2W\,U_z + U\,W_z\right)\,dz \\
&= \rho\pi\,a^2\left(2\dot{U} + U\,U_x + 2W\,U_z\right)\,dz
\end{aligned} \tag{7.75}$$

Hence it is neither the Eulerian acceleration nor the Lagrangian acceleration that is involved! The correction $U\,W_z$ to the Lagrangian acceleration is known as the "axial divergence term."

In the following we present three applications of the Rainey equations.

7.5.1   Second-Order Loads on a Fixed Bottom-Mounted Vertical Cylinder

For a bottom-mounted cylinder, the horizontal load in regular waves is obtained as

$$F_x = \rho\pi\,a^2\left[\int_{-h}^{0}(2\dot{U} + U\,U_x + 2W\,U_z)\,dz + 2\eta\,\dot{U}(0,0,0,t)\right] \tag{7.76}$$

giving (in deep water)

$$F_x = -2\,\rho\,\pi\,a^2\,g\,A\,\sin\omega t - \frac{5}{4}\rho\,\pi\,a^2\,g\,k\,A^2\,\sin 2\omega t \tag{7.77}$$

As seen in Section 7.3 this provides a good approximation of the 2ω component for $ka < 0.1$. Beyond this value, free surface effects (the contribution from the second-order potential $\varphi_D^{(2)}$) cannot be neglected.

In a bichromatic wave system, with free surface elevation

$$\eta(x,t) = A_1\,\cos(k_1 x - \omega_1 t) + A_2\,\cos(k_2 x - \omega_2 t) \tag{7.78}$$

Rainey's equations give the following second-order horizontal load at the difference frequency $\omega_1 - \omega_2$:

$$F_-^{(2)}(t) = \rho\,\pi\,a^2\,A_1\,A_2\,(k_1 - k_2)\left[\frac{3}{2}\frac{\omega_1\omega_2}{k_1+k_2} - g\right]\sin(\omega_1 - \omega_2)\,t \tag{7.79}$$

where deep water is again assumed, that is, $k_1 h > 3$ and $k_2 h > 3$, and the contribution (7.64) from the second-order incident potential has been discarded.

As already commented, Newman's approximation only provides the real part of the QTF. It may be wondered whether the Rainey contribution (7.79) properly approximates the imaginary part of the QTF. In the case of a vertical cylinder, it is easy to calculate exactly the difference frequency QTF. This was done by Molin & Chen (2002).

Figures 7.24 through 7.26 show the real and imaginary parts of the "exact" QTF, compared with Rainey (equation (7.79)) and with Newman's approximation taken here as $\sqrt{f_d(\omega_1)\,f_d(\omega_2)}$. The QTFs are made nondimensional with $\rho\,g\,a\,A_1\,A_2$ (or $2\rho\,g\,a\,A_1\,A_2$ in the case of (7.79)). The 3 figures are for 3 different values of $k_1 a$: 0.1, 0.2, and 0.4, and the QTFs are shown vs $(k_1 - k_2)/k_1$.

It can be seen that there is a fairly good agreement between the exact QTFs and their approximations, as provided by Newman for the real part and by Rainey for the imaginary part. As $k_1 - k_2$ increases from zero, the imaginary part quickly becomes dominant over the real part at the lowest $k_1 a$ value.

Take, for instance, a cylinder of diameter 20 m in a sea state with peak periods around 14 s. Then $ka \sim 0.2$. From figure 7.25 it can be seen that the imaginary part of the QTF dominates when $\Delta k/k \simeq 2\,\Delta\omega/\omega$ exceeds 0.2, that would be for beat periods lower than 140 s! Newman's approximations are therefore ill suited to the analysis of the slow-drift motion of semi-submersible platforms and a load model based on Rainey's equations is a better alternative.

7.5.2   Wave Drift Force on a Truncated Vertical Column Free to Heave

It has been explained earlier that wave drift forces in regular waves are related to the radiated and diffracted wave systems by the structure. Since neither the Morison nor the Rainey equations account for free surface effects, it would seem like that they cannot predict any drift force. This has just been confirmed in the fixed vertical cylinder case. However when the structure is free to respond to the incoming waves, and some

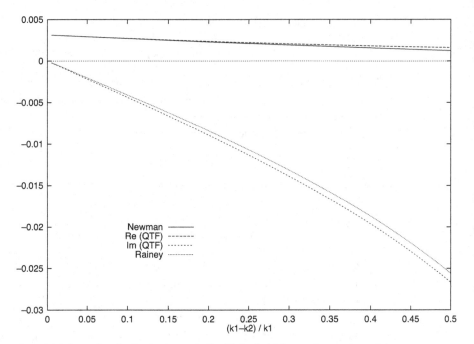

Figure 7.24 Second-order load on a vertical cylinder in bichromatic sea. $ka = 0.1$.

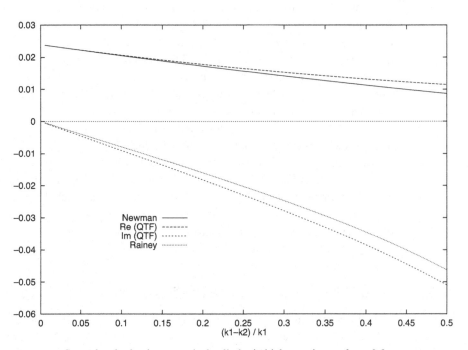

Figure 7.25 Second-order load on a vertical cylinder in bichromatic sea. $ka = 0.2$.

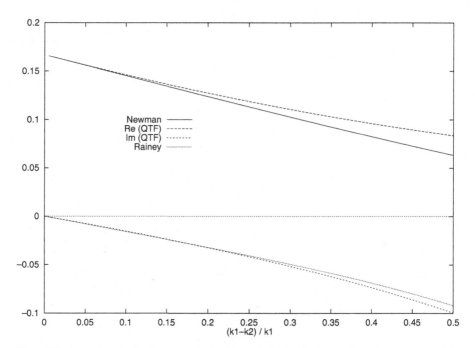

Figure 7.26 Second-order load on a vertical cylinder in bichromatic sea. $ka = 0.4$.

external damping is introduced in its equation of motion, the Morison and Rainey equations do predict some nonzero time-averaged horizontal load.

Let us consider a slender and truncated vertical cylinder, of radius a and draft d, restrained in surge and pitch, and free to heave. An additional linear damping, in the form $\zeta \, B_C$ is introduced in the heave equation of motion, with B_C the critical damping. Physically this damping may be due to viscous effects or to some PTO (Power Take Off).

Following the Hooft method introduced in Chapter 6, the vertical excitation load at the foot of the column, in a regular wave of frequency ω and amplitude A, is well approximated by

$$F_z(t) = \rho g \, A \, (\pi a^2 - 2ka^3) \, e^{-kd} \, \cos \omega t \tag{7.80}$$

(still assuming $ka \ll 1$ and deep water).

The mass M plus vertical added mass M_a of the column is $M + M_a \simeq \rho \pi a^2 d + 2 \rho a^3$ and the vertical stiffness is $K = \rho g \pi a^2$. The heave response is then

$$Z(t) = A \, \Re \left\{ z_0 \, e^{-i\omega t} \right\} \tag{7.81}$$

with z_0 the complex RAO

$$z_0 = \frac{\rho g \, (\pi a^2 - 2k \, a^3) \, e^{-kd}}{-\omega^2 (M + M_a) - i \, \zeta \, \omega \, B_C + K} \tag{7.82}$$

As an application case we take a radius of 4 m, a draft of 24 m and 5% of critical damping. Figure 7.27 shows the heave RAO from equation (7.82) and calculated with the diffraction-radiation software COREV. The agreement is quite good.

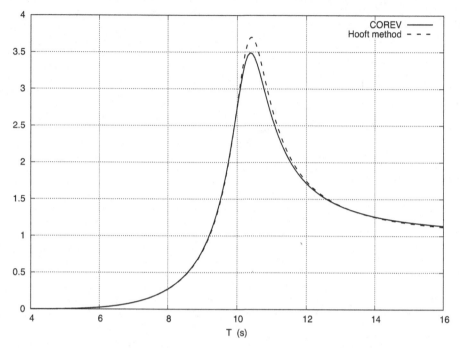

Figure 7.27 Floating column. Heave RAO at 5% critical damping.

From the inertia term in the Morison equation the local horizontal force is obtained as

$$dF_x = 2 \rho \pi a^2 \, \Phi_{xt}(0,z,t) \, dz \tag{7.83}$$

and the z integration is to be performed from $-d + Z(t)$ up to $\eta(0,t)$. The nonzero time averaged value results from the integration between $-d + Z(t)$ and $-d$, leading to the second-order drift force:

$$F_{dx\,\text{Morison}} = \rho g \pi A^2 k a^2 e^{-kd} \, \Im\{z_0\} \tag{7.84}$$

According to Rainey's equations, an end load must be added at the foot of the column, given by

$$F_{xe} = \rho \pi a^2 U \left(W - \dot{Z}\right) \tag{7.85}$$

with $U = \Phi_x(0,-d,t)$, $W = \Phi_z(0,-d,t)$. Its time-averaged value is

$$\overline{F_{xe}} = -\frac{1}{2} \rho g \pi A^2 k a^2 e^{-kd} \, \Im\{z_0\} \tag{7.86}$$

so that the drift force becomes

$$F_{dx\,\text{Rainey}} = \frac{1}{2} \rho g \pi A^2 k a^2 e^{-kd} \, \Im\{z_0\} \tag{7.87}$$

half the Morison value.

Figure 7.28 shows the nondimensional drift force $F_{dx}/(\rho g a A^2)$ calculated with COREV, and obtained from equation (7.84) and from equation (7.87). In the low

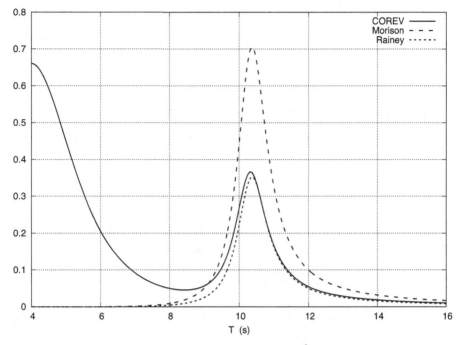

Figure 7.28 Floating column. Horizontal drift force $F_{dx}/(\rho g a A^2)$.

period range, COREV predicts a significant drift force, due to diffraction effects, which both Morison and Rainey fail to reproduce, as expected. However at the heave resonant period, a peak appears for all 3 methods and it is clear that the Morison value is about twice too high.

The drift force is usually calculated from equation (7.19), which, in this case of infinite depth and zero incidence, reduces to

$$F_{dx} = \rho g A^2 \frac{1}{k} \left[\frac{1}{2\pi} \int_0^{2\pi} H(\theta)\, H^*(\theta)\, \cos\theta\, d\theta + \Re(H(0)) \right] \qquad (7.88)$$

with $H(\theta)$ the Kochin function.

The heaving column creates a flux $Q = -\pi a^2\, \dot{Z}(t)$ at its foot which, owing to its deep draft and small diameter, can be viewed, from the far field, as a submerged pulsating source. From Wehausen & Laitone (1960, eq. (13.17''')), the resulting radiated velocity potential is

$$\Phi_R = -\Re \left\{ \frac{\pi}{2} k\, a^2\, A\, z_0\, \omega\, e^{k(z-d)}\, H_0(kR)\, e^{-i\omega t} \right\} \qquad (7.89)$$

with H_0 the Hankel function. The Kochin function is then

$$H(\theta) = -i \frac{\pi}{2}\, k^2 a^2\, z_0\, e^{-kd} \qquad (7.90)$$

and the drift force is, from equation (7.88),

$$F_{dx} = \frac{1}{2}\, \rho g\, \pi\, A^2\, k\, a^2\, e^{-kd}\, \Im\{z_0\} \qquad (7.91)$$

in agreement with Rainey!

It is somewhat puzzling that Rainey, which is waveless, correctly predicts the drift force at the heave natural frequency. What is clear from this exercise is that the Morison equation fails to properly reproduce the second-order loads and that the Rainey equations offer a valuable alternative.

7.5.3   Second-Order Vertical Loads on a Submerged Circular Cylinder

Finally, consider the two-dimensional case of a fixed circular cylinder under the free surface, in infinite depth. In a regular wave, application of Rainey's equations immediately gives the vertical drift force

$$\overline{F}_{dz} = 2\,\rho\,\pi\,a^2\,g\,k^2\,A^2\,e^{-2kd} \tag{7.92}$$

(per unit length) where d is the submergence and a the radius. According to Ogilvie (1963) this is a good approximation when $ka \ll 1$ and $2kd \gg 1$.

In a bichromatic wave system, the following load at the difference frequency $\omega_1 - \omega_2$ (with $\omega_1 > \omega_2$) is obtained

$$F_z^{(2)} = 2\,\rho\,\pi\,a^2\,A_1\,A_2\,\omega_1\left[\omega_2\,(k_1 + k_2)\,e^{-(k_1+k_2)d} + (\omega_1 - \omega_2)\,(k_1 - k_2)\,e^{-(k_1-k_2)d}\right]$$
$$\cos(\omega_1 - \omega_2)\,t \tag{7.93}$$

It is to be noted that, at variance with the horizontal load, the vertical QTF is here real (no component in $\sin(\omega_1 - \omega_2)\,t$). The vertical drift force has nothing to do with radiated or diffracted waves, and is just the result of the decay of the wave action with submergence: as a result, the $-1/2\,\rho\,V^2$ term in the Bernoulli equation creates an upward suction force.

7.6   Second-Order Loads in Irregular Waves

When the QTFs have been derived for all couples (ω_i, ω_j) in the incoming wave system, the second-order loads can be expressed.

For the sake of simplicity, we assume long-crested seas. To first-order the free surface elevation of the incoming waves is written as

$$\eta^{(1)}(x,y,t) = \Re\left\{\sum_i A_i\,e^{\,i(k_i x\cos\beta + k_i y\sin\beta - \omega_i t + \theta_i)}\right\} \tag{7.94}$$

with β the angle of propagation.

The second-order loads decompose into two components, at the sum and difference frequencies:

• Sum frequency

$$\overrightarrow{F}_+^{(2)}(t) = \Re\left\{\sum_i\sum_j A_i\,A_j\,\overrightarrow{f}_+^{(2)}(\omega_i,\omega_j,\beta)\,e^{\,i[-(\omega_i + \omega_j)t + \theta_i + \theta_j]}\right\} \tag{7.95}$$

- Difference frequency

$$\vec{F}_-^{(2)}(t) = \Re\left\{\sum_i \sum_j A_i \, A_j \, \vec{f}_-^{(2)}(\omega_i,\omega_j,\beta)\, e^{\,i[-(\omega_i-\omega_j)t+\theta_i-\theta_j]}\right\} \quad (7.96)$$

with $\vec{f}_+^{(2)}(\omega_i,\omega_j,\beta)$ and $\vec{f}_-^{(2)}(\omega_i,\omega_j,\beta)$ the QTFs. Here the reference point is taken at $x = y = 0$.

When the proposed Newman approximation (7.61) is used, the difference frequency second-order loads take the form:

$$F_{-k}^{(2)}(t) = \left\{\left[\sum_i A_i \, \sqrt{|f_{dk}(\omega_i,\beta)|}\, \cos(-\omega_i t + \theta_i)\right]^2 \right.$$
$$\left. + \left[\sum_i A_i \, \sqrt{|f_{dk}(\omega_i,\beta)|}\, \sin(-\omega_i t + \theta_i)\right]^2 \right\}\, \text{sign}(f_{dk}(\beta)) \quad (7.97)$$

This is only valid for the second-order loads in surge and sway for which the drift force keeps a constant sign when the frequency varies. It also applies to yaw provided that the reference point is momentarily taken some distance away from amidships (so that its sign does not change with frequency). The advantage of (7.97) is that it consists in two simple summations instead of a double summation, offering reduced CPU cost in time domain simulations. It has much similarity with the wave envelope (or pseudo-envelope) squared (see Chapter 2), showing that the slowly varying force follows the wave envelope.

In the frequency domain, the spectra of the sum and difference frequency second-order loads are obtained as:

- Sum frequency

$$S_{F_{+k}^{(2)}}(\Omega) = 8 \int_0^{\Omega/2} S(\omega)\, S(\Omega-\omega)\, \|f_{+k}^{(2)}(\omega,\Omega-\omega,\beta)\|^2\, d\omega \quad (7.98)$$

- Difference frequency

$$S_{F_{-k}^{(2)}}(\Omega) = 8 \int_0^\infty S(\omega)\, S(\omega+\Omega)\, \|f_{-k}^{(2)}(\omega,\omega+\Omega,\beta)\|^2\, d\omega \quad (7.99)$$

When approximation (7.61) is used, equation (7.99) can be rewritten as

$$S_{F_{-k}^{(2)}}(\Omega) = 8 \int_0^\infty S(\omega)\, S(\omega+\Omega)\, f_{dk}(\omega,\beta)\, f_{dk}(\omega+\Omega,\beta)\, d\omega \quad (7.100)$$

Figure 7.29 shows the spectral density of the second-order difference frequency loads in surge for the N'Kossa barge, in head seas, from approximation (7.100) and from (7.99) with the QTFs computed with Hydrostar (neglecting the free surface integral). In the limit $\Omega \to 0$ the two curves converge to the same value, as expected. However they are in a ratio about 2 from $\Omega = 0.1$ rad/s, showing (in this particular case) that Newman approximation may not be applied to stiff moorings.

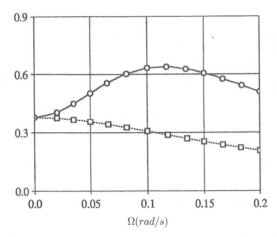

Figure 7.29 N'Kossa barge. Spectrum of the second-order difference frequency exciting load in surge, vs the difference frequency Ω (in rad/s). JONSWAP wave spectrum, $H_S = 3.9$ m, $T_P = 15.5$ s, $\gamma = 2$ (from Chen, 1994).

Mean Wave Drift Force in Irregular Waves

In an irregular sea state, the mean drift force is simply obtained from summation over all Airy components. For a long-crested sea state this gives

$$F_d = \sum_i A_i^2 \, f_d(\omega_i, \beta) = 2 \int_0^\infty S(\omega) \, f_d(\omega, \beta) \, d\omega \tag{7.101}$$

and for a short-crested sea state

$$F_d = 2 \int_0^\infty \int_0^{2\pi} S(\omega, \beta) \, f_d(\omega, \beta) \, d\beta \, d\omega \tag{7.102}$$

When the normalized drift force is nearly constant with the wave frequency over the wave spectrum range, F_d can be approximated as

$$F_d \simeq 2 \, \overline{f_d} \int_0^\infty S(\omega) \, d\omega = \frac{H_S^2}{8} \, \overline{f_d} \tag{7.103}$$

As far as the mean wave loads are concerned an irregular sea state is therefore "equivalent" to a regular wave of amplitude $H_S/(2\sqrt{2})$.

(In the previous equations F_d (resp. f_d) denotes any component of the wave drift force.)

7.7   Mooring Behavior

Mooring design is an important issue in offshore and marine engineering. Mooring systems must keep the floating bodies in place, sometimes within a limited range around a reference position (for instance, in the case of LNG transfer).

The horizontal motion of moored structures decomposes into three components:

- a mean offset under the time-averaged loads due to wind, waves, and current;

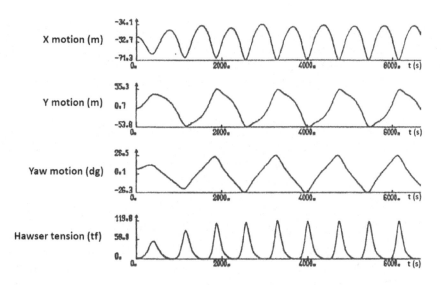

Figure 7.30 Numerical simulation of the fishtailing motion of a bow-moored tanker (from Molin, 1981).

- the first-order response to the sea state, taking place at the wave frequencies;
- the slow-drift or low-frequency response, induced by the second-order wave loads and the time-varying wind loads. This slow-drift response takes place at the natural frequencies of the mechanical system composed by the body and its mooring. In the case of ships or FPSOs at single point moorings a dynamic response, known as *fishtailing*, can appear under the action of steady wind and/or current only (see Figure 7.30). This behavior is parent to the *flutter* instability of wings or bridge decks, well-known in aerodynamics (see Section 5.2.2 in Chapter 5)[5].

7.7.1   Mean Wave, Current, and Wind Loads

Waves

When the wave spectrum $S(\omega)$ and the normalized drift forces $f_d(\omega)$ are known, the mean wave loads are obtained from equation (7.101) above.

As seen above, in Section 7.2.4, drift forces are affected by current superposition. This can be accounted for when the wave drift damping matrix is known, or through Aranha's formula. A procedure to account for this effect in time domain simulation models is proposed in Section 7.7.3.

---

[5] According to the stability criterion exposed in Section 5.2.2 on flutter, it would seem like there can be no fishtailing instability when a ship is moored by a bow hawser. Instability results from the fact that a third degree of freedom comes into play, namely surge, as the figure shows.

Current

Current mostly contributes to the mooring loads via viscous effects, friction and drag. Drag is usually dominant, except in the case of streamlined bodies (ship hulls) in head current where the boundary layers remain attached. For ship hulls the reference model is the semi-infinite flat plate, for which the frictional force is written as

$$F_f = \frac{1}{2} \rho C_f(\text{Re})\, S\, U_C^2 \tag{7.104}$$

with Re the Reynolds number based on the plate or ship length L: $\text{Re} = U_C\, L/\nu$ and S the wetted area of the hull. In the laminar regime ($\text{Re} < 10^5$) the frictional coefficient C_f is given by the Blasius formula

$$C_f = \frac{1.328}{\sqrt{\text{Re}}} \tag{7.105}$$

It is only in the case of model tests at very small scale that the laminar regime is fulfilled over the whole length of the hull. At full scale, and for most experimental models, the boundary layer is mainly turbulent and the ITTC formula can be used:

$$C_f = \frac{0.075}{(\log_{10}\text{Re} - 2)^2} \tag{7.106}$$

The frictional force is usually multiplied by a form coefficient $1+k$, with k typically between 0.2 and 0.4.

As written above, this only applies to well-streamlined ship hulls. In the case of barges or blunt FPSOs, the flow massively separates at the stern, inducing drag. As soon as the current comes at an angle, other effects come into play such as lift (at small headings). For some classes of ships (VLCCs, LNG-carriers), there are documented current load coefficients resulting from systematic towing tests in wave tanks (OCIMF, 1977). Note that no reference is being made to the Reynolds number in the OCIMF coefficients. Palo (1983) presents a compilation of available references for mean wind and current loads where it appears that they may widely disagree (see also Palo, 1986).

An important factor is the depth over draft ratio: current loads strongly increase as the underkeel clearance is reduced. In shallow locations, the current profile may be strongly biased over the depth and some averaged velocity is usually considered.

An alternative to towing is to do tests in a wind tunnel, with the problem that there will be a boundary layer at the floor that is not realistic. Another route, widely followed nowadays and more and more reliable, is to do calculations with a CFD code.

Wind

Wind loads show much similarity to current loads. A distinction is that, due to boundary layer effects, the mean wind profile is strongly sheared over the height of the superstructures whereas, in deep water, the current profile by the free surface is close to vertical. Wind loads are usually assessed through wind tunnel tests or CFD computations. Some reference coefficients are provided by OCIMF for VLCCs and by the Society of International Gas Tanker and Terminal Operators (2007) for LNG carriers.

7.7.2   Slow-Drift Motion

For the sake of simplicity, consider the simple case of a ship moored in head seas. Its surge motion decomposes into a mean offset $x^{(0)}$, a wave frequency component $x^{(1)}(t)$, and a low-frequency component $X^{(2)}(t)$, which is taken to obey the equation

$$(M + M_a) \ddot{X}^{(2)} + B \dot{X}^{(2)} + K X^{(2)} = F_-^{(2)}(t) \tag{7.107}$$

with M the mass of the ship, M_a the added mass at zero frequency, B a linear damping, K the mooring stiffness, and $F_-^{(2)}(t)$ the low-frequency hydrodynamic load.

The spectrum of the slow-drift response is then obtained as

$$S_{X^{(2)}}(\Omega) = \frac{S_{F^{(2)}_-}(\Omega)}{[-(M + M_a) \Omega^2 + K]^2 + B^2 \Omega^2} \tag{7.108}$$

from which the variance (the standard deviation squared) follows as

$$\sigma^2_{X^{(2)}} = \int_0^\infty \frac{S_{F^{(2)}_-}(\Omega)}{[-(M + M_a) \Omega^2 + K]^2 + B^2 \Omega^2} \, d\Omega \tag{7.109}$$

When the damping is low, most of the response takes place in a small neighborhood of the natural frequency $\Omega_0 = \sqrt{K/(M + M_a)}$. The variance can be approximated as

$$\sigma^2_{X^{(2)}} \simeq S_{F^{(2)}_-}(\Omega_0) \int_0^\infty \frac{d\Omega}{[-(M + M_a) \Omega^2 + K]^2 + B^2 \Omega^2} \simeq \frac{\pi \, S_{F^{(2)}_-}(\Omega_0)}{2 \, B \, (M + M_a) \, \Omega_0^2} \tag{7.110}$$

From this equation it is clear that the amplitude of the low-frequency motion, even though being driven by second-order loads, small in magnitude, can be large when the natural frequency Ω_0 and the damping B are low.

In linear sea-keeping radiation, damping plays the major role in mitigating resonance, with some additional contribution from viscous effects, for instance in the case of ship roll. At the low slow-drift frequencies, radiation damping is usually negligible and other energy dissipation processes come into play. The main damping components have been identified as **wave drift damping**, viscous damping at the hull (friction and/or flow separation), and damping due to the mooring lines.

Some damping effect also arises from the wind loads. It can easily be taken into account by subtracting the slow-drift velocity from the wind velocity while computing the wind loads.

7.7.3   Wave Drift Damping

Historically the wave drift damping effect was first identified by Wichers & van Sluijs in 1979. By doing surge extinction tests on a tanker model, in still water and in regular waves, they observed that the surge damping increased in waves, at a rate proportional to the wave amplitude squared and dependent on the wave period (see Figure 7.31).

In some sense, the wave drift damping results from the inconsistency in the way second-order theory is being applied. The underlying assumption to the perturbation scheme is that first-order quantities are small (of order ε) and second-order quantities

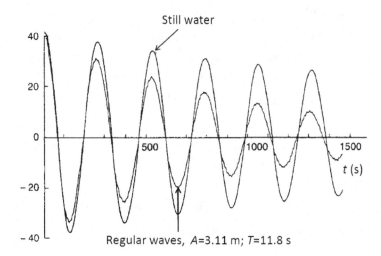

Figure 7.31 Surge decay test of a VLCC in still water and in regular waves (from Wichers, 1988).

even smaller (of order ε^2). The low-frequency second-order loads are actually smaller than the first-order loads, but they drive a response that is larger than the first-order response.

In the model equation of the slow-drift response

$$(M + M_a)\,\ddot{X}^{(2)} + B\,\dot{X}^{(2)} + K\,X^{(2)} = F_-^{(2)}(t) \tag{7.111}$$

a way to try and improve things is to make the second-order load $F_-^{(2)}$ dependent on the slow-drift position $X^{(2)}$ and slow-drift velocity $\dot{X}^{(2)}$:

$$(M + M_a)\,\ddot{X}^{(2)} + B\,\dot{X}^{(2)} + K\,X^{(2)} = F_-^{(2)}(X^{(2)}, \dot{X}^{(2)}, t) \tag{7.112}$$

or, through a Taylor series development limited to the very first terms:

$$(M + M_a)\,\ddot{X}^{(2)} + B\,\dot{X}^{(2)} + K\,X^{(2)} = F_-^{(2)}(0,0,t) + X^{(2)}\,\frac{\partial F_-^{(2)}(0,0,t)}{\partial X^{(2)}}$$

$$+ \dot{X}^{(2)}\,\frac{\partial F_-^{(2)}(0,0,t)}{\partial \dot{X}^{(2)}} \tag{7.113}$$

Separating time-averaged values from fluctuating parts, one gets

$$(M + M_a)\,\ddot{X}^{(2)} + \left[B - \overline{\frac{\partial F_-^{(2)}(0,0,t)}{\partial \dot{X}^{(2)}}}\right]\dot{X}^{(2)} + \left[K - \overline{\frac{\partial F_-^{(2)}(0,0,t)}{\partial X^{(2)}}}\right]X^{(2)}$$

$$= F_-^{(2)}(0,0,t) + X^{(2)}\,\frac{\partial F_-'^{(2)}(0,0,t)}{\partial X^{(2)}} + \dot{X}^{(2)}\,\frac{\partial F_-'^{(2)}(0,0,t)}{\partial \dot{X}^{(2)}} \tag{7.114}$$

The interpretation is that the damping and the stiffness are modified, together with the excitation.

The possible modification of the stiffness does not seem to be relevant to the horizontal components of the slow-drift motion, but it plays an important role in the vertical components for such structures as large semis (or submarines), which have natural frequencies in heave, roll, and pitch well below the wave frequencies. For a submerged cylinder the vertical drift force is strongly dependent on the submergence (see equation (7.92)). A consequence is that submarines at low forward speed may become unstable under waves. Likewise, the natural frequencies in roll and pitch of large semis may notably decrease in high waves.

When the time scale of the slow-drift response is much longer than the wave periods, the slow-drift velocity $\dot{X}^{(2)}$ varies little over a few consecutive wave cycles and it can be interpreted as the opposite of a current superimposed onto the waves. The sensitivity of the wave drift force to current has been described in Section 7.2.4 and the concept of wave drift damping has been introduced there: when the slow-drift velocity of the structure is $(\dot{X}^{(2)}, \dot{Y}^{(2)})$, the normalized drift force is written as

$$\overrightarrow{f_d}(\omega, \beta, \dot{X}^{(2)}, \dot{Y}^{(2)}) = \overrightarrow{f_d}(\omega, \beta, 0, 0) - \begin{pmatrix} b_{d11}(\omega, \beta) & b_{d12}(\omega, \beta) \\ b_{d21}(\omega, \beta) & b_{d22}(\omega, \beta) \end{pmatrix} \begin{pmatrix} \dot{X}^{(2)} \\ \dot{Y}^{(2)} \end{pmatrix} + O(\tau^2)$$

(7.115)

where the b_{dij} terms can be obtained by running a diffraction-radiation code with slow forward speed or approximated through Aranha's formulas (7.34) to (7.37).

Unfortunately neither standard diffraction-radiation codes nor Aranha's formulas provide information on the wave-drift damping in yaw[6].

In irregular seas, the wave drift damping is obtained by summation over all Airy components of the wave signal. For a long-crested sea state of spectrum $S(\omega)$ and direction β, this leads to

$$-\frac{\overline{\partial F_-^{(2)}(0,0,t)}}{\partial \dot{X}^{(2)}} = -\sum_j A_j^2 \frac{\partial f_d(\omega_j)}{\partial \dot{X}^{(2)}} = -2 \int_0^\infty S(\omega) \frac{\partial f_d(\omega)}{\partial \dot{X}^{(2)}} \, d\omega$$

$$= 2 \int_0^\infty S(\omega) \, b_{d11}(\omega) \, d\omega$$

(7.116)

where $b_{d11}(\omega)$ is the normalized wave drift damping ($B_{dij} = A^2 b_{dij}$).

$$B_{dij}(\beta) = 2 \int_0^\infty S(\omega) \, b_{dij}(\omega, \beta) \, d\omega$$

(7.117)

When Aranha's formula (7.34) is used and the wave spectrum is known analytically, the differentiation of the drift force can be avoided through integration by parts. For instance, in deep water and head waves ($\beta = 0$), one gets for B_{d11}:

$$B_{d11} = \frac{2}{g} \int_0^\infty \omega \, f_{d1}(\omega, 0) \left(2 S(\omega) - \omega \frac{\partial S(\omega)}{\partial \omega} \right) d\omega$$

(7.118)

In time domain simulation models, a way to incorporate the wave drift damping effect and the enhancement of the slow-drift force with current is to apply the

---

[6] This problem was considered by Finne & Grue (1998) who proposed a theoretical model accounting for a slow-drift velocity in yaw. They present numerical results in the case of an FPSO and conclude that the wave drift damping in yaw is comparable to the viscous damping.

following Newman-type formulation:

$$F_{-k}^{(2)}(t) = \left\{ \left[\sum_i A_i \sqrt{|\tilde{f}_{dk}(\omega_i, \beta)|} \, \cos D_i(t) \right]^2 \right.$$

$$\left. + \left[\sum_i A_i \sqrt{|\tilde{f}_{dk}(\omega_i, \beta)|} \, \sin D_i(t) \right]^2 \right\} \text{sign}(f_{dk}(\beta)) \qquad (7.119)$$

where

$$D_i(t) = k_i \, (X^{(2)}(t) - U_c \, t) \, \cos\beta + k_i \, (Y^{(2)}(t) - V_C \, t) \, \sin\beta - \omega_i \, t + \theta_i \qquad (7.120)$$

and

$$\tilde{f}_{dk} = \left(1 - \frac{4\,\omega_i \, U_L}{g} \right) f_{dk} \left(\omega_i - \frac{\omega_i^2 \, U_L}{g}, \beta + \frac{2\,\omega_i \, U_T}{g} \right) \qquad (7.121)$$

with

$$U_L = (\dot{X}^{(2)}(t) - U_C) \, \cos\beta + (\dot{Y}^{(2)}(t) - V_C) \, \sin\beta \qquad (7.122)$$

$$U_T = (\dot{X}^{(2)}(t) - U_C) \, \sin\beta - (\dot{Y}^{(2)}(t) - V_C) \, \cos\beta \qquad (7.123)$$

Here the wave elevation is given in a coordinate system (x, y) traveling with the current.

7.7.4   Damping Due to the Mooring Lines

It has been established that mooring lines (and risers) significantly contribute to slow-drift damping. This is due to their friction on the sea floor, to internal energy dissipation, and to hydrodynamic drag forces upon them. Here we are only considering the last contribution.

Through model tests Huse & Matsumoto (1989) produced some evidence of the significance of mooring line damping. They also showed that the slow-drift damping effect is enhanced by the first-order response of the mooring lines (see Figure 7.32).

An estimate, and a better understanding, of mooring line damping can be obtained through the following procedure. For the sake of simplicity, consider the case of a floating body retained by only one catenary line, in the vertical plane $y = 0$. The motion of the body consists in a linear response $(x_F^{(1)}(t), z_F^{(1)}(t))$ at the fairlead and a slow-drift motion $X^{(2)}(t)$. As a result, the line responds dynamically and drag forces take place along the line, dissipating energy to which the slow-drift damping can be related through some simplifying assumptions:

1. The line responds linearly to its top excitation. Let $\alpha(\omega, s)$ (resp. $\gamma(\omega, s)$) be the transfer functions relating the local transversal motion of the line to the top x motion $x_F(t)$ (resp. top z motion $z_F(t)$). In a regular wave of unit amplitude and frequency ω, the local transverse velocity $v(s, t)$ (with respect to still fluid) is given by

$$v(s, t) = \Re \left\{ -i\omega \left(\alpha(\omega, s) \, a_{xF}(\omega) + \gamma(\omega, s) \, a_{zF}(\omega) \right) e^{-i\omega t} \right\} \qquad (7.124)$$

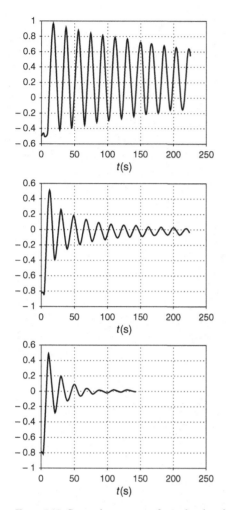

Figure 7.32 Surge decay tests of a tanker in still water. With aerial mooring lines (top), with catenary mooring lines (middle), with catenary mooring lines and forced wave frequency motion of the turret (bottom) (from Huse & Matsumoto, 1989).

where $(a_{xF}(\omega), a_{zF}(\omega))$ are the RAOs of $(x_F(t), z_F(t))$ and s is the curvilinear abscissa.

2. The line response to the top slow-drift motion is quasi-static:

$$V^{(2)}(s,t) = \alpha(0,s)\,\dot{X}^{(2)}(t) \tag{7.125}$$

3. The first-order transverse velocities of the line are much larger than the second-order velocity

$$|v^{(1)}(s,t)| \gg |V^{(2)}(s,t)| \tag{7.126}$$

4. The first- and second-order motions are uncorrelated:

$$E\left\{X^{(2)}(t)\, x_C^{(1)}(t)\right\} = E\left\{X^{(2)}(t)\, z_C^{(1)}(t)\right\} \equiv 0 \tag{7.127}$$

5. The line oscillates in a fluid at rest.

The local drag force applied on a line element ds is given by:

$$dF = -\frac{1}{2}\, C_D\, \rho\, D\, \left|v^{(1)}(s,t) + V^{(2)}(s,t)\right|\, \left(v^{(1)}(s,t) + V^{(2)}(s,t)\right) ds \tag{7.128}$$

(High local K_C numbers are assumed here, so the dependent flow formulation applies.)

The slow-drift velocity $V^{(2)}$ being assumed to be low compared to $v^{(1)}$ a Taylor expansion is applied to $|v^{(1)} + V^{(2)}|\, (v^{(1)} + V^{(2)})$, giving:

$$|v + V|\, (v + V) = v\, |v| + 2\, |v|\, V + O(V^2) \tag{7.129}$$

where the superscripts $^{(1)}$ and $^{(2)}$ have been omitted, for simplicity.

Only the second term in the right-hand side contributes to the damping of the slow-drift response. The associated dissipated power is obtained by:

$$P = \frac{1}{T}\, \int_{t_0}^{t_0+T} dt\, \int_0^l C_D\, \rho\, D\, \left|v(s,t)\right|\, V^2(s,t)\, ds \tag{7.130}$$

This can be simplified owing to the fact that the two time scales are different and that $v(s,t)$ is a Gaussian process:

$$\int_{t_0}^{t_0+T} |v(s,t)|\, V^2(s,t)\, dt \simeq \overline{|v(s,t)|}\, \int_{t_0}^{t_0+T} V^2(s,t)\, dt$$

$$\simeq \sqrt{\frac{2}{\pi}}\, \sigma_v(s)\, \int_{t_0}^{t_0+T} V^2(s,t)\, dt \tag{7.131}$$

where the bar $\overline{|v|}$ means time-averaged value.

Finally, the averaged power P is written as:

$$P = \sqrt{\frac{2}{\pi}}\, \rho \int_0^l C_D\, D\, \sigma_v(s)\, \alpha^2(0,s)\, ds\, \left\{\frac{1}{T}\int_{t_0}^{t_0+T} \dot{X}^{(2)^2}\, dt\right\} \tag{7.132}$$

This means that the slow-drift damping is given by:

$$B_{a11} = \sqrt{\frac{2}{\pi}}\, \rho \int_0^l C_D\, D\, \sigma_v(s)\, \alpha^2(0,s)\, ds \tag{7.133}$$

where

$$\sigma_v^2(s) = \int_0^\infty \left\|\alpha(\omega,s)\, a_{xF}^{(1)}(\omega) + \gamma(\omega,s)\, a_{zF}^{(1)}(\omega)\right\|^2\, \omega^2\, S(\omega)\, d\omega \tag{7.134}$$

with $S(\omega)$ the wave spectrum.

In some catenary line configurations, a small horizontal displacement at the fairlead will induce large transversal displacement of the lower part of the line, meaning a high value of $\alpha(\omega,s)$, which appears in the final expression of the slow-drift damping through its third power. This makes it clear why the induced damping can be large.

The developments above also clarify the damping enhancement due to the first-order response, which introduces a multiplicative factor $2\,|v^{(1)}|/|V^{(2)}|$ compared to the slow-drift motion alone.

The procedure above can easily be extended to the case of a line at some angle to the slow-drift velocity and generalized to a complete mooring system. A practical difficulty is the derivation of the transfer functions $(\alpha(\omega,s),\beta(\omega,s),\gamma(\omega,s))$. When the wave frequencies are low compared to the natural frequencies of the line dynamics, a quasi-static approach can be used.

Drag loads on risers and TLP tethers also contribute to slow-drift damping.

A blunt way to account for mooring line damping, in time domain simulation models, is to solve jointly the slow-drift response and the dynamics of the mooring lines. This means several hundred degrees of freedom, long time simulations (to obtain a sufficient number of low frequency cycles), and small time steps (to properly reproduce the dynamics of the lines). With current computer resources this is feasible.

7.7.5 Viscous Damping at the Hull

It is by far the most difficult damping component to properly evaluate. This is due to the variety of geometric shapes (barges, tankers, semi-submersible platforms, etc.), to the different forms of viscous loads (friction, drag), and to the fact that several flows coexist at different frequencies and different K_C numbers, from which the low-frequency damping effect must be extracted.

Two generic cases are considered here:

- A tanker moored to a single point mooring, undergoing surge motion. Due to the relatively streamlined hull shape, the viscous damping mainly originates from friction.
- A barge, or a pontoon of a semi, in beam waves. This is parent to the two-dimensional case of a rectangular shape in combined wave frequency and low-frequency flows. The main difference, with the previous case of mooring line damping, is that the K_C numbers are low: the dependent flow field formulation may not be applied.

Surge Viscous Damping of Ship Hulls

As in the mean current load case, reference can be made to the equivalent flat plate. It is important here to understand that a flow that periodically reverses is in no way related to a steady current. In the latter case, a boundary layer develops from the leading edge, the local friction coefficient depends on the local Reynolds number $\mathrm{Re}(x) = U x/\nu$ with x the downstream coordinate. In the periodic flow case, the local friction coefficient depends on the oscillatory Reynolds number $\mathrm{Re}_o = A^2\omega/\nu$ with $U = A\omega\cos\omega t$ the plate velocity.

In the laminar regime, the frictional stress is obtained as (Stokes, 1851)

$$\tau = -\rho A \omega \sqrt{\omega\nu}\,\cos\left(\omega t + \frac{\pi}{4}\right) \qquad (7.135)$$

45 degrees out of phase with (minus) the velocity. To express energy dissipation an additional $\sqrt{2}/2$ coefficient must be applied, giving the "effective" frictional coefficient

$$C_f = \sqrt{\frac{2}{\text{Re}_o}} \tag{7.136}$$

The surge frictional damping is then

$$B_{v11} = \frac{\sqrt{2}}{2} \rho \sqrt{\omega \nu} \, S \tag{7.137}$$

with S the hull or flat plate area.

In the case of smooth flat plates, these theoretical results are well verified by experiments, as long as the flow remains laminar. Transition to turbulence appears at oscillatory Reynolds numbers around $3 \cdot 10^5$.

In the case of a FPSO undergoing surge oscillations with a typical amplitude of 10 m and a typical frequency of 0.005 Hz, the oscillatory Reynolds number is equal to $6 \cdot 10^6$: the flow is turbulent. In model tests at a 1:50 scale, following Froude scaling law, the oscillatory Reynolds number is divided by $50^{3/2}$, so drops down to 10^4: the flow is laminar.

In fact, damping values derived from extinction tests are in fair agreement with theoretical values, when multiplied with a corrective coefficient $1 + k$, where k is comprised between 0.5 and 1. It is likely that these relatively high values of the k coefficient (compared to the values typically encountered in forward speed resistance) are due to flow separation at the aft and stern.

In the case of the oscillatory flat plate in the laminar regime, the nonlinear convective terms disappear from the Navier–Stokes equations, which become linear. This means that if the plate is undergoing a bichromatic motion, that is two sinusoidal signals superimposed, the associated viscous loads are additive. In other words, the low-frequency frictional damping is the same whether the low-frequency motion coexists with the wave motion or not. It is tempting to write that this result holds for the FPSO model, whether the model oscillates in waves or in still water. This is well verified in model tests since the difference between the measured dampings, in still water and in waves, is interpreted as the wave drift damping and that this experimental wave drift damping compares well to numerical predictions based on potential flow theory.

The situation is quite different at full scale.

For the oscillatory flat plate in the turbulent regime, at oscillatory Reynolds number higher than about $6 \cdot 10^5$, different expressions of the frictional coefficient can be found in the literature. For instance:

$$C_f = 0.09 \, \text{Re}_o^{-0.2} \qquad \text{(Jonsson, 1966)} \tag{7.138}$$

or

$$C_f = 0.024 \, \text{Re}_o^{-0.123} \qquad \text{(Justesen, 1988)} \tag{7.139}$$

The phase shift between the frictional stress and the velocity is apparently less than in the laminar regime, and can be assumed to be nil.

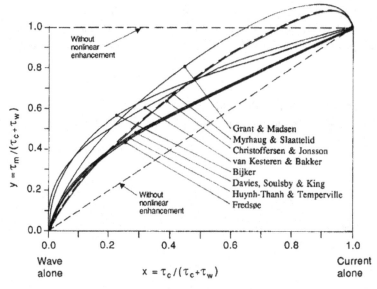

Figure 7.33 Frictional stress on the sea floor in wave plus current. Mean stress $\tau_m/(\tau_c + \tau_w)$ in wave plus current vs mean stress $\tau_c/(\tau_c + \tau_w)$ in current alone (from Soulsby *et al.*, 1993).

Moving now to the case of a low-frequency motion combined with a wave frequency motion, the main implication of the turbulent regime is that the frictional loads are no longer additive. There is extensive literature on the parent problem of the wave and current boundary layer on the sea floor. From this literature, it appears that the mean frictional stress, in wave plus current, is much greater than in current alone.

Figure 7.33, taken from Soulsby *et al.* (1993), shows the mean shear stress τ_m in waves plus current compared to the mean shear stress τ_c in current alone, from different theoretical models. The shear stresses τ_m and τ_c are normalized by the sum $\tau_c + \tau_w$ of the mean stress in current τ_c and the stress amplitude τ_w under regular waves. These results are for particular values of the ratios z_0/h and A/z_0, where h is the water depth, z_0 the roughness height, and A the amplitude of the water motion at the bottom (here $z_0/h \simeq z_0/A \simeq 10^{-4}$, meaning a nearly smooth bottom). There is some scatter in the results from the different numerical models but they all agree on the fact that the mean stress increases under waves, particularly when $\tau_c/(\tau_c + \tau_w)$ is low, that is, when the wave flow is dominant.

These results do not really provide a basis for drawing any definitive and quantified conclusions relative to our problem: as argued at the beginning of this paragraph, there is quite a difference between the boundary layers under current and under a low-frequency sinusoidal flow. Still it might be expected that the same effect should be observed, that is, at full scale, the low-frequency viscous damping should be enhanced under waves, compared to its still water value.

Under the assumption that the wave flow is dominant over the low-frequency flow, we can propose an estimate of the low-frequency viscous damping, following a method similar to that used for the damping due to the mooring lines. We again refer to the flat plate model, undergoing a combined motion of velocity $u(t) + U(t)$, $u(t)$ standing for the wave frequency velocity and $U(t)$ for the slow-drift velocity. We assume the wave frequency velocity $u(t)$ to be dominant: $|u| \gg |U|$. We write the frictional load as

$$dF = \frac{1}{2} \rho \, C_f(\text{re}_o) \, (u + U) \, |u + U| \, dS \tag{7.140}$$

re_o being the oscillatory Reynolds number associated with $u(t)$.

Following the same approximations as in the previous case, we write

$$(u + U) \, |u + U| \simeq u \, |u| + 2 \, |u| \, U + O(U^2) \tag{7.141}$$

$$\frac{1}{T} \int_{t_0}^{t_0+T} |u(t)| \, U^2(t) \, dt \simeq \frac{1}{T} \, \overline{|u|} \int_{t_0}^{t_0+T} U^2(t) \, dt \tag{7.142}$$

Applying again stochastic linearization

$$\overline{|u|} \simeq \sqrt{\frac{2}{\pi}} \, \sigma_u, \tag{7.143}$$

the low-frequency damping is derived as

$$dB = \sqrt{\frac{2}{\pi}} \, \rho \, C_f(\text{re}_o) \, \sigma_u \, dS \tag{7.144}$$

To move from the oscillatory flat plane to the ship hull undergoing slow-drift motion in waves, we have the problem that the standard deviation σ_u of the wave frequency velocity varies along the hull (in fact it should be the standard deviation of the relative velocity of the hull with respect to the fluid). Moreover, this relative velocity has two tangential components. A gross estimate consists in considering only the incident wave velocity, writing

$$\sigma_u = \sqrt{m_2} = \sqrt{m_0} \, \sqrt{\frac{m_2}{m_0}} = \frac{H_S}{4} \, \frac{2\pi}{T_2} \tag{7.145}$$

Likewise, the oscillatory Reynolds number is taken from the mean wave amplitude and the mean wave particle velocity:

$$\overline{A} = \sqrt{\frac{\pi}{2}} \, \sqrt{m_0} \tag{7.146}$$

$$\overline{A\omega} = \sqrt{\frac{\pi}{2}} \, \sqrt{m_2} \tag{7.147}$$

giving $\text{re}_o = \overline{A} \, \overline{A\omega} / \nu$:

$$\text{re}_o = \overline{A} \, \overline{A\omega} / \nu = \frac{\pi^2}{16} \, \frac{H_S^2}{\nu \, T_2} \tag{7.148}$$

Finally, the low-frequency viscous damping is obtained as

$$B = \sqrt{\frac{\pi}{2}} \rho \, C_f(\text{re}_o) \, \frac{H_S}{T_2} \, S \tag{7.149}$$

with, from Justesen's formula:

$$C_f = 0.024 \, \text{re}_o^{-0.123} \qquad (7.150)$$

The validity of these results cannot be guaranteed.

It can be observed that the hulls of FPSOs are far from being smooth, being covered with various protections for boat landing, for the SCRs and for the mooring lines, with anodes in some cases, that all will contribute to viscous dissipation. This is at variance with LNG carriers, which are extremely streamlined and where only friction comes into play.

Viscous Damping of Semi-Submersible Platforms

We now consider the second generic case, which is the individual geometric elements of a semi-submersible platform or a TLP, in beam waves. We use a two-dimensional idealization, in a plane perpendicular to the column axis (therefore horizontal) or to the pontoon axis (therefore vertical).

From a kinematic point of view, it is equivalent to consider a strip of the column or pontoon in forced motion in a fluid at rest, if we accept that the flow due to the incoming waves is locally uniform. This motion consists in a slow-drift part, horizontal, and a wave frequency part, with horizontal and vertical components.

This problem differs from the viscous loads on the mooring lines because the considered geometric elements have dimensions of the same order as the relative motion amplitude with respect to the fluid. As a result, in the case of rounded shapes, it may be that no separation of the boundary layers occurs, or limited separation with the vortices emitted being reabsorbed at each cycle, not being convected away.

In Chapter 4 some experimental data have been shown on the drag coefficients for circular or square shapes, in low K_C periodic flows. In the case of the circle, the flow remains attached up to some K_C value that depends on the Stokes parameter $\beta = D^2/\nu T$. Considering the circular column of a semi, at full scale the diameter D is around 20 m, and the period T 10 s or 100 s, respectively for the wave or slow-drift flow. This means β numbers of the order of 10^6 or 10^7 and turbulent boundary layers. Experiments suggest that at such high β numbers separation is delayed to K_C numbers around 4 or 5 and that, for smooth walls, drag coefficients up to separation are very low: values less than 0.1 have been reported at β numbers around 10^5 (see Figure 4.22). Probably much lower values would be found at $\beta = 10^6$ or $\beta = 10^7$.

Practically this means that, at full scale, not much damping is expected to arise from circular columns as long as the K_C numbers are below 5, that is, for amplitudes of motion less than about one diameter. It must be borne in mind that things might go differently at model scale, where the β number is divided by the scale factor to the power 3/2: on one side the separation threshold is lowered (to $K_C \simeq 2$ for $\beta \simeq 1000$), on the other side drag coefficients are increased.

When the cylinder has sharp angles, the oscillatory flow separates and experimental drag coefficients show a steady increase as the K_C number goes down to zero. In the case of a square, values as high as 4 or 5 are attained with a zero K_C number (Bearman *et al.*, 1985). See Figure 4.23.

Figure 7.34 Experimental model. The loads are measured on the middle section.

This is for sinusoidal flow. When two oscillatory flows coexist, at distant frequencies, it is easy to convince oneself, through experiments, that the dependent flow-field formulation of the drag force cannot be applied to the lower frequency component. Figures 7.35 and 7.36 refer to experiments carried out on a square model, 0.35 m in side, undergoing unidirectional bichromatic forced motion (see Figure 7.34). The fast motion has a period equal to 1.75 s and a K_C number equal to 1. For the slow motion component they are $T = 10$ s and $K_C = 3$. Figure 7.35 shows the product $(v+V)|v+V|$ vs time, with v the high-frequency component and V the low-frequency component; the velocity amplitudes are in a ratio of about 2 to 1. Figure 7.36 shows the drag force derived from the measurements, obtained by subtracting the inertia force (opposite to the acceleration) from the total measured force. It is clear that there is no proportionality between the two time series: in the second one it looks as if the low-frequency component has completely disappeared.

In Martigny *et al.* (1994), where these experiments are reported, an attempt is made at applying so-called wake models, such as those being used to study the on-bottom stability of pipes and flowlines (see Chapter 4). It is found that wake effects do lead to a decrease of the low-frequency damping, however not to the extent shown in Figure 7.36.

The alternative is to use the independent flow field formulation of the drag load:

$$dF_D = \left\{ \frac{1}{2}\rho\, C_{Dlf}\, D\, V(t)\, |V(t)| + \frac{1}{2}\rho\, C_{Dwf}\, D\, v(t)\, |v(t)| \right\} dl \qquad (7.151)$$

where the problem now is to evaluate how much the low-frequency drag coefficient C_{Dlf} is sensitive to the coexisting wave frequency flow.

Figure 7.37 shows some results obtained for different types of high-frequency motion. What is plotted in the figure is not the drag coefficient C_{Dlf} but the linear damping coefficient C_{blf} related to C_{Dlf} through

$$C_{blf} = \frac{2\,K_C}{3\,\pi^2}\, C_{Dlf} \qquad (7.152)$$

The slow motion has a period equal to 10 s and a K_C number varying from 2 to 4; the fast motion has a period equal to 1.75 s and a k_C number equal to 1.5. Different types of fast motion are achieved: horizontal, vertical, circular, and elliptic (vertical

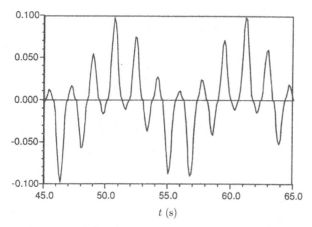

Figure 7.35 Square cylinder undergoing bichromatic forced motion in still water. Product $(v + V) |v + V|$ vs time.

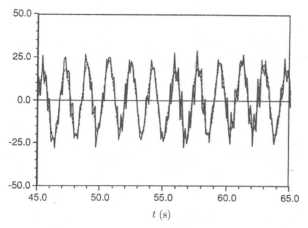

Figure 7.36 Square cylinder undergoing bichromatic forced motion in still water. Measured drag force.

amplitude divided by two). It can be seen that, in the case of the horizontal motion, the damping coefficients are hardly changed but that, in the other cases, they are significantly increased. At any rate the high-frequency component always increases the low-frequency damping, suggesting that the independent flow field formulation is on the conservative side.

7.7.6 Relative Importance of the Different Damping Components

The previous section has shown the difficulties that arise in properly evaluating the different components of the slow-drift damping, particularly the viscous component. Still it may be wondered whether all components equally contribute and whether some can a priori be ignored.

This is a difficult exercise in the general case. In the following we investigate their sensitivity to the significant wave height of the sea state, the spectral shape and

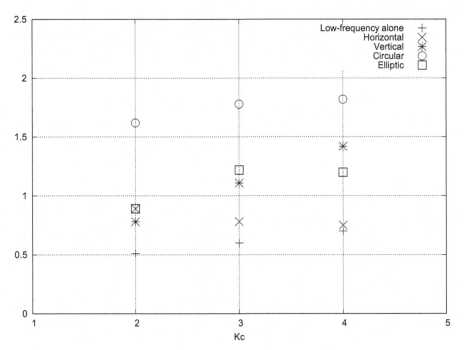

Figure 7.37 Square cylinder undergoing bichromatic forced motion in still water. Damping coefficient of the slow motion vs the K_C value of the low-frequency motion, for different types of high-frequency motion.

the mean period T_2 being kept the same. We assume the idealized one dof system considered previously:

$$(M + M_a) \, \ddot{X}^{(2)} + B(X^{(2)}, \dot{X}^{(2)}) \, \dot{X}^{(2)} + K \, X^{(2)} = F_-^{(2)}(t) \tag{7.153}$$

In the damping term we add the following components:

Viscous damping
We assume that it results from drag forces, written following the independent flow field formulation

$$F_D = -\frac{1}{2} \rho \left(\sum_i C_{Di} \, S_i \right) |\dot{X}^{(2)}| \, \dot{X}^{(2)} \tag{7.154}$$

Under the assumption that $X^{(2)}(t)$ be a Gaussian process, we can write

$$|\dot{X}^{(2)}| \, \dot{X}^{(2)} \simeq \sqrt{\frac{8}{\pi}} \, \sigma_{\dot{X}^{(2)}} \, \dot{X}^{(2)} \tag{7.155}$$

and, under the assumption of a narrow-banded signal

$$|\dot{X}^{(2)}| \, \dot{X}^{(2)} \simeq \sqrt{\frac{8}{\pi}} \, \Omega_0 \, \sigma_{X^{(2)}} \, \dot{X}^{(2)} \tag{7.156}$$

with Ω_0 the natural frequency.

Therefore the viscous damping takes the form:

$$B_v = \Omega_0 \ \sigma_{X^{(2)}} \ b_v \tag{7.157}$$

Damping due to the mooring lines
Following the formulation proposed in Section 7.7.4, this damping component varies linearly with the significant waveheight:

$$B_{ml} = H_S \ b_{ml}(T_2) \tag{7.158}$$

Wave drift damping
The wave drift damping is quadratic with H_S:

$$B_d = H_S^2 \ b_d(T_2) \tag{7.159}$$

As a result the equation of motion becomes

$$(M + M_a) \ \ddot{X}^{(2)} + \left(\Omega_0 \ \sigma_{X^{(2)}} \ b_v + H_S \ b_{ml} + H_S^2 \ b_d\right) \dot{X}^{(2)} + K \ X^{(2)} = H_S^2 \ f_-^{(2)}(t) \tag{7.160}$$

where the excitation load also varies quadratically with H_S.

The variance of the slow-drift response is then obtained as

$$\sigma_{X^{(2)}}^2 = \frac{\pi \ H_S^4 \ S_{f_-^{(2)}}(\Omega_0)}{2 \left[\Omega_0 \ \sigma_{X^{(2)}} \ b_v + H_S \ b_{ml}(T_2) + H_S^2 \ b_d(T_2)\right] (M + M_a) \ \Omega_0^2} \tag{7.161}$$

From this equation some conclusions can be drawn:

- The relative contribution of the viscous damping depends on the natural frequency of the system. It becomes negligible when the natural frequency drops to zero.
- The viscous damping is the same whatever the sea state, whereas mooring line damping varies linearly with H_S and wave drift damping varies quadratically. As H_S increases the wave drift damping becomes dominant and the variance of the response is asymptotically obtained as

$$\sigma_{X^{(2)}}^2 = \frac{\pi \ H_S^2 \ S_{f_-^{(2)}}(\Omega_0)}{2 \ b_d(T_2) \ (M + M_a) \ \Omega_0^2} \tag{7.162}$$

meaning that the standard deviation $\sigma_{X^{(2)}}$ varies **linearly** with H_S.

Conversely, when H_S goes to zero, the standard deviation is obtained from

$$\sigma_{X^{(2)}}^3 = \frac{\pi \ H_S^4 \ S_{f_-^{(2)}}(\Omega_0)}{2 \ b_v \ (M + M_a) \ \Omega_0^3} \tag{7.163}$$

again close to linear with H_S (to the power 4/3).

From this exercise it may be concluded that a good estimate of wave drift damping is critical in North Sea conditions whereas, in less severe areas, such as the gulf of Guinea, viscous and mooring line dampings probably play the major role.

7.8   Time Domain Simulation of Slow-Drift Motion

Even though, as just seen above, methods working in the frequency domain can be applied to a certain extent, it is customary to tackle the slow-drift motion of moored structures through time domain simulations. This method offers many advantages, for instance simulating moorings with nonlinear restoring forces, complex environmental conditions with waves, wind, and current of different directions, nonlinear damping effects, the sudden failure of a mooring line, etc.

In most numerical models, only the horizontal components (surge, sway, and yaw) of the slow-drift motion are integrated in time. The wave frequency response is simply added up at every time step, from pre-calculated RAOs and from the wave history.

In a fixed coordinate system Oxy the position of the structure is defined by the coordinates $X_G^{(2)}, Y_G^{(2)}$ of its center of gravity and its heading $\psi^{(2)}$. In this coordinate system the equations of its slow-drift motion are written as:

$$M \frac{d^2 X_G^{(2)}}{dt^2} = F_x \tag{7.164}$$

$$M \frac{d^2 Y_G^{(2)}}{dt^2} = F_y \tag{7.165}$$

$$\frac{d}{dt}\left(I \frac{d\psi^{(2)}}{dt}\right) = C_z \tag{7.166}$$

M being the mass and I the yaw inertia which, strictly, varies in time via the first-order angular response. In practice, one generally considers that the inertia I remains constant. The right-hand sides represent the external loads to take into account, which are:

- the low-frequency second-order wave loads;
- the wind loads;
- the restoring loads from the mooring system (plus, possibly, the risers);
- the hydrodynamic loads other than from the waves, accounting for the current and for the slow-drift velocities and accelerations of the structure.

All these loads must be formulated in such a way that the damping effects, as previously described, are adequately accounted for.

Incoming waves

The wave conditions may be specified from an experimental record of the free surface elevation (when exploiting results from model tests for instance, more rarely from in situ measurements). More frequently they are specified as a wave spectrum $S(\omega)$. The usual technique is to divide a restricted frequency range (for instance from $\omega_P/2$ to $5\,\omega_P/2$ where ω_P is the peak frequency) into N sub-intervals of equal width $\Delta\omega$. The carrier frequencies ω_i can then be taken in the middle of each interval, offering possible computational advantages for the second-order wave loads through FFT techniques, but with the drawback that the wave signal repeats itself, with a periodicity $2\pi/\Delta\omega$ that must be longer than the duration of the simulation. An alternative is to pick up the carrier frequencies at random within each sub-interval. This avoids repetition, and also that all $\omega_i - \omega_j$ be multiples of $\Delta\omega$.

To first-order the incoming wave elevation is then given by

$$\eta_I^{(1)}(x,y,t) = \sum_{i=1}^{N} A_i \, \cos[k_i(x\cos\beta + y\sin\beta) - \omega_i t + \theta_i] \tag{7.167}$$

where the amplitudes are obtained from

$$A_i^2 = 2\,S(\omega_i)\,\Delta\omega \tag{7.168}$$

and the wave numbers k_i from the dispersion equation

$$\omega_i^2 = g\,k_i\,\tanh k_i h \tag{7.169}$$

In (7.167) the phase angles θ_i are also picked up at random within the interval $[0\ 2\pi]$.

The choice of the number of wave components, besides avoiding repetition of the wave signal, is dictated by the need to properly reproduce the spectra of the second-order loads in the vicinity of the natural frequencies. This criterion leads to N being much higher than the ratio $\omega_P/(\zeta\,\Omega_0)$ where Ω_0 is the lowest natural frequency (in surge, sway or yaw) and ζ the associated damping ratio. With a typical 5% damping value and a frequency ratio equal to 20 (a natural period of 200 s with a T_P equal to 10 s), this means $N \gg 400$!

Prior to running the time domain simulations, a Hydrodynamic Database (HDB) must be constituted by running systematically a diffraction-radiation code for a large number of wave directions and frequencies, storing RAOs and drift forces $f_d(\omega_i, \beta_k)$ (and possibly QTFs as $f_-^{(2)}(\omega_i, \omega_j, \beta_k)$). These HDBs may be sea state dependent, for instance if nonlinear damping terms are introduced in the first-order equations of motion.[7]

Input of second-order loads

When dealing with the horizontal components of the slow-drift motion, it is usual to resort to some Newman approximation, that is to say, express the slowly varying wave loads from the knowledge of the drift force $f_d(\omega, \beta)$ in regular waves. At each time step the loads are introduced as, for instance, equation (7.97). It is just a matter of interpolating in the HDB to account for the instantaneous heading and for the carrier frequencies. The limitations of Newman approximations have been outlined, in the cases of stiff moorings and/or shallow water depth.

It is necessary to introduce the wave drift damping effects. This can be done in two different ways:

---

[7] In most diffraction-radiation codes, the elementary diffraction and radiation potentials are stored panel-wise, enabling recomputation of the RAOs and of the drift forces sea state by sea state; this also enables to vary the mechanical properties such as the position of the center of gravity or the inertias.

1. by using the formulation proposed at the end of Section 7.7.3 on the wave drift damping;
2. by introducing, in the equation of motions, an averaged linear damping given by

$$B_{dij}(\beta) = 2 \int_0^{\infty} S(\omega) \, b_{dij}(\omega, \beta) \, d\omega \qquad (7.170)$$

where the b_{dij} terms are obtained through Aranha's formula or previously calculated with a diffraction-radiation code accounting for forward speed.

Even though, in both cases, one would use Aranha's formula, these two methods are not equivalent: in the second case only the mean value of the second-order loads is corrected for the effects of the slow-drift velocity; in the first case it is the instantaneous value. The maxima of the resulting motions can therefore be different.

Wind loads
In a body-fitted coordinate system, the wind loads are usually expressed as

$$F_{WX} = \frac{1}{2}\rho_a \, C_{dWX}(\alpha_W) \, S_W \, V_W^2 \qquad (7.171)$$

$$F_{WY} = \frac{1}{2}\rho_a \, C_{dWY}(\alpha_W) \, S_W \, V_W^2 \qquad (7.172)$$

$$C_{W\psi} = \frac{1}{2}\rho_a \, C_{dW\psi}(\alpha_W) \, S_W \, L_W \, V_W^2 \qquad (7.173)$$

where S_W is a reference area, L_W a reference length, and $C_{dWX}, C_{dWY}, C_{dW\psi}$ force and moment coefficients dependent on the wind incidence α_W, and V_W the wind velocity (usually taken 10 m above MSL). These coefficients need to be determined beforehand through wind tunnel tests, or CFD calculations, or from the literature.

Wind speed fluctuations cover a wide frequency domain and also participate in the low-frequency excitation of moored structures. It is easy, from a given wind spectrum $S_W(f)$, to generate a wind time signal in a similar way as for the wave elevation:

$$V_W(t) = \overline{W} + \sum_{i=1}^{N_W} v_i \, \cos(2\pi f_i t + \psi_i) \qquad (7.174)$$

where

$$v_i^2 = 2 \, S_W(f_i) \, \Delta f \qquad (7.175)$$

The wind velocity and direction may also be input from in situ measurements, for instance to simulate squalls (see Figure 2.13).

In the load equations (7.171), (7.172), (7.173), the slow body velocity may be subtracted from the wind velocity. Some damping effect is thus introduced for the surge and sway components.

Mooring loads
The mooring system may be represented globally, as a restoring force vs the offset, or line by line. The second way is obviously preferable to have access to the peak mooring loads, or to simulate a line failure.

The line response can be considered to be quasi-static. The tension at the fairlead (and the line inclinations) may then be derived through the catenary equations, from the position $(x_F(t), y_F(t), z_F(t))$ of the fairlead (adding up 3 dof low frequency and 6 dof wave frequency components). However the lines may respond dynamically to the wave frequency motion of the fairleads. This will affect the tension peaks. It has also been emphasized that the wave response of the lines enhances the slow-drift damping from the mooring lines. Finally, the lines may vibrate under vortex shedding.

Whether the dynamic line response may be important or not can be assessed prior to the simulations, by checking whether the line resonant frequencies lie in the wave frequency range. As a rule of thumb, the natural frequencies can be approximated by

$$\omega_n = \frac{n\pi}{l}\sqrt{\frac{T}{m}} \tag{7.176}$$

with n the mode number, l the length of the line, T the mean tension along the line, and m the mass plus added mass per unit length.

One way to account for the dynamic line effects (excluding a possible VIV response) is to jointly solve, in the time domain, the wave frequency of the lines and the low-frequency motion of the floater. In this way the mooring line damping is naturally introduced. Some software offers this capability, but the CPU cost is greatly increased.

Another way is to linearize the line response and then, following the procedure introduced in Section 7.7.4, derive damping values directly introduced into the equations of motion.

Hydrodynamic loads

Here we are concerned with the hydrodynamic loads, other than the second-order wave loads, due to the current and to the slow-drift velocities and accelerations in surge, sway, and yaw.

For a fixed structure, the current loads result from a combination of friction, drag, lift, and Munk moment (in yaw) effects. They are usually expressed in a similar way as the wind loads:

$$F_{CX} = \frac{1}{2}\rho_w \, C_{dCX}(\alpha_C) \, S_C \, V_C^2 \tag{7.177}$$

$$F_{CY} = \frac{1}{2}\rho_w \, C_{dCY}(\alpha_C) \, S_C \, V_C^2 \tag{7.178}$$

$$C_{C\psi} = \frac{1}{2}\rho_w \, C_{dC\psi}(\alpha_C) \, S_C \, L_C \, V_C^2 \tag{7.179}$$

For a moving structure, it is customary to use the same load expressions with the relative current, subtracting the slow-drift velocity. This procedure is legitimate for the wind loads but it is questionable for the hydrodynamic loads: in the case of wind, the wind velocity is much higher than the slow-drift velocities, the relative flow does not reverse, and the reduced velocity $V_W T/L$ (with V_W the wind velocity, T the slow-drift period, and L the horizontal scale) is very high, justifying a quasi-static approach. For the hydrodynamic loads, the current and slow-drift velocities are comparable, and the reduced velocity $V_C T/L$ is of order one: strictly the loads cannot be related to the instantaneous kinematics, the past flow history should be accounted for.

Accounting for the relative current velocity introduces damping effects in surge and sway but not in yaw. For shiplike bodies at single point moorings yaw damping is critical since, depending on the amount of damping, fishtailing instability may occur or not. The usual procedure is to integrate drag loads along the hull, via a strip theory, where, locally, the current, sway and yaw-induced velocities are combined. Different formulations have been proposed, for example by Molin & Bureau (1980), Wichers (1988), and others.

Finally, inertia loads, opposite to the accelerations, must be introduced, usually from the matrix of added masses and inertias at zero frequency. This can be done following maneuvering theory (Norrbin, 1970).

In model tests, it is frequently observed that the natural frequencies of the systems vary somewhat with the sea states, differing from their still water values. This implies that the low frequency added masses deviate from their theoretical values. It can be due to viscous effects and to the coupling with the wave frequency motion: Figure 7.38, from the same experiments as reported in Section 7.7.5, shows the sensitivity of the low frequency added mass coefficient to the type of high-frequency motion superimposed.

The sensitivity of the low-frequency inertia loads to the wave frequency component may also be due to purely potential effects. It results from the same inconsistency as leads to the concept of wave drift damping: the second-order motion is not small

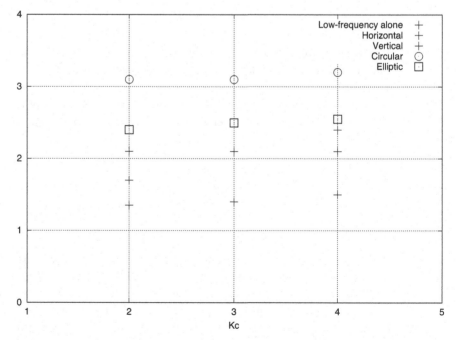

Figure 7.38 Square cylinder undergoing bichromatic forced motion in still water. Added mass coefficient of the slow motion vs the K_C value of the low-frequency motion, for different types of high-frequency motion.

compared to the first-order response. The concept of **wave drift added mass** was introduced by Kinoshita *et al.* (2002). A complete theory and some comparisons with experimental results are provided by Yoshida *et al.* (2005). Significant effects are found for an array of four cylinders.

Duration and exploitation of the simulations

The choice of the time stepping scheme presents no difficulty and standard methods such as Runge–Kutta or Predictor-Corrector or Newmark can be used. As written earlier, usually only the slow-drift components are integrated in time, with the wave frequency response superimposed at every time step.

The duration of the simulations must be long enough to allow a statistical exploitation. It can be shown (Pinkster & Wichers, 1987) that the expected relative value of the standard deviation of the variance of the response strongly depends on the damping ratio, and is given by

$$\sigma'_{\sigma^2} = \frac{1}{\sqrt{2\pi \, \zeta \, N_C}} \tag{7.180}$$

with ζ the damping ratio and N_C the number of low-frequency cycles.

This equation means that, if one desires for example an accuracy of 20% on the variance (i.e., 10% on the standard deviation), and that the damping ratio is 5%, 80 cycles are required. Many more are necessary to obtain a representative distribution of maxima. (It can be noted that the same problem arises in model testing.)

It is customary to perform a large number of simulations, each one 3 full-scale hours long, varying the wave history.

To derive design values, several strategies are possible.

The first one consists in fitting the distribution of the tension peaks with a statistical law, theoretical or empirical, and deducing the extreme value associated with the target probability level. It is important to realize that the second-order forces, quadratic with respect to the wave signal, are not Gaussian processes. When the Newman approximation in geometric mean is applied, one easily establishes that they follow the exponential law:

$$p(F_{-j}^{(2)}) = \frac{1}{F_{dj}} \, e^{-F_{-j}^{(2)}/F_{dj}} \qquad\qquad F_{-j}^{(2)} \geq 0 \tag{7.181}$$

with F_{dj} the mean drift force:

$$F_{dj} = 2 \int_0^\infty S(\omega) \, f_{dj}(\omega) \, d\omega \tag{7.182}$$

When the mechanical system governing the response is linear, it can be shown that the response becomes Gaussian asymptotically again when the damping ratio goes to zero, therefore that the maxima follow Rayleigh's law. In the case of a nonlinear damping there are theoretical models (for a state-of-the-art survey see Alliot, 1990).

The hypothesis of a linear mechanical system is rarely verified. It has been seen that the viscous forces on the hull are rather quadratic. The restoring laws of the catenary mooring systems are generally quite nonlinear, especially in the vicinity of their operating points in extreme conditions. And there remains the problem of how to combine

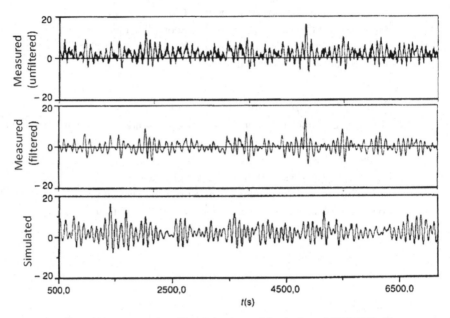

Figure 7.39 Slow-drift sway motion of a semi-submersible platform (NEKTON). Top: measured response during model tests. Middle: same signal, low-pass filtered. Bottom: simulated sway motion (with a different wave history).

the low-frequency response and the wave frequency response. In these cases there is hardly any other solution than to exploit directly the maxima obtained during the numerical simulations, in order to calibrate a predefined law of the Weibull type or other.

Figure 7.39 shows an example of comparison between experimental and numerical slow-drift motion.

7.9   Low-Frequency Motions in Heave, Roll, and Pitch

Semi-submersible platform are usually designed in such a way that their natural periods in heave, roll, and pitch be above the energetic wave periods. Typical values are 25 s to 30 s for heave and 30 s to 50 s for roll and pitch. At such periods the first-order wave loads are nil. Excitation results from the second-order wave loads (and from the time-varying wind loads).

Similar situations are encountered with Spar platforms, with submarines at low immersion (see Figure 7.40), with gravity platforms in towing and installation phases, etc.

Compared to the case of horizontal low-frequency motions, an important difference is that the natural periods are not low enough to justify the use of Newman approximations in the derivation of the second-order loads. One should calculate the complete QTF matrix. In the case of not too massive structures, the use of approximations based, for instance, on the Rainey equations can be an alternative.

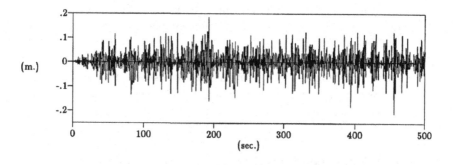

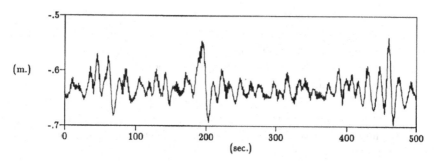

Figure 7.40 Numerical simulation of the heave motion of a submarine under waves (model test scale). Top: wave elevation; bottom: heave response (from Boudet, 1992).

When the excitation loads have been obtained, the next issue, as usual with resonant systems, is the damping mechanisms.

With regard to wave drift damping, it does not exist in heave. On the other hand, the second-order mean wave loads are sensitive to immersion. This results in a change in vertical stiffness, generally a decrease: The structure is all the more attracted to the free surface the closer it is.

The case of roll and pitch is more complex, since they result from local loads both vertical and horizontal. As with heave, the sensitivity of the mean second-order loads to immersion can result in a decrease of the stiffnesses. This phenomenon has often been observed for large semi-submersible platforms: Given an initial inclination in waves, the raised pontoon is more attracted toward the free surface than the lowered pontoon, hence possibly a loss of stability if the metacentric height is insufficient. On the other hand, the horizontal loads that apply near the free surface are undoubtedly sensitive to a local horizontal velocity, induced by the slow rotations, hence a possible wave drift damping effect. This aspect of the problem has apparently never been explored.

Concerning mooring line damping, there is a lack of experimental or numerical references. The effect should be weaker than in the horizontal case, since the catenary shape of the line is usually less sensitive to a vertical displacement of the fairlead than to a horizontal displacement.

Most of the damping then results from viscous effects, flow separation at the pontoons. The difficulties in properly formulating the associated loads are the same as in the horizontal case.

Second-order roll resonance has been observed for some FPSOs with very low metacentric heights, and roll natural frequencies lower than the wave frequencies. See, for instance, Rezende & Chen (2007). It can be observed that such low-frequency roll responses may also be due to parametric instabilities (see Chapter 8).

7.10   High-Frequency Second-Order Response (Springing)

In ship hydrodynamics **springing** is defined as a continual hull girder vibration due to wave loading. Springing may occur under forward speed, when the wave encounter frequencies meet the resonant hull frequencies. Second-order wave excitation has also been advocated. More transient vibratory responses, usually due to bow slamming, are known as **whipping** or **ringing**.

In offshore hydrodynamics, by definition, springing is a resonant response due to sum frequency second-order wave excitation. Other types of vibratory responses, due to higher-order wave loads, or to wave impacts, are defined as ringing. Ringing is considered in Chapter 8.

Springing is a resonant behavior specific to TLPs, which typically have natural periods in heave, roll, and pitch in the range 2 s to 4 s. Springing is therefore likely to occur in sea states with peak periods from 4 s to 8 s. Even though there is relatively little energy at such peak periods, these are relatively frequent sea states and springing must be taken into account to assess the fatigue life of the tethers.

The procedure to follow is similar to that applied to the slow-drift behavior.

The first step is the determination of the QTFs. As already explained, there is a significant difference between sum frequency QTFs and difference frequency QTFs, which is that, for the former, the contribution of the second-order diffraction potential (at frequency $\omega_i + \omega_j$) cannot be neglected. It has been seen in Section 7.3 that, for a vertical cylinder, it is only at very low wave numbers ($ka < 0.1$) that approximations such as Rainey's equations can be justified. The $f_+^{(2)}$ QTFs must be computed for all couples (ω_i, ω_j) such that $\omega_i + \omega_j$ be equal to a resonant frequency in heave, roll, or pitch. This calculation supposes very fine meshes of the hull and of the free surface, which entail fairly dissuasive calculation costs. Some diffraction-radiation codes offer this capability.

The QTFs once calculated, and the wave spectrum given, one can deduce the spectrum of the second-order loads $S_{F_+^{(2)}}(\Omega)$ by applying equation (7.98), and then move on to calculating the response, in the frequency domain. Once again this raises the problem of identifying and estimating the damping mechanisms.

As in the case of low-frequency response, the damping has various origins: radiation damping, viscous damping due to friction and separation at sharp edges, structural damping in the tendons, in the foundations, dissipation in the soil, etc. Each component amounting to a few per thousand of the critical damping!

For TLPs, the main contribution to the hydrodynamic damping comes from flow separation at the base of the columns and at the pontoons, provided that their edges be sufficiently sharp. The amplitude of the resonant motion, compared to the diameter of the columns, is extremely low (a few centimeters to compare with 20 or 30 m!). Model tests, and theoretical or numerical investigations, have shown that the damping force due to separation is no longer quadratic, but linear, with the oscillatory velocity, the reason for this behavior being the wake effects (see, for example, Huse & Muren, 1987; Huse, 1990, and Huse & Utnes, 1994). The viscous damping is also very sensitive to the superposition of a current.

Note that, unlike in the case of slow-drift motion, there is no equivalent here of the wave drift damping, simply because the initial hypotheses for the development in series of perturbation are (for once) well verified: the second-order response is actually small compared to the first-order one!

Finally, there remains the nontrivial problem of determining the statistical distribution of the maxima of the high-frequency motion. For TLPs there is also the problem of how to combine the linear and second-order responses to derive the peak loads in the tendons, and calculate the fatigue life. The reader is referred to the specialized literature (Naess, 1992; Winterstein *et al.*, 1994; see also Naess & Moan, 2013).

7.11   References

ALLIOT J-M. 1990. Statistique des forces et des mouvements de dérive lente des structures flottantes amarrées. Etude bibliographique, IFP report 38 076 (in French).

ARANHA J.A.P. 1991. Wave groups and slow motion of an ocean structure, in *Proc. 6th Int. Workshop Water Waves & Floating Bodies*, 5–8 (www.iwwwfb.org).

ARANHA J.A.P. 1994. A formula for 'wave damping' in the drift of a floating body, *J. Fluid Mech.*, **275**, 147–155.

ARANHA J.A.P. 1996. Second-order horizontal steady forces and moment on a floating body with small forward speed, *J. Fluid Mech.*, **313**, 39–54.

BATTJES J.A., BAKKENES H.J., JANSSEN T.T., VAN DONGEREN A.R. 2004. Shoaling of subharmonic gravity waves, *J. Geophys. Res.*, **109**, C02009.

BEARMAN P.W., DOWNIE M.J., GRAHAM J.M.R., OBASAJU E.D. 1985. Forces on cylinders in viscous oscillatory flow at low Keulegan-Carpenter numbers, *J. Fluid Mech.*, **154**, 337–356.

BOUDET L. 1992. Etude numérique et expérimentale du comportement d'un submersible sous houle irrégulière : analyse de la réponse en pilonnement à basse fréquence, PhD thesis, Paris VI University (in French).

CHEN X.B. 1988. Etude des réponses du second-ordre d'une structure soumise à une houle aléatoire, PhD thesis, Nantes University (in French).

CHEN X.B. 1994. Approximation on the quadratic transfer function of low frequency loads, in *Proc. 7th Int. Conf. Behaviour of Offshore Structures, BOSS'94*, Vol. 2, 289–302, C. Chryssostomidis Editor, Pergamon.

CHEN X.B. 2007. Middle-field formulation for the computation of wave-drift loads, *J. Eng. Math.*, **59**, 61–82.

Chen X.B., Rezende F., Malenica Š., Fournier J.R. 2007. Advanced hydrodynamic analysis of LNG terminals, in *Proc. 10th International Symposium on Practical Design of Ships and Other Floating Structures*. PRADS2007, Houston.

Clark P.J., Malenica Š., Molin B. 1993. An heuristic approach to wave drift damping, *Applied Ocean Research*, **15**, 53–55.

Dev A.K. 1996. Viscous effects in drift forces on semi-submersibles, PhD thesis, TU Delft.

Eatock Taylor R., Hung S.M. 1987. Second order diffraction forces on a vertical cylinder in regular waves, *Applied Ocean Res.*, **9**, 19–30.

Emmerhoff O.J., Sclavounos P.D. 1992. The slow drift motion of arrays of vertical cylinders, *J. Fluid Mech.*, **242**, 31–50.

Finne S., Grue J. 1998. On the complete radiation-diffraction problem and wave-drift damping of marine bodies in the yaw mode of motion, *J. Fluid Mech.*, **357**, 289–320.

Huse E. 1977. Wave induced mean forces on platforms in direction opposite to wave propagation, *Norwegian Maritime Research*.

Huse E. 1990. Resonant heave damping of tension leg platforms, in *Proc. 22nd Offshore Technology Conf.*, paper 6317.

Huse E., Matsumoto K. 1989. Mooring line damping due to first- and second-order vessel motion, Marintek report 511151.00-1.

Huse E., Muren P. 1987. Drag in oscillatory flow interpreted from wake considerations, in *Proc. 19th Offshore Technology Conf.*, paper 5370.

Huse E., Utnes T. 1994. Springing damping of tension leg platforms, in *Proc. 26th Offshore Technology Conf.*, paper 7446.

Jonsson I.G. 1966. Wave boundary layers and friction factors, in *Proc. 10th Conf. Coastal Eng.*, Vol. 1, 127–148, Tokyo.

Justesen P. 1988. Prediction of turbulent oscillatory flow over rough beds, *Coastal Eng.*, **12**, 257–284.

Kinoshita T., Bao W., Yoshida M., Ishibashi K. 2002. Wave drift added mass of a cylinder array free to respond to the incident waves, in *Proc. 19th Offshore Mechanics and Arctic Eng. Conf.*, OMAE paper 28442.

Ledoux A., Molin B., Delhommeau G., Remy F. 2006. A Lagally formulation of the wave drift force, in *Proc. 21st Int. Workshop Water Waves & Floating Bodies*, Loughborough (www.iwwwfb.org).

Liu Y.N., Molin B., Kimmoun O., Remy F., Rouault M.-C. 2011. Experimental and numerical study of the effect of variable bathymetry on the slow-drift wave response of floating bodies, *Appl. Ocean Res.*, **33**, 199–207.

Madsen O.S. 1986. Hydrodynamic forces on circular cylinders, *Appl. Ocean Res.*, **8**, 151–155.

Malenica Š., Clark P.J., Molin B. 1995. Wave and current forces on a vertical cylinder free to surge and sway, *Appl. Ocean Res.*, **17**, 79–90.

Malenica Š., Chen X-B. 1997. L'influence d'un courant sur les surélévations de la surface libre autour d'un corps flottant, in *Actes des 6èmes Journées de l'Hydrodynamique*, 177–188 (in French; http://website.ec-nantes.fr/actesjh/).

Martigny D., Molin B., Scolan Y-M. 1994. Slow-drift viscous damping of TLP pontoons, in *Proc. 13th Offshore Mechanics & Arctic Eng. Conf., OMAE*, Houston.

Maruo H. 1960. The drift of a body floating in waves, *J. Ship Research*, **4**, 1–10.

MATSUI T., YEOB L.S., KIMITOSHI S. 1991. Hydrodynamic forces on a vertical cylinder in current and waves, *J. Soc. Nav. Arch. Japan*, **170**, 277–287.

MEI C.C., BENMOUSSA C. 1984. Long period oscillations induced by short-wave groups over an uneven bottom, *J. Fluid Mech.*, **139**, 219–235.

MOE G. (Ed.) 1993. Vertical Resonant Motions of TLP's (VRMTLP), Final Report. NTH report R-1-93.

MOLIN B. 1979. Second-order diffraction loads upon three-dimensional bodies, *Applied Ocean Res.*, **1**, 197–202.

MOLIN B. 1981. Un modèle de comportement des navires amarrés sur point unique, PhD thesis, ENSM Nantes (in French).

MOLIN B. 1994. Second-order hydrodynamics applied to moored structures. A state-of-the-art survey, *Ship Technol. Res.*, **41**, 59–84.

MOLIN B., BUREAU G. 1980. A simulation model for the dynamic behavior of tankers moored to single point moorings, in *Proc. International Symposium on Ocean Engineering Ship Handling*, SSPA, Gothenburg.

MOLIN B., CHEN X.B. 2002. Approximations of the low frequency second-order wave loads: Newman vs Rainey, in *Proc. 17th Int. Workshop Water Waves & Floating Bodies*, Cambridge (www.iwwwfb.org).

MOLIN B., FAUVEAU V. 1984. Effect of wave directionality on second-order loads induced by the set-down, *Applied Ocean Res.*, **6**, 66–72.

MOLIN B., HAIRAULT J.-P. 1983. On second-order motion and vertical drift forces for three-dimensional bodies in regular waves, in *Proc. Int. Workshop on Ship and Platform Motions*, Berkeley.

NAESS, A. 1992. Prediction of extremes related to the second-order sum-frequency response of a TLP, in *Proc. 2nd Intl. Offshore Polar Eng. Conf., ISOPE*.

NAESS A., MOAN T. 2013. *Stochastic Dynamics of Marine Structures*. Cambridge University Press.

NEWMAN J.N. 1967. The drift force and moment on ships in waves, *J. Ship Research*, **11**, 51–60.

NEWMAN J.N. 1974. Second order, slowly varying forces on vessels in irregular waves, in *Proc. Int. Symp. Dynamics Marine Vehicles & Structures*, 182–186.

NEWMAN J.N. 1990. Second-harmonic wave diffraction at large depth, *J. Fluid Mech.*, **213**, 59–70.

NORRBIN N.H. 1970. Theory and observation on the use of a mathematical model for ship maneuvering in deep and confined waters, in *Proc. 18th Intern. Symposium on Naval Hydrodynamics*, Pasadena.

OGILVIE T.F. 1963. First- and second-order forces on a cylinder submerged under a free surface, *J. Fluid Mech.*, **16**, 451–472.

OIL COMPANIES INTERNATIONAL MARINE FORUM 1977. *Prediction of wind and current loads on VLCCs*, Witherby & Co. Ed.

PALO P.A. 1983. Steady wind- and current-induced loads on moored vessels, in *Proc. 15th Offshore Technology Conf.*, paper 4530.

PALO P.A. 1986. Full-scale vessel current loads data and the impact on design methodologies and similitude, in *Proc. 18th Offshore Technology Conf.*, paper 5205.

PINKSTER J.A. 1975. low frequency phenomena associated with vessels moored at sea, *Society Petroleum Engineers Journal*, December, 487–494.

PINKSTER J.A. 1980. Low frequency second order wave exciting forces on floating structures, Ph.D. Dissertation, Delft.

PINKSTER J.A., VAN OORTMERSSEN G. 1977. Computation of the first- and second-order wave forces on oscillating bodies in regular waves, in *Proc. 2nd Int. Conf. Numerical Ship Hydrodynamics*, 136–156.

PINKSTER J.A., WICHERS J.E.W. 1987. The statistical properties of low frequency motions of nonlinearly moored tankers, in *Proc. 19th Offshore Technology Conf.*, paper 5457.

RAINEY R.C.T. 1989. A new equation for calculating wave loads on offshore structures, *J. Fluid Mech.*, **204**, 295–324.

RAINEY R.C.T. 1995. Slender-body expressions for the wave load on offshore structures, *Proc. R. Soc. Lond. A*, **450**, 391–416.

REZENDE F.C., CHEN X.B. 2007. Second-order roll motions for FPSOs operating in severes environmental conditions, in *Proc. 2007 Offshore Technology Conference*, Houston, OTC paper 18906.

ROUAULT M.-C., MOLIN B., LEDOUX A. 2007. A Lagally formulation of the second-order slowly-varying drift forces, in *Actes des 11èmes Journées de l'Hydrodynamique*, Brest (http://website.ec-nantes.fr/actesjh/).

SCHÄFFER H.A. 1993. Infragravity waves induced by short-wave groups, *J. Fluid. Mech.*, **247**, 551–588.

SCOLAN Y-M. 1989. Contribution à l'étude des non-linéarités de surface libre. Deux cas d'application : clapotis dans un bassin rectangulaire ; diffraction au second ordre sur un groupe de cylindres verticaux, PhD thesis, Paris VI University (in French).

SHAO Y.L., FALTINSEN O.M. 2010. Use of body-fixed coordinate system in analysis of weakly nonlinear wave–body problems, *Applied Ocean Res.*, **32**, 20–33.

Society of International Gas Tanker & Terminal Operators Ltd. 2007. Prediction of wind loads on large liquefied gas carriers.

SOULSBY R.L., HAMM L., KLOPMAN G., MYRHAUG D., SIMONS R.R., THOMAS G.P. 1993. Wave-current interaction within and outside the bottom boundary layer, *Coastal Engineering*, **21**, 41–69.

STOKES G.G. 1851. On the effect of the internal friction of fluids on the motion of pendulums, *Transaction of the Cambridge Philosophical Society*, **9**, 8–106.

VAN DONGEREN A.P., VAN NOORLOOS J., STEENHAUER K., BATTJES J., JANSSEN T., RENIERS A. 2004. Shoaling and shoreline dissipation of subharmonic waves, in *Proc. 29th Intl. Conf. Coastal Engng*, 1225–1237.

VAN NOORLOOS J.C. 2003. Energy transfer between short wave groups and bound long waves on a plane slope, M.Sc. thesis, TU Delft.

WEHAUSEN J.V., LAITONE E.V. 1960. Surface waves, in *Handbuch der Physik*, **9**, Springer Verlag.

WICHERS J.E.W. 1988. A simulation model for a single point moored tanker, Publication No. 797, Maritime Research Institute Netherlands (MARIN).

WICHERS J.E.W., VAN SLUIJS M.F. 1979. The influence of waves on the low frequency hydrodynamic coefficients of moored vessels, in *Proc. 11th Offshore Technology Conf.*, paper 3625.

WINTERSTEIN S.R., UDE T.C., KLEIVEN G. 1994. Springing and slow-drift responses: predicted extremes and fatigue vs. simulation, in *Proc. 7th Intl. Conf. on the Behaviour of Offshore Structures, BOSS'94*, Vol. 3, 1–15.

YOSHIDA M., KINOSHITA T., BAO W. 2005. Nonlinear hydrodynamic forces on an accelerated body in waves, *J. Offshore Mech. Arc. Eng.*, **127**, 17–30.

8 Large Bodies: Other Nonlinear Effects

8.1 Higher-Order Diffraction Loads

Interest in these loads is historically associated with the TLP concept, and with some deep water GBS platforms in the North Sea. Recently, there has been renewed attention associated with the wave loads upon the masts of offshore wind turbines.

In Chapter 7, it was mentioned that TLPs are sensitive to sum frequency second-order loads (so-called **springing** loads), which may induce resonant responses in heave, roll, and pitch. Springing is of concern for the fatigue life of the tethers. The natural periods in heave, roll, and pitch are typically in the range 2–4 s; it is in sea states with peak periods of about twice that the springing behavior occurs.

Model tests on the Heidrun TLP, in North Sea design sea states, that is with peak periods around 15 s, revealed high-frequency fluctuating loads in the tethers, at about 4 to 5 times the wave frequencies. This suggests nonlinearities much higher than second-order coming into play. At variance with springing, which consists in a more or less steady response, this behavior, new at the time, baptized **ringing**,[1] appears sporadically under very high or very steep waves, not necessarily breaking nor impacting. Similar phenomena were observed in model tests on the deep water Draugen GBS.

As an illustration, we present, in Figures 8.1 and 8.2, some time series from model tests on a simplified GBS platform, obtained within the frame of a research project dedicated to springing and ringing (Scolan *et al.*, 1997). The experimental model has been shown in Figure 1.19: it consists in a vertical cylinder, 46 cm in diameter, linked via a steel blade to a rigid and heavy base, lying on the basin floor. The water depth is 2.9 m. The natural period of the bending mode can be adjusted by varying the blade stiffness and/or by loading the cylinder with weights. An external device (consisting in a perforated plate in a bucket filled with water) can be used to vary the damping of the system, the initial damping being very low (0.25% of critical damping).

Figure 8.1 (top) shows the time trace of the free surface elevation at the cylinder location during calibration tests. The signal input into the wavemaker results in a group of high and steep waves, as would occur in a design sea state. The highest crest to trough value is slightly above 35 cm, the mean wave period is about 2 s. At 1:60 scale, they would be 21 m and 15.5 s, the cylinder diameter being 27.6 m.

---

[1] The coinages springing and ringing have long been used to describe the elastic response of ship hulls, under wave loads and slamming.

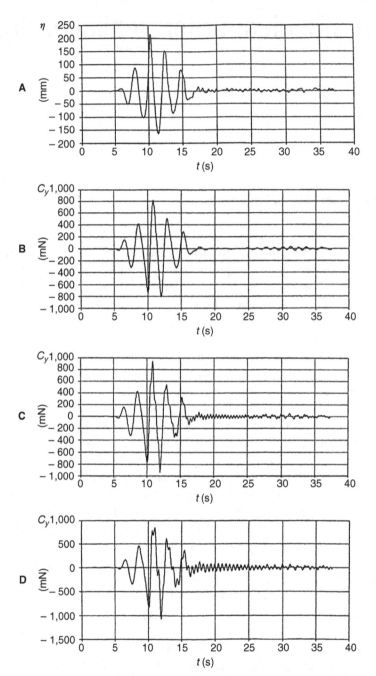

Figure 8.1 Model tests on a vibrating cylinder. Free surface elevation (A). Bending moment at foot for natural periods 0.23 s (B), 0.44 s (C), and 0.56 s (D).

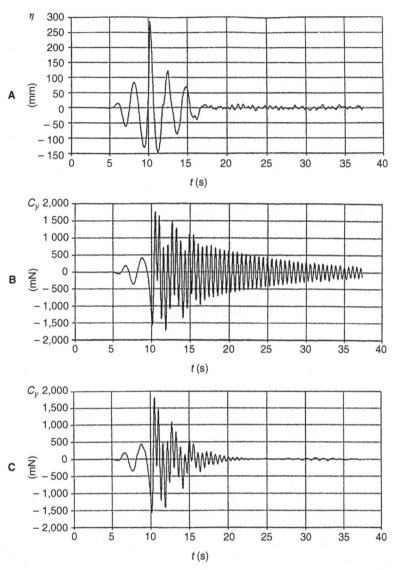

Figure 8.2 Model tests on a vibrating cylinder. Free surface elevation (A). Bending moment at foot for natural periods 0.56 s and damping ratios 0.7% (B), and 3.2% (C).

The other 3 plots in Figure 8.1 show the bending moment in the flexible blade during the passage of the wave group, for 3 different values of the natural period of the bending mode: 0.23 s, 0.44 s, and 0.56 s, which would be 1.8 s, 3.4 s, and 4.3 s at full scale. In the stiffest case no dynamic response can be observed, the system can be considered as perfectly rigid. In the following plots, a high-frequency component appears, superimposed onto the quasi-linear response. Depending upon the phase shift between the two components, a decrease or an increase of the peak value results.

Figure 8.2 shows another wave packet, and the measured bending moments, at the same natural period of 0.56 s, for two values of the damping ratio: 0.7% and 3.2%. In both cases the very first peaks are the same: Contrary to springing, damping is not a critical parameter in ringing.

In these tests, as in those carried out on the Heidrun model, the Keulegan–Carpenter numbers are sufficiently low and the Stokes parameter (β) is sufficiently high to ascertain that the flow remains attached to the cylinder wall: Viscous effects do not come into play. Even though the waves produced are rather steep, ringing events are observed in nonbreaking waves, suggesting that nonlinear diffraction effects are at hand. The order of nonlinearity is presumably equal to the ratio between the ringing and the wave frequencies, around 4 or 5. Fourth- or fifth-order diffraction theory seems to be out of reach; however, some work has been done on third-order diffraction theory, briefly summarized in the following.

8.1.1   Third-Order Diffraction Theory

As at second-order, the starting point is the Stokes expansion, now pursued to third-order. The velocity potential is developed as

$$\Phi = \Phi^{(1)} + \Phi^{(2)} + \Phi^{(3)} + \dots \tag{8.1}$$

where $\Phi^{(1)}$ is of order ε, $\Phi^{(2)}$ of order ε^2, $\Phi^{(3)}$ of order ε^3, the "small" parameter ε being identified with the wave steepness kA.

In a regular wave of frequency ω, the successive potentials $\Phi^{(1)}$, $\Phi^{(2)}$, and $\Phi^{(3)}$ have the following time dependences:

$$\Phi^{(1)} = \mathcal{R}\left\{\varphi^{(1)} e^{-i\omega t}\right\} \qquad \Phi^{(2)} = \varphi_0^{(2)} + \mathcal{R}\left\{\varphi^{(2)} e^{-2i\omega t}\right\} \tag{8.2}$$

$$\Phi^{(3)} = \mathcal{R}\left\{\varphi_1^{(3)} e^{-i\omega t}\right\} + \mathcal{R}\left\{\varphi^{(3)} e^{-3i\omega t}\right\} \tag{8.3}$$

The same time dependence follows for the load components $\vec{F}^{(1)}$, $\vec{F}^{(2)}$, and $\vec{F}^{(3)}$. When one is concerned only with the loads taking place at the frequency 3ω the potentials $\varphi_0^{(2)}$ and $\varphi_1^{(3)}$ can be discarded. The load component at the triple frequency $\vec{F}^{(3)} = \mathcal{R}\{\vec{f}^{(3)} \exp(-3i\omega t)\}$ divides into 3 terms:

$$\vec{f}^{(3)} = \vec{f}_1^{(3)} + \vec{f}_2^{(3)} + \vec{f}_3^{(3)} \tag{8.4}$$

where

$$\vec{f}_1^{(3)} = -\frac{i\omega}{8g} \int_{\Gamma_0} \left[\varphi^{(1)} \nabla\varphi^{(1)} \cdot \nabla\varphi^{(1)} + v^2 \left(\varphi^{(1)}\right)^3\right] \vec{n_0} \, d\Gamma \tag{8.5}$$

$$\vec{f}_2^{(3)} = -\frac{1}{2}\rho \iint_{S_{B_0}} \nabla\varphi^{(1)} \cdot \nabla\varphi^{(2)} \, \vec{n_0} \, dS - \rho v \int_{\Gamma_0} \varphi^{(1)} \, \varphi^{(2)} \, \vec{n_0} \, d\Gamma \tag{8.6}$$

$$\vec{f}_3^{(3)} = 3i\omega \rho \iint_{S_{B_0}} \varphi^{(3)} \, \vec{n_0} \, dS \tag{8.7}$$

with $v = \omega^2/g$. As in the previous chapters S_{B_0} is the mean wetted hull and Γ_0 is the mean waterline. Here the body is assumed to be fixed and wall-sided at the waterline.

In the particular case of a vertical cylinder, standing on the sea floor, the first-order potential $\varphi^{(1)}$ is known analytically (see Chapter 6). The second-order potential $\varphi^{(2)}$ can also be derived semi-analytically in the whole fluid domain, in particular, at the free surface (e.g., see Chau & Eatock Taylor, 1992, or Malenica & Molin, 1995). By applying Haskind–Hanaoka theorem in the same way as was done in Chapter 7 for the second-order diffraction problem, it is possible to evaluate the $\overrightarrow{f_3}^{(3)}$ component, through an integral over the whole free surface. This integral involves the first- and second-order potentials $\varphi^{(1)}$, $\varphi^{(2)}$, and the linear radiation potential at the frequency 3ω.

Numerical results obtained by Malenica & Molin (1995) (see also Malenica, 1994) are reproduced in Figure 8.3, which shows the real and imaginary parts of the different contributions to the third-order wave load in surge, in the particular case $kh = 8$. They are plotted for $ka \le 0.25$ (with a the cylinder radius), which is the range of practical interest for ringing. In this deep water case the second- and third-order incident potentials are nil (the third-order effect is just a modification of the wave number, see Chapter 3). It can be seen that the contribution $f_3^{(3)}$ from the third-order diffraction potential $\varphi_D^{(3)}$ cannot be neglected when ka is larger than about 0.05. This contribution is associated with the emission, from the cylinder, of waves at 3 times the frequency of the incoming waves, therefore of wave lengths 9 times shorter.

Other numerical results can be found in Teng & Kato (2002) who solve the first- and second-order problems numerically, and derive the third-order loads following the same variant of the Haskind–Hanaoka theorem. On the cylinder case they confirm Malenica's results. They give results for the heave force and pitch moment, and also consider the cases of a truncated cylinder and a hemisphere.

In both Malenica & Molin (1995) and Teng & Kato (2002) comparisons are shown with experimental values of the third-order loads, with reasonably good agreement.

8.1.2   Numerical Wave Tanks

With the progress of computer resources, it has become possible to solve, in time and space, the Navier–Stokes or Euler equations in a finite domain around the considered body, with appropriate conditions at the boundaries of the numerical domain (to let the waves get in and get out). Nowadays, many researchers use commercial, or open source, CFD solvers and an extensive academic community has developed around OpenFOAM.

Historically the first fully nonlinear wave tanks relied on potential flow theory and were restricted to the two-dimensional case of a vertical strip, the waves being generated at one end (usually with a numerical wavemaker) and absorbed at the other end (usually through dissipative terms in the free surface boundary conditions). Early developments were based on the so-called MEL (*Mixed Euler-Lagrange*) method introduced by Longuet-Higgins & Cokelet (1976) to simulate overturning waves. In

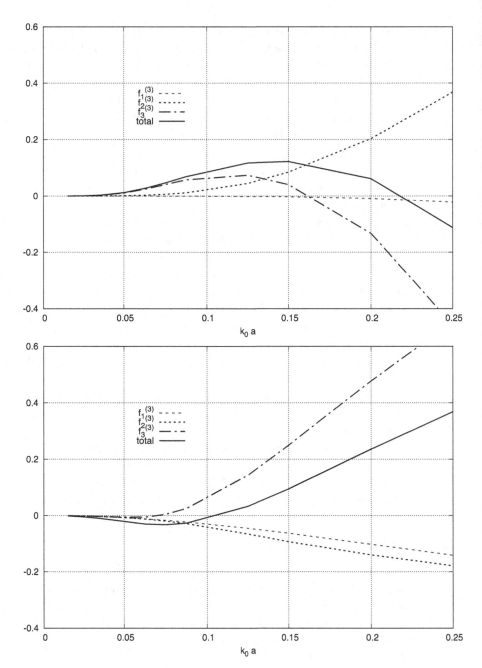

Figure 8.3 Bottom-mounted vertical cylinder. Different components to the third-order horizontal load (normalized by $\rho g A^3$). Real (top) and imaginary (bottom) parts. $kh = 8$ (from Malenica & Molin, 1995) (Cambridge University Press, reprinted with permission).

this method markers are distributed on the free surface, and the free surface conditions can then be written in the form

$$\frac{dX_i}{dt} = u(X_i, Z_i, t) \qquad \frac{dZ_i}{dt} = w(X_i, Z_i, t) \qquad \frac{d\Phi_i}{dt} = -g\,z + \frac{1}{2}\,(\nabla\Phi_i)^2 \quad (8.8)$$

with $(u, w) = \nabla\Phi_i$, Φ_i being the velocity potential associated with the marker i in (X_i, Z_i).

At each time step, the free surface elevation, and the velocity potential there, being known, the Laplace equation is solved (through the integral equation method) to derive the normal derivative of the potential at the free surface (so-called Dirichlet to Neumann problem). The tangential component being estimated through finite differentiation, the equations (8.8) can be integrated to the next time step.

Following the same method, some three-dimensional wave tanks were developed, one early application being wave interaction with ships at forward speed (e.g., see Nakos, 1990). Ferrant (1998) tackled the vertical cylinder case with his code XWAVE. Figure 8.4 shows the wave field around the cylinder at one instant, in the case $ka = 0.2$, $kA = 0.12$, $kh = 8$. The elevations of the diffracted waves are exaggerated three-fold compared to the incoming wave. Short waves emanating from the cylinder, at the double and triple wave frequencies, are clearly visible.

After a steady state has been reached, harmonics can be extracted through Fourier analysis, and compared with second- and third-order theories. Figure 8.5 shows the third-harmonic load components derived from XWAVE simulations, compared to the values obtained by Malenica (1994), in the case $kh = 8$. In the XWAVE computations the wave steepness has been taken very low, that is, $kA = 0.06$ (about 15% of limiting steepness). It can be seen that the agreement is excellent over the whole range of ka values. The two numerical models are mutually validated.

When the incoming wave amplitude is gradually increased in the numerical wave tank, the sensitivity to the wave steepness of the double and triple frequency loads can be investigated. This was done by Ferrant (1998) who, in the case $ka = 0.245$, $kh \simeq 10$, shows the evolution of the different harmonics up to $kA = 0.2$. Comparisons are shown with harmonics extracted from small-scale model tests by Huseby & Grue (2000). A slight decay of the normalized triple load component $f^{(3)}/(\rho g A^3)$ with the steepness is observed (about 25% over the kA range), in good agreement with the experimental values.

A drawback of the MEL method is that it cannot cope with wave breaking. The simulation stops when the plunging waves reenter the free surface. A remedy (used in XWAVE) is to assign the free surface markers to move only vertically. Overturning is then ruled out but the physics are biased.[2]

With the markers moving only vertically, the free surface conditions appear in the so-called Zakharov form (from Zakharov, 1968):

$$\eta_t + \nabla\eta \cdot \nabla\widetilde{\phi} - \widetilde{w}\,(1 + \nabla\eta \cdot \nabla\eta) = 0 \qquad (8.9)$$

---

[2] Some tricks can be applied where, from a breaking criterion, some localized (in time and space) pressure is applied at the free surface to dissipate the proper amount of energy.

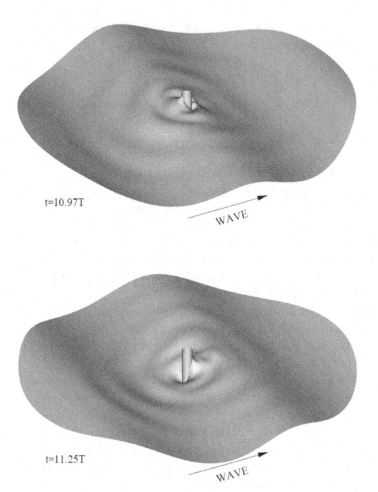

t=10.97T

WAVE

t=11.25T

WAVE

Figure 8.4 Nonlinear wave interaction with a vertical cylinder (from Ferrant *et al.*, 1999) (Courtesy of WIT Press from the book [Advances in Fluid Mechanics, Nonlinear Water Wave Interaction]).

$$\widetilde{\phi}_t + g\,\eta + \frac{1}{2}\left(\nabla\widetilde{\phi}\right)^2 - \frac{1}{2}\,\widetilde{w}^2\left(1 + \nabla\eta \cdot \nabla\eta\right) = 0 \qquad (8.10)$$

where $\nabla = [\partial/\partial x, [\partial/\partial y]$ is the horizontal gradient, and $\widetilde{\phi}(x,y,\eta,t)$, $\widetilde{w}(x,y,\eta,t)$ are the velocity potential and vertical velocity at the free surface.

This is the form used by most people dealing with nonlinear wave propagation and nonlinear wave–body interaction (with the restriction that the body must be fixed and wall-sided at the waterline). To advance η and $\widetilde{\phi}$ in time, at each time step a Dirichlet to Neumann problem must be solved to obtain the vertical velocity $\widetilde{w}$ at the free surface. Besides the integral equation method, other routes can be followed such as the HOS method (for simple geometries) or the so-called extended Boussinesq models (e.g., see Bingham *et al.*, 2009), which rely on polynomial approximations of the potential and vertical velocity over the depth.

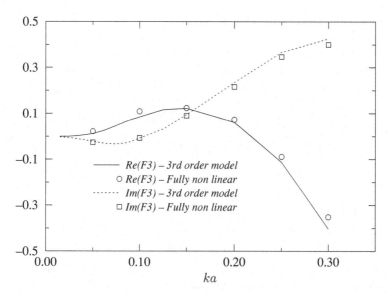

Figure 8.5 Triple frequency wave load on a vertical cylinder. Comparison between asymptotic and numerical values (from Ferrant *et al.*, 1999) (Courtesy of WIT Press from the book [Advances in Fluid Mechanics, Nonlinear Water Wave Interaction]).

8.1.3   Simplified Formulations of the Ringing Loads

Third-order diffraction loads have so far been obtained only for simple axisymmetric geometries in regular waves. Their worth is mostly to serve as a reference for simplified formulations that can be applied in more realistic sea states, in focused seas or in irregular waves.

Simplified theoretical models have been proposed, usually based on the assumptions of small perturbation of the incoming wavefield and quasi-rigid free surface. These models do not account for the diffracted waves at super- and sub-harmonics. It has been seen in Chapter 6 that the inertia term in the Morison equation provides a good approximation to the diffraction loads when ka (k the wave number and a the radius) is lower than 0.5. Likewise, at second-order, Rainey's equations (see Section 7.5.1) provide a good estimate when ka is less than 0.10–0.15. In deep water the wave number k_2 of the second-order free waves being $4k$, this means $k_2a \leq 0.5$, which sounds consistent. Faltinsen, Newman, and Vinje (1995) have extended Rainey's equations to third-order, still neglecting free surface effects. Comparisons between their formulation and exact results from the third-order diffraction theory of Malenica & Molin show that the domain of validity of the FNV theory is similarly restricted to $ka \leq 0.05$, or $k_3a \leq 0.5$, k_3 being the wave number of the third-order free waves at frequency 3ω (equal to $9k$ in deep water).

By chance the Rainey and FNV formulations provide a reasonably good order of magnitude of the second- and third-order loads. It is the phase angles that are wrong (by close to 180 degrees). This phase information is of primary importance

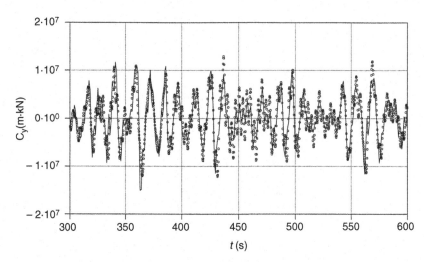

Figure 8.6 Bending moment at the base of the Clarom cylinder in irregular waves, from experiments (symbols) and computations (from Scolan *et al.*, 1997).

when one needs to add up linear and nonlinear load components. Unfortunately, at the present time, there is no alternative formulation. Rainey and FNV equations offer the advantage that they are applicable in irregular seas, where the additional difficulty is to formulate the incident kinematics up to the instantaneous free surface. This brings us back to the problems addressed in Chapter 3 on wave crest kinematics, leading to the stretching models. Numerical tests, reported in Scolan *et al.* (1997), show that, in the end, the kinematic model matters as much as the load model. Figure 8.6, taken from that paper, shows the best fit obtained between measured and calculated bending moments, in a steep irregular wave case: the overall agreement is good, but some experimental peaks are missed, or reproduced with a slight time shift.

Recently the ringing problem has received renewed attention associated with the development of offshore wind farms. An important difference with offshore oil structures is that the diameters are smaller (less than 10 m), so the flow is likely to separate in large waves. This is seen as the so-called secondary load cycle observed in experiments but unpredicted by potential flow theory (Grue 2002). Another difference is that the water depths are shallower, implying the presence of non-negligible high-frequency components (absent in deep water) in the incoming waves. Slamming loads under breaking waves, more likely to occur at reduced depths, are also an issue. Again, many experimental campaigns have been carried out, with rigid and flexible cylinders, in constant and variable bathymetry, in regular, irregular, and transient waves (e.g., see Kristiansen & Faltinsen, 2017). Another difference with the old ringing problem is that, over the past twenty years, CFD has made much progress, with the consequence that CFD software has become the main investigation tool (e.g., see Liu *et al.*, 2019), at the expense of potential flow models (and of analytical work).

8.2   Third-Order Wave–Frequency Effects

It has been seen, in Chapter 3, that the third-order correction to the velocity potential of a regular wave, in deep water, only consists in a modification of the wave length: the initial wave number $k = \omega^2/g$ becomes $k' = k\,(1 - k^2 A^2)$.

The case of a bichromatic wave system was also considered, and it was obtained that the cross-interaction between the two wave components results in a further modification of the wave numbers. In the particular case when the two wave components have the same frequency ω, the first one traveling along the Ox axis, and the second one at an angle β, with the first-order free surface elevation given by

$$\eta^{(1)}(x,y,t) = A_1 \cos(kx - \omega t) + A_2 \cos(kx \cos\beta + ky \sin\beta - \omega t) \qquad (8.11)$$

the modification of the wave number of the first component is obtained as

$$k_1^{(2)} = \frac{1}{2} k^3 A_1^2 f(0) + k^3 A_2^2 f(\beta) = -k^3 A_1^2 + k^3 A_2^2 f(\beta) \qquad (8.12)$$

where deep water is still assumed and the function f is a complicated expression given in equation (3.148) and shown in Figure 3.17. The function f is close to a straight line from -2 to $+2$ when the angle β goes from 0 to 180 degrees: when β is larger than about 90 degrees the third-order cross-interaction tends to increase the wave numbers, that is, decrease the wave lengths.

Similar phenomena occur when one wave component is not a plane wave, for instance the combined reflected and radiated waves from a structure. To illustrate the consequences of the third-order interaction between the two wave systems, consider a simple body, a vertical plate, shown in Figure 8.7, that underwent extensive model tests at BGO-FIRST. The plate is 1.2 m wide, bottom-mounted (at a depth of 3 m), and is attached to one of the sidewalls. By mirror symmetry the setup is equivalent to a 2.4 m wide plate in the middle of a 32 m wide tank. The plate is located 19 m from the wavemaker.

Figure 8.7 The 1.2 m long plate at BGO-FIRST.

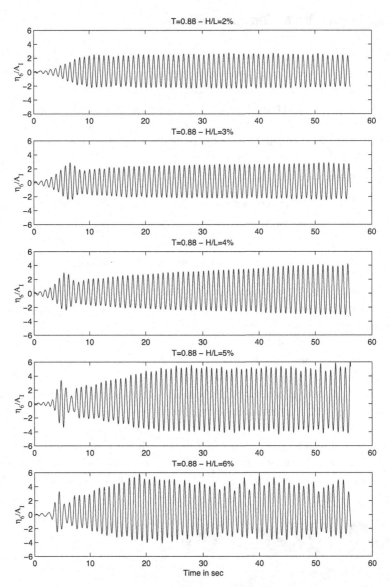

Figure 8.8 Time series of the normalized free surface elevation at the plate-wall corner. Wave period: 0.88 s. Steepnesses $2A_I/L$ from 2% (top) to 6% (bottom) (from Molin *et al.*, 2005) (Cambridge University Press, reprinted with permission).

The tests are run in regular waves, with wave lengths from 1.2 m to 3 m and steepnesses $2A_I/L$ from 2% to 6%. Figure 8.8 shows, in the 1.2 m wave length case, time series of the free surface elevations in the plate-wall corner, for the different steepnesses. The time series are shown from the arrival of the wave front, and they are normalized by the amplitude A_I of the incoming waves (as measured during calibration tests). If the physics were linear the same plot would be obtained 5 times. It can be seen that only in the lowest steepness case a steady state is attained, with a periodic signal oscillating roughly between -2 and $+2$, as predicted by linear theory. From the 3% steepness, the signal amplitude increases with time, at a rate that turns out to be proportional to the square of the steepness. In the 4% and 5% cases, it looks as if a steady state has been attained after many cycles, with the signal amplitude being finally about twice the linear value. At 6% steepness the signal becomes somewhat erratic after the first ramp: This is associated with local breakings taking place in the basin.

These results are taken from Molin *et al.* (2005). It is argued there that the observed phenomena are actually due to the third-order interaction between the incoming waves and the wave system reflected by the plate: this interaction tends to decrease the wave length of the incoming waves, ahead of the plate, where the reflected wave system is the strongest, inducing focusing effects in a similar way as a local shoal would do.

In Molin *et al.* (2005), a parabolic equation is proposed to model the evolution of the incoming wave system as it progresses toward the plate. A steady state is assumed, with the incident velocity potential written in the form

$$\varphi_I = \frac{-\mathrm{i}\, A(\varepsilon^2 x, \varepsilon y)\, g}{\omega}\, \mathrm{e}^{\mathrm{i}\, k\,(1-\varepsilon^2)\, x}\, \mathrm{e}^{[k+\varepsilon^2 k_1^{(2)}(\varepsilon^2 x, \varepsilon y)]z} \tag{8.13}$$

with ε the wave steepness here defined as $k\, A_I$, k the wave number and A_I the incoming wave amplitude away from the plate. In this equation the wave amplitude A is made a complex quantity, slowly varying in space due to tertiary interactions with the reflected wave system from the plate. This reflected wave system is locally identified to a plane wave of amplitude A_R and direction β_R. The complex amplitude A is shown to obey the parabolic equation

$$2\,\mathrm{i}\, k\, A_x + A_{yy} + 2\, k^4 \left[A_R^2\, f(\beta_R) + A_I^2 - \|A\|^2 \right] A = 0 \tag{8.14}$$

where $f(\beta)$ is the interaction function given in equation (3.148).

Equation (8.14) is similar to the parabolic approximation of the mild slope equation (e.g., see Dingemans, 1997, or Radder, 1979), with the difference that here the forcing term is not due to shoaling but to the tertiary interaction with the reflected wave field.

The linear diffraction problem is first solved with the incoming waves unmodified ($A \equiv A_I$). Then the parabolic equation (8.14) is solved numerically starting from some distance l ahead of the plate (here the location of the wavemakers). This provides an updated incoming wave system reaching the plate, whereafter the diffraction problem is solved again, then the parabolic equation, etc., until convergence is reached, if ever.

Figure 8.9 illustrates the evolution of the RAO of the free surface elevation along the plate, at the 5% steepness. Convergence is reached in a few iterations and good agreement is obtained with the experimental RAOs, in spite of the crude approximations inherent to the parabolic model. It is striking that RAO values higher than 4 are obtained at the wall (in $y = 0$) when linear theory predicts values less than 2.

More comparisons, with plates of different lengths, at different facilities, are shown in Molin *et al.* (2006, 2010, 2014). In these papers conclusive comparisons are also made with simulations from a numerical wave tank, based on the extended Boussinesq equations, developed by Jamois (2005). An important parameter is shown to be the size of the *interaction area* where the interactions between incoming and reflected

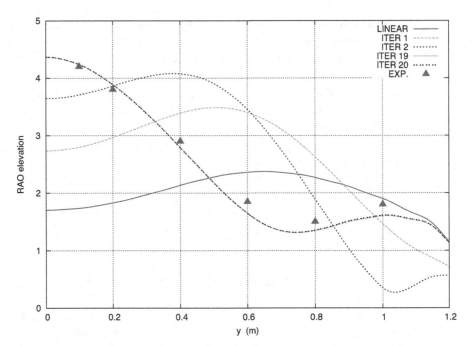

Figure 8.9 Evolution of the RAO of the free surface elevation along the plate through the iterations. Period 0.88 s. Steepness $2\,A_I/L = 5\%$.

waves take place. It is easy to show through running the parabolic model that, depending on the distance from the wavemaker to the plate, different RAOs are obtained. The finite width of the tank may also come into play. These issues are addressed in Chatjigeorgiou & Molin (2013) and Molin *et al.* (2014).

Similar runups occur in the case of ships or barges in beam (or close to beam) waves (see Figure 6.29 in Chapter 6). In fact, our first experience with the phenomenon was while running tests on the roll response of barges at resonance, where the barge model started shipping water and nearly sank. These runups occur in irregular waves as well (e.g., see Zhao *et al.*, 2019) and they are believed to have contributed to the loss of the Prestige tanker following enhanced wave loads amidships.

The parabolic model has been extended to the case of multichromatic seas by Ouled Housseine *et al.* (2020). Good agreement is reported therein with results from experiments in irregular seas.

In the case of semi-submersible platforms or TLPs, the reflected/radiated wave field is much weaker, and it takes large amplitude waves, and columns of large diameters, to see any effect. However unexpectedly high free surface elevations may be observed as Figure 6.28 shows. Further investigations are presented in Molin *et al.* (2014) where it is found that, albeit present in steep regular waves, the third-order runup effects are hardly noticeable in irregular waves.

Of more concern is the case of reflective coastal structures (seawalls, harbor entrances, etc.), many wave lengths long. In Molin *et al.* (2014) some tests are reported on an idealized harbor entrance (a gap through a vertical wall). The measured free surface elevations in the gap and along the wall are unrelated with the predictions of linear theory, and chaotic behaviors are quickly observed in most cases.

8.3   Parametric Instabilities

Parametric instabilities are resonant motions taking place at subharmonics of the excitation. In marine transportation a well-known case is parametric roll that affects some classes of ships.

As an introduction, consider the case of a floating body assumed to be small compared to the wave lengths, a buoy, for instance, small enough that the free surface may be considered as flat around the waterline, its elevation $\eta(t)$ varying only with time. Be $z(t)$ the heave motion of the buoy.

Consider now that the buoy, following some perturbation, takes an inclination in roll (the incoming waves propagating in the Ox direction). The pressure being quasi-hydrostatic from the free surface, an estimate can be made of the restoring moment in roll, accounting for the changes in immersed volume and position of the center of buoyancy. It can be written as a time-varying hydrostatic stiffness

$$K_{44}(t) = \rho g \left[\mathsf{V}(t) \overline{GB}(t) + I_{YY} \right] \tag{8.15}$$

where it is assumed that the buoy is wall-sided at the waterline, so that the waterplane inertia I_{YY} does not change.

The immersed volume V varies according to

$$\mathsf{V}(t) = \mathsf{V}_0 + [\eta(t) - z(t)] \, S_{F_0} \tag{8.16}$$

and the algebraic distance from center of gravity to center of buoyancy:

$$\overline{GB}(t) = \overline{GB}_0 - \frac{[\eta(t) - z(t)] \, S_{F_0}}{\mathsf{V}_0} \, z_B \tag{8.17}$$

Inserting (8.16) and (8.17) into (8.15), and keeping leading order terms, we get

$$K_{44}(t) = K_{44_0} - K_{33} \, z_G \, [\eta(t) - z(t)] \tag{8.18}$$

where z_G is the vertical coordinate of the center of gravity with respect to the free surface.

The roll equation of motion is finally obtained as

$$(I + I_a) \, \ddot{\alpha} + \left(K_{44_0} - K_{33} \, z_G \, [\eta(t) - z(t)] \right) \alpha = 0 \tag{8.19}$$

In regular waves of frequency ω it takes the form

$$\ddot{\alpha} + (\omega_R^2 + \omega^2 \, f \, \cos \omega t) \, \alpha = 0 \tag{8.20}$$

with $\omega_R = \sqrt{K_{44}/(I + I_a)}$ the roll natural frequency.

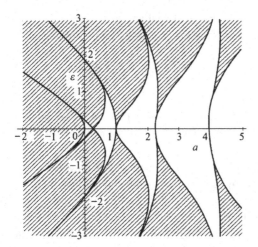

Figure 8.10 Stability (white) and instability (hatched) regions of the Mathieu equation.

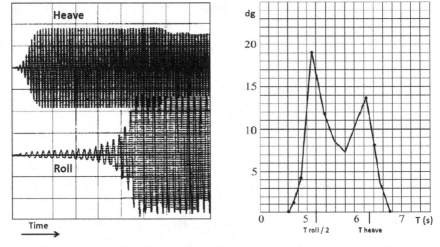

Figure 8.11 Wave response of a cylindrical buoy. Time series of the heave and roll motions (left). Roll response amplitude in 0.45 m amplitude waves (right).

Writing $\omega t = \tau$ equation (8.20) is transformed into

$$\frac{d^2\alpha}{d\tau^2} + (a + 2\varepsilon \cos \tau)\alpha = 0 \qquad (8.21)$$

where $a = \omega_R^2/\omega^2$, $\varepsilon = f/2$. Equation (8.21) is known as the Mathieu equation. As Figure 8.10 shows, the Mathieu equation has unstable solutions (growing exponentially in time) for some regions of the plane (a, ε). When the forcing ε is weak, instability zones are located in the neighborhoods of the values $a = n^2/4$, with n an integer, more particularly by $a = 1/4$. This means a wave frequency ω equal to twice the resonant frequency ω_R.

An example of parametric roll is shown in Figure 8.11, taken from Chantrel (1984). The left part shows the time traces of the heave and roll motions of a cylindrical buoy, of diameter and draft both equal to 8 m, in a regular wave of amplitude 0.45 m and period 6.56 s. The tests are done on a scale 1:18. The natural periods in heave and roll are respectively 6.3 s and 10.4 s. The figure shows a slow appearance of the roll motion, then a quick increase and a leveling at an amplitude of about 8 degrees. The right plot shows that the roll response has two peaks, the first one when the wave period is half the roll natural period, the second one when the wave period coincides with the heave natural period. At the first peak the parameter $a = \omega_R^2/\omega^2$ is exactly 1/4. At the second peak its value is 0.37, but the heave response is very high (RAO close to 3 at this wave amplitude – see Figure 6.25 in Chapter 6).

When some linear damping is introduced in the Mathieu equation, it is found that the instability occurs when the forcing parameter ε exceeds some threshold, then the unstable motion grows without bounds. With a quadratic damping, as results from flow separation, a balance finally arises between production and dissipation of energy, and a steady state may be reached, as Figure 8.11 shows.

Increasing the damping is a good way to avoid parametric instabilities: in the case of the cylindrical buoy, the parametric roll was eliminated by adding a disc at the base. One should also avoid that natural frequencies of different degrees of freedom be in a ratio 1:2.

In the tests with the buoy, other unstable behaviors, combining heave and pitch, appeared when the wave frequency was the sum of the natural frequencies in heave and pitch, both of them oscillating at their own natural frequencies. Similar behaviors, combining heave and pitch, have been reported for Spars (Dern, 1972; Haslum & Faltinsen, 1999; Koo *et al.*, 2004).

In naval engineering, a great deal of literature has been devoted to parametric roll (or "slow roll") that affects some ships such as containerships, particularly in head or following waves, when the wave encounter frequency is twice the roll natural frequency (e.g., see Blocki, 1980). Good agreement with experiments has been obtained with weakly nonlinear numerical models, combining nonlinear hydrostatics and nonlinear Froude–Krylov loads with linear diffraction and radiation (e.g., see Malenica *et al.*, 2006 or Park *et al.*, 2013). By "nonlinear hydrostatics and nonlinear Froude–Krylov loads," it is meant that the hydrostatic and linear hydrodynamic pressures (due to the incoming waves) are integrated up to the instantaneous incident free surface (see Figure 8.12).

8.4 Hydrodynamic Impact and Water Entry

There are many situations, in offshore and marine operations, where phenomena such as wave impacts and slamming take place: steep waves hitting masts of wind turbines, ship bow slamming, sea-launching of underwater equipment, free-fall lifeboats, sloshing in tanks, etc. All these phenomena have in common a fast water entry, communicating to the fluid particles accelerations much larger than gravity, of very short duration, and very high pressures, localized in time and space.

Historically the first theoretical works on water entry, by von Karman and Wagner, in the early 1930s, were concerned with the sea-landing of seaplanes. Due to the slenderness of their floats, a two-dimensional idealization was appropriate.

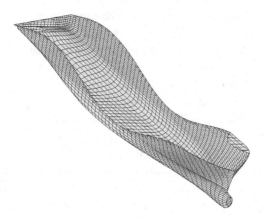

Figure 8.12 Instantaneous wetted hull of a containership for parametric roll analysis (from Malenica *et al.*, 2006).

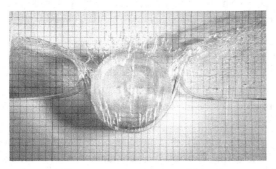

Figure 8.13 Water entry of a circular cylinder (courtesy M. Greenhow).

Figure 8.13 shows the water entry of a circular cylinder at high speed. The moment when the picture was taken is a bit beyond the initial stage since the cylinder is already half immersed. Noticeable are the jets and the rise-ups of the free surface at either side. In the asymptotic theory presented below the jets are ignored: they carry negligible mass and negligible momentum, but not negligible kinetic energy! In fact, at the initial penetration stage, half the kinetic energy goes into the jets, the other half into the bulk of the fluid. This is of no concern when the impact loads are derived from momentum considerations.

8.4.1   Wagner Theory

The problem is illustrated in Figure 8.14: a two-dimensional symmetric body enters an initially flat free surface $y = 0$[3] at time $t = 0$, with a velocity $U(t)$[4] possibly varying in time. The figure refers to a time when the body has penetrated the initial free surface by a small distance $h(t) = \int_0^t U(\tau)\,d\tau$. Two jets are ejected but, as argued above, they are cut off from the fluid domain and replaced by two end points, the **spray roots**, C_1

---

[3] In this section, Oy is the vertical axis.
[4] We take the vertical velocity positive downward.

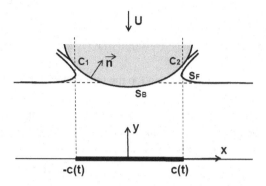

Figure 8.14 Water entry of a rounded symmetric shape and idealization following Wagner

and C_2 in the figure. Because of mass conservation, due to the area occupied by the body in the fluid, the free surface has risen up on both sides and the spray roots are located at an altitude somewhat higher than $y = 0$.

In the absence of sharp edges, viscosity plays a negligible role: in a laminar regime, the thickness of the boundary layer is of order $\sqrt{\nu t}$ (with ν the kinematic viscosity), quite negligible. Assuming incompressibility is more a matter of concern since the induced fluid velocities may be very high, close to sound velocity especially when the water is aerated (see Figure A.1). A major consequence of compressibility is that there is an upper bound to the pressure, the **acoustic pressure**, given by $\rho c_s V$, with c_s the sound velocity and V the flow velocity.

Nevertheless we assume that incompressibility (and irrotationality) holds and we revert to potential flow theory. The velocity potential $\Phi(x, y, t)$ verifies the following boundary value problem:

$$\Delta\Phi = 0 \qquad \text{in the fluid domain} \qquad (8.22)$$

$$\nabla\Phi \rightarrow 0 \qquad x^2 + y^2 \rightarrow \infty \qquad (8.23)$$

$$\nabla\Phi \cdot \vec{n} = -U\, n_y \qquad \text{on the wetted hull } S_B \qquad (8.24)$$

$$g\,\eta + \Phi_t + \frac{1}{2}(\nabla\Phi)^2 = 0 \qquad \text{on the free surface } S_F \qquad (8.25)$$

$$\Phi_y - \eta_x \Phi_x = \eta_t \qquad \text{on the free surface } S_F \qquad (8.26)$$

The hydrodynamic load is obtained from integrating the pressure as

$$\vec{F} = -\rho \int_{S_B} \left[g\,y + \Phi_t + \frac{1}{2}(\nabla\Phi)^2 \right] \vec{n}\, ds \qquad (8.27)$$

Taking advantage of the fact that the free surface is a material surface at atmospheric pressure, this can be rewritten as

$$\vec{F} = -\rho g \int_{S_B} y\, \vec{n}\, ds - \rho \frac{d}{dt} \left[\iint_{S_B \cup S_F} \Phi\, \vec{n}\, ds \right] \qquad (8.28)$$

It is possible to numerically solve this problem in time and space, using a numerical method such as described in Section 8.1.2, with a special treatment applied to the jets. See for instance Zhao & Faltinsen (1993) or Battistin & Iafrati (2003). These numerical models are helpful to provide reference solutions to assess the validity of the approximate models described below.

The following simplifying assumptions are now introduced:

1. The penetration $h(t)$, and the free surface elevation $\eta(x,t)$, are small compared to the wetted length.
2. The normal vectors $\vec{n}$, to the wetted body, and to the free surface, are close to vertical.

   These two assumptions apply to relatively flat bodies in the early entry stage. They allow simplification of the kinematic conditions, at the body and at the free surface, by transferring them to the axis $y = 0$. They become

$$\Phi_y(x,0,t) = -U(t) \qquad -c(t) \le x \le c(t) \tag{8.29}$$

$$\eta_t(x,t) = \Phi_y(x,0,t) \qquad |x| > c(t) \tag{8.30}$$

3. Fluid accelerations are much higher than gravity. This assumption applies at times t small compared to U/g.

Then the hydrostatic term in the free surface dynamic condition can be discarded. The quadratic term is discarded as well in spite of the high fluid velocities in the jets. The dynamic condition then reduces to $\Phi_t = 0$ applied in $y = 0$ or, equivalently

$$\Phi(x,0,t) = 0 \qquad |x| > c(t) \tag{8.31}$$

The vertical impact load is now given by

$$F_y = -\rho \frac{d}{dt} \left(\int_{-c(t)}^{c(t)} \Phi(x,0,t)\,dx \right) = \frac{d}{dt}[U(t)\,M_a(t)] \tag{8.32}$$

with M_a the vertical added mass, which varies in time due to the varying wetted length.

The solution to the boundary value problem is well known: It is the flow around a flat plate, of length $2c(t)$, in unbounded fluid (by symmetry), moving at velocity $U(t)$ in its normal direction. It can be derived from the flow around a circular cylinder by conformal mapping. The complex potential $f(z,t) = \Phi(x,y,t) + i\,\psi(x,y,t)$ (where $z = x + iy$ and ψ is the stream function) is given by

$$f(z) = -i\,U(t)\,\sqrt{z^2 - c^2(t)} + i\,U(t)\,z \tag{8.33}$$

The velocity potential below the plate (at $y = 0^-$) is then

$$\Phi(x,0^-,t) = -U(t)\,\sqrt{c^2(t) - x^2} \tag{8.34}$$

and the vertical load is obtained as

$$F_y = \rho \frac{d}{dt}\left[U(t) \int_{-c(t)}^{c(t)} \sqrt{c^2(t) - x^2}\,dx \right] = \rho \frac{\pi}{2} \frac{d}{dt}\left[U(t)\,c^2(t) \right] \tag{8.35}$$

The last problem to address is the determination of the wetted length $2c(t)$. In the so-called von Karman approximation it is taken from the intersection of the body with the axis $y = 0$, neglecting the rise-up of the free surface. To derive $c(t)$ we write that, at the spray roots C_1 and C_2, the body shape being given as $y = f(x)$, the free surface elevation $\eta(c(t),t)$ is equal to $f(c(t)) - h(t)$, $h(t)$ being the penetration height.

The vertical velocity of the free surface is obtained from

$$\eta_t(x,t) = \Phi_y(x,0,t) = U(t)\,\frac{x}{\sqrt{x^2 - c^2(t)}} - U(t) \tag{8.36}$$

and its elevation is then

$$\eta(x,t) = \int_0^t U(\tau) \, \frac{x}{\sqrt{x^2 - c^2(\tau)}} \, d\tau - h(t) \tag{8.37}$$

The half-wetted length $c(t)$ therefore satisfies the integral equation

$$f\big(c(t)\big) = \int_0^t U(\tau) \, \frac{c(t)}{\sqrt{c^2(t) - c^2(\tau)}} \, d\tau \tag{8.38}$$

In the case where the entry velocity is constant, the solution is

$$t = \frac{2}{\pi U} \int_0^c \frac{f(x)}{\sqrt{c^2 - x^2}} \, dx \tag{8.39}$$

giving the instant of contact t as a function of the half-wetted length c.

The determination of the wetted length is drastically simplified when one uses the so-called **displacement potential**, which is the time integral of the velocity potential (Korobkin & Pukhnachov, 1988). It is easy then to establish that the penetration height h and the half-wetted length c verify the following relationship

$$h(t) = \frac{2}{\pi} \int_0^c \frac{f(x)}{\sqrt{c^2 - x^2}} \, dx = \frac{2}{\pi} \int_0^{\pi/2} f(c \, \sin \theta) \, d\theta \tag{8.40}$$

The superiority of the displacement potential is related to the fact that the hydrodynamic load at time t does not depend on the past history of the flow, as a result of the simplified free surface condition $\Phi = 0$.

In the case of a circular cylinder of radius a, as a first approximation $f(x)$ is given by

$$f(x) = \frac{x^2}{2a} \tag{8.41}$$

that is a parabola of identical curvature. For a steady entry velocity U one gets

$$t = \frac{c^2}{4 U a} \quad \text{that is} \quad c(t) = 2 \sqrt{U a t} \tag{8.42}$$

This is $\sqrt{2}$ times the value obtained through the von Karman approach.
The impact load is given by

$$F_y = \frac{\pi}{2} \rho \frac{d}{dt} \big[4 U^2 a t \big] = 2 \pi \rho a U^2 \tag{8.43}$$

twice the von Karman value.
When the entry velocity varies in time an additional term appears:

$$F_y = 2 \pi \rho a U^2 + 2 \rho \pi a \dot{U} h(t) \tag{8.44}$$

In the case of a wedge with a deadrise angle β, meaning $f(x) = x \tan \beta$, the half-wetted length is obtained as

$$c(t) = \frac{\pi}{2} U t \cot \beta \tag{8.45}$$

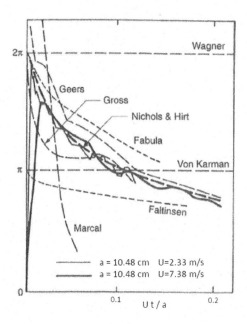

Figure 8.15 Impact load $F/(\rho U^2 a)$ on a circular cylinder, from measurements and from theoretical/numerical models (from Cointe & Armand, 1987).

and the impact load is

$$F_y = \frac{\pi^3}{4} \rho U^3 t \cot^2 \beta \tag{8.46}$$

$(\pi/2)^2 \simeq 2.5$ times the von Karman value!

These two cases are representative of the water entry of rounded shapes (with curvature a at the impact point) or symmetric angular shapes. In the first case, the impact load jumps instantly from zero to the finite value $2\pi\rho a U^2$; in the second case it increases linearly with time.

These are asymptotic results that apply, in the case of the rounded shape, to very low penetration ($Ut/a \ll 1$) and very high Froude number ($U/\sqrt{ga} \gg 1$). Figure 8.15 shows measured and calculated impact loads (normalized by $\rho U^2 a$) on a circular cylinder, as a function of the nondimensional time Ut/a. It can be seen that there is a scatter in the experimental values, and that the Wagner value 2π is attained only in the very first moment when the penetration length is less than a few percent of the radius.

Experimental scatter is due to difficulties in measuring a load with a very stiff model. There is also an effect of air cushioning that reduces the impact load, especially in the case of wedges of very small deadrise angles.

8.4.2   Local Pressures

There are many cases when local loads or pressures are needed. For instance containment systems in LNG tanks must be able to withstand local pressures due to sloshing. The problem is that, when the pressures are derived from the potential $-U(t)\sqrt{c^2(t) - x^2}$ through the Bernoulli equation, infinite values are obtained at the spray roots in $\pm c(t)$. The pressure is still integrable and the results obtained above for the impact load hold, but the pressure values by the spray roots are unphysical.

This is due to the fact that the potential $-U(t)\sqrt{c^2(t) - x^2}$ is the flow around a flat plate: this flow turns around the edges with infinite local velocity, whereas, in the water entry case, the flow continues tangentially as a jet. This deficiency can be remedied by matching this "outer" flow, valid some distance away from the spray roots, with a local "inner" jet flow. The mathematical technique, known as matched asymptotic expansions (e.g. see van Dyke, 1964), was first applied by Cointe & Armand (1987). They obtained that the jet has the following features:

- Thickness:

$$\delta = \frac{\pi}{8} U^2 \frac{c}{\dot{c}^2} \tag{8.47}$$

where $\dot{c} = dc/dt$.
- Flow velocity inside the jet

$$V_{\text{jet}} = 2\,\dot{c} \tag{8.48}$$

The pressure along the wetted length, jet included, is obtained as (see also Zhao & Faltinsen, 1993)

$$p = \rho U \frac{c\,\dot{c}}{\sqrt{c^2 - x^2}} - \rho U \frac{c\,\dot{c}}{\sqrt{2c\,(c - x)}} + 2\rho\dot{c}^2 \frac{\sqrt{\tau}}{(1 + \sqrt{\tau})^2} \quad \text{for } 1 \le \tau < \infty \tag{8.49}$$

$$p = 2\rho\dot{c}^2 \frac{\sqrt{\tau}}{(1 + \sqrt{\tau})^2} \qquad\qquad\qquad \text{for } 0 < \tau \le 1 \tag{8.50}$$

where x and τ are linked through

$$x = c + \frac{1}{\pi}\delta\left(-\ln\tau - 4\sqrt{\tau} - \tau + 5\right) \tag{8.51}$$

The value $\tau = 1$ corresponds to $x = c$. When τ goes from 1 to 0, x goes from c to $+\infty$, along the jet. When τ increases from the value 1, x decreases until it is equal to 0 for some value of τ.

The pressure peak is attained at the spray root ($x = c$) and is equal to $\rho\dot{c}^2/2$. It must be borne in mind that this is for an incompressible fluid. As mentioned above, compressibility leads to an upper bound of the pressure, equal to $\rho c_s V$, with c_s the sound velocity.

Figure 8.16 shows, in the rounded case, the pressure coefficient $C_p = p/(1/2\,\rho U^2)$ from equations (8.49), (8.50), for three successive values of the submergence Ut/a. The values from the outer solution (8.34), singular in $x = c$, are also shown for reference.

Figure 8.17 shows the pressure coefficients for 3 wedges of deadrise angles 5, 10, and 15 degrees. They are shown vs the vertical coordinate referred to the initial free surface in nondimensional form, that is $y/(Ut)$. Zhao & Faltinsen (1993) used a fully

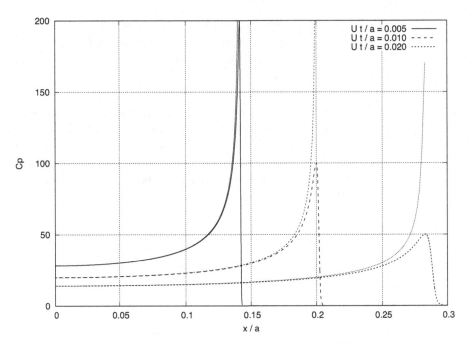

Figure 8.16 Water entry of a circular cylinder. Pressure coefficients at different times.

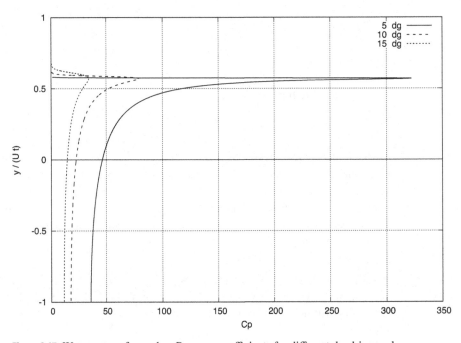

Figure 8.17 Water entry of a wedge. Pressure coefficients for different deadrise angles.

nonlinear time domain approach (neglecting gravity) to simulate the water entry of wedges and obtained pressure maps very close to the Wagner asymptotic approach, for deadrise angles less than 20 degrees.

Experimental validation is another matter: as impact pressures are very localized in time and space, numerous and very tiny sensors, sampled at very high frequencies (typically 20 kHz), must be used. Pressure sensors moreover have their own drawbacks such as being expensive and sensitive to temperature. Rigidity of the wall and of the sensor settings is also an issue as it may affect the measurements. Hydroelastic issues in slamming are discussed by Faltinsen (2000).

8.4.3　More Elaborate Models

Wagner theory represents the leading order approximation of the fully nonlinear problem, with respect to the small parameter defined as the ratio between the penetration depth and the characteristic horizontal size of the contact region. Different models, which can be classified as weakly nonlinear, have been proposed in the literature. Among them the generalized von Karman model (GVKM), the generalized Wagner model (GWM), and the modified Logvinovich model (MLM) appear to be the most often used in practice.

Modified Logvinovich Model

The Wagner solution being taken as a first-order approximation, a second-order approximation may be derived through a Taylor development applied to the velocity potential from $y = 0$ down to $y = -h(t) + f(x)$:

$$\Phi(x, f - h, t) \simeq \Phi(x, 0, t) + (f - h)\,\Phi_y(x, 0, t) = \Phi(x, 0, t) + (f - h)\,\dot{h} \qquad (8.52)$$

At the same time the nonlinear Bernoulli equation is developed to second-order (Korobkin, 2005):

$$p = -\rho\,\dot{h}^2 \left[\frac{\dot{c}}{\dot{h}} \frac{c}{\sqrt{c^2 - x^2}} - \frac{1}{2} \frac{1}{1 + f_x^2} \left(\frac{c^2}{c^2 - x^2} + f_x^2 \right) + \frac{\ddot{h}}{\dot{h}} \left(\sqrt{c^2 - x^2} + f - h \right) \right]$$

$$(8.53)$$

Even though this expression looks complicated, its numerical implementation is straightforward. It can be noted that the pressure in (8.53) is not integrable. This can be remedied by fitting (8.53) with a local jet flow at the contact point, or by simply setting to zero the pressure when negative. In fact, physically the total pressure cannot be less than the vapor pressure otherwise cavitation occurs. Cavitation does occur in some water entry events.

Generalized von Karman and Wagner Models

In these models, the body boundary condition is applied at the actual wetted position of the body. The free surface is still considered as flat and the homogeneous Dirichlet

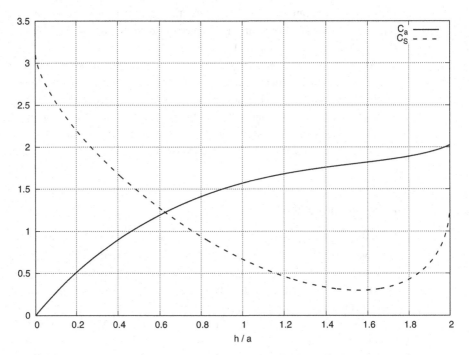

Figure 8.18 Generalized von Karman model applied to a circular cylinder. Added mass coefficient $C_a = M_a/(\rho a^2)$ and slamming coefficient $C_S = a\, dC_a/dh$ vs h/a with a the radius and h the penetration.

condition $\Phi = 0$ is applied there. In the GVKM model, the contact points are found from the intersection of the body contour with the initial free surface. In the GWM approach a numerical variant of the Wagner condition is applied.

Figure 8.18 shows the application of the Generalized von Karman model to a circular cylinder. The analytic expression of the added mass given by Greenhow & Li (1987) has been used. As already mentioned the initial value of the slamming coefficient, at $h = 0$, is π, against 2π with the Wagner model. Physically, when the submergence exceeds 20% or 30% of the diameter, separation may occur (see Figure 8.13), followed by a cavity formation.

Figure 8.19 shows a comparison between experimental measurements and calculations of the water entry load upon a ship section. The experimental results are taken from Zhao *et al.* (1996). The ship section is in free fall, and enters water at a velocity of about 2.8 m/s. Both the MLM and GWM give numerical predictions in fair agreement with the experiments. However, the GWM does a better job in predicting local pressures. On the other hand, the MLM is easier to implement and much cheaper computationally.

The MLM and GWM can be extended to the case of asymmetric impacts, for instance in the case of an inclined ship section. One difficulty there is the determination of the contact points. See de Lauzon *et al.* (2015) for instance.

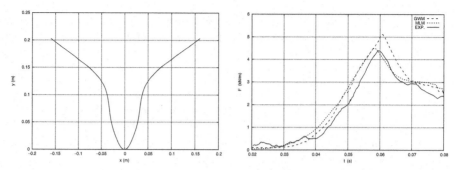

Figure 8.19 Water entry of a ship section. Ship section shape (left). Vertical load from experiments and from the MLM and GWM methods (right) (courtesy Bureau Veritas).

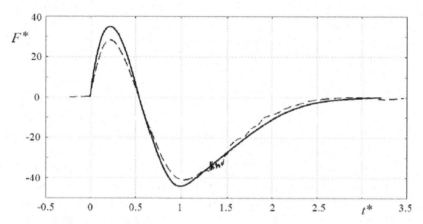

Figure 8.20 Nondimensional hydrodynamic load on a wedge during entry and exit stages. Solid line theoretical, dashed line CFD (from Korobkin *et al.*, 2014a).

8.4.4 Water Exit

In some applications, for instance ship whipping, or when lowering objects through the free surface, both entry and exit stages must be considered in the computations of the hydrodynamic loads. The downward force that occurs during the exit stage can be of the same order as the upward impact load and, in general, lasts longer (Korobkin, 2013). The simplest model of water exit is based on the von Karman approach where the wetted part of the body is taken as the part below the mean water level. This approach provides a good approximation of the magnitude of the load but it under-predicts its duration. A more elaborate approach is provided by Korobkin (2013) and Korobkin *et al.* (2014b) where the flat disk approximation is used for bodies of small deadrise angle undergoing high accelerations. A comparison of the model with numerical results from a CFD code is provided in Figure 8.20.

In this figure, it can be observed that the downward exit force is greater and lasts longer than the entry load. A similar phenomenon occurs in wave-in-deck events, when waves hit platform decks from underneath: the upward load is followed by a downward suction force as the wave retreats and the wetted area decreases in time or moves along the deck (e.g., see Baarholm & Faltinsen, 2004).

8.4.5   Wave Impacts

Within von Karman and Wagner theories, where gravity is neglected, the water entry of a rounded body into a flat free surface and the impact of a rounded bulk of fluid upon a flat solid surface are equivalent. When both are rounded, within the Wagner approximation, the impact load is given by

$$F_y = 2\,\rho\,\pi\,R\,U^2 \tag{8.54}$$

with $R = R_b\,R_w/(R_b + R_w)$, R_b the radius of curvature of the body, R_w the radius of curvature of the fluid. This is for two-dimensional bodies and constant velocity has been assumed.

In coastal engineering, there is extensive literature on the impact of breaking waves upon vertical piles. The horizontal impact load is usually expressed as

$$F(t) = \frac{1}{2}\rho\,\lambda\,\eta\,C_s(U\,t/D)\,D\,U^2 \tag{8.55}$$

Here η is the elevation of the wave crest, $\lambda\,\eta$ is the height of the impact area, a the radius, and $D = 2a$ the diameter. Due to aeration, the density ρ may be quite inferior to the standard value 1025 kg/m^3.

The slamming coefficient C_s depends on the penetration distance Ut relative to the diameter. Different formulations can be found in the literature. Goda *et al.* (1966) have proposed $C_s = \pi\,(1 - Ut/a)$, the reference value, at the initial instant, being the von Karman model. Based on experiments by Campbell & Weynberg, DNVGL (2017) gives

$$C_s = 5.15\left[\frac{D}{D + 19\,s} + \frac{0.107\,s}{D}\right] \tag{8.56}$$

where $s = Ut$. The Generalized von Karman approximation has been shown in Figure 8.18.

Another C_s shape (WO) has been proposed by Wienke & Oumeraci (2005), following large scale experiments in the large wave flume (GWK) of the Coastal Research Center (FZK) in Hanover, with a cylinder 0.70 m in diameter.

These different slamming coefficients are shown in Figure 8.21. There is a wide scatter although the DNVGL and WO coefficients are somewhat close during the initial penetration stage. When the penetration is higher than 0.2 diameter the WO coefficient is set to zero while the DNVGL coefficient levels at $C_s = 0.8$. This accounts for the drag load that follows the impact force when the cylinder slice is fully submerged. In the case of horizontal tubular elements impacted by waves, DNVGL proposes a combination of buoyancy, slamming, inertia, and drag components.

Wave impacts against seawalls have also been widely studied. Similar impacts occur in LNG tanks of LNG carriers. In low filling conditions (on the ship's return trip) the liquid motion in the tank may resemble a hydraulic bore traveling back and

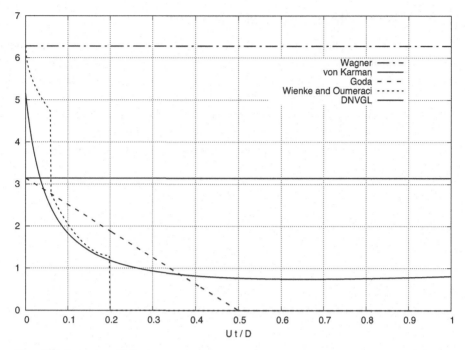

Figure 8.21 Impact coefficient.

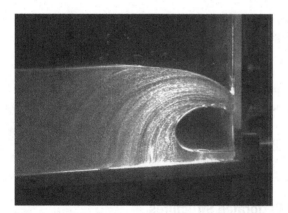

Figure 8.22 Shallow water wave impact upon a vertical wall (courtesy O. Kimmoun).

forth. Figure 8.22 shows a shallow water impact with an enclosed air pocket. Due to the air compressibility, the resulting pressures at the wall show some oscillations after the initial peak (see Figure 9.22 in Chapter 9). The highest pressures are attained when the wave front is nearly vertical as it approaches the wall, resulting in high local vertical accelerations and an upward jet. This type of impact is known as **flip-through** (Lugni et al., 2006).

8.4.6   Water Entry of Three-Dimensional Bodies

The three-dimensional shapes equivalent to the cylinder and to the wedge are the sphere and the cone. Unlike in two dimensions, even though these geometries are simple, there are no analytical solutions to the Wagner problem (there are analytical solutions for some particular cases such as elliptic paraboloids – see Scolan & Korobkin, 2001). There are approximate Wagner theories and fully numerical results from nonlinear potential flow models, or from CFD codes.

When a von Karman approach is adopted, the following entry loads are obtained[5]:

- Cone:

$$F_z = 4 \rho \cot^3 \beta \, U^4 t^2 \qquad (8.57)$$

with β the deadrise angle.

- Sphere:

$$F_z = 2^{5/2} \rho \, a^{3/2} U^{5/2} t^{1/2} \qquad (8.58)$$

with a the radius of the sphere.

Approximate Wagner theories for the cone and sphere can be found in Faltinsen & Zhao (1997).

For the cone they obtain

$$F_z = 4 \left(\frac{4}{\pi}\right)^3 \rho \cot^3 \beta \, U^4 t^2 \qquad (8.59)$$

that is about 2.06 times the von Karman value.

For the sphere they get

$$F_z \simeq 6 \sqrt{3} \, \rho \, a^{3/2} U^{5/2} t^{1/2} \qquad (8.60)$$

about 1.84 times the von Karman value.

These results have been confirmed by Battistin & Iafrati (2003) who use a purely numerical method.

A tentative conclusion is that, in three dimensions, when a von Karman approach is used, a systematic factor 2 correction should be applied.

8.5   Hydrodynamics of Perforated Structures

Perforated seawalls (Figure 8.23) are widely used in coastal and harbor engineering. Introduced by Jarlan (1961), they are known to reduce wave reflection, dissipate energy, and improve harbor tranquility. They have inspired designers of the early GBS platforms in the North Sea (for instance, the Ekofisk reservoir), even though it was not clearly assessed at the time how much the openings would reduce the wave loading. Other cases of perforated, or ventilated, structures are mud mats, roofs of suction

---

[5] Here constant entry velocity U is assumed.

Figure 8.23 Jarlan walls at Dieppe harbor (credits D. Lajoie).

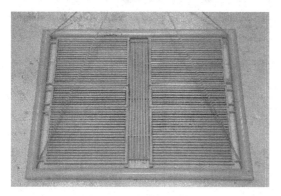

Figure 8.24 Experimental model of hatch cover (from Molin & Nielsen 2004).

anchors, or some protection covers used to shelter subsea equipment (Figure 8.24). Vertical screens, used in wave tanks as wave absorbers, perforated breakwaters, and nets of fishfarms, are other examples. The openings may be circular, rectangular, or just horizontal or vertical slots (Figure 8.25), with quasi-identical performances. The main discriminating parameter, as will be seen further, is the porosity, or open-area ratio, τ, that is the area of the openings divided by the total area.

An essential assumption made in the theoretical model proposed here is that the flow separates through the openings. This will always be the case when the openings have sharp angles. When they are rounded, separation will also take place provided that the local Keulegan–Carpenter number (that is the amplitude of flow motion though the openings related to their widths) be large enough. As a result, there occurs locally a drag load, applied to the solid part, proportional to the square of the local relative velocity.

Next, the succession of solid and open parts is idealized as a porous wall where the traversing velocity, and pressure drop, are continuous. From the parent case of channel flow through a diaphragm (see Figure 8.26), the pressure drop is written as

$$\Delta p = \frac{1 - \tau}{2 \, \mu \, \tau^2} \, \rho \, v \, |v| \tag{8.61}$$

Figure 8.25 The Lucciana jetty of Monaco Hercule harbor when under construction at La Ciotat (credits C. Colmard).

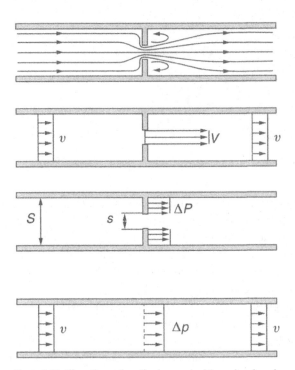

Figure 8.26 Flow through a diaphragm inside a circular pipe: streamlines, velocity profiles, pressure differential, and porous membrane idealization.

where Δp is the (locally averaged) pressure drop, v the (locally averaged) traversing velocity, τ the open-area ratio, and μ a discharge coefficient depending on the shape of the openings (typically $\mu \simeq 0.5$ for sharp openings, and $\mu \sim 1$ for rounded ones).

According to an ideal fluid theory only inertia loads should appear, related to the flow acceleration. Again by reference to the flow through two-dimensional or circular diaphragms, it can be proved that inertia effects become nil in the limit when the density of openings goes to infinity, notwithstanding the value of the open-area ratio (Molin, 2011; Molin & Remy, 2015). In the following, for the sake of simplicity, we first make the assumption that the openings are numerous enough that inertia effects can be ignored. We also assume the wall thickness to be negligible.

Finally, what may sound like a contradiction, we assume potential flow theory to hold in the fluid domains at either side of the porous wall. This assumption is actually consistent with the previous assumptions of quasi-infinite number of openings: The emitted vortical structures remain confined in a layer of thickness scaling with the opening size and they quickly disintegrate, as in the case of air flow through honeycombs in a wind tunnel. The limitations of the proposed model are then about the same as for the equivalent solid body (i.e., they will depend on the value of a global Keulegan–Carpenter number based on the flow amplitude and on the global dimensions).

The boundary conditions taken at the perforated wall are therefore:

• Pressure drop:

$$\Delta p = \frac{1 - \tau}{2 \mu \tau^2} \rho \, v_{\mathrm{rel}} \, |v_{\mathrm{rel}}| \tag{8.62}$$

with v_{rel} the relative velocity in the normal direction:

$$v_{\mathrm{rel}} = \left(\vec{V}_{l,r} - \vec{U} \right) \cdot \vec{n} \tag{8.63}$$

$\vec{V}_l$ the fluid velocity on the left-hand side, $\vec{V}_r$ the fluid velocity on the right-hand side, $\vec{U}$ the local velocity of the porous wall, and $\vec{n}$ the normal vector.

• Mass conservation:

$$\vec{V}_l \cdot \vec{n} = \vec{V}_r \cdot \vec{n} \tag{8.64}$$

The pressure is obtained through the Bernoulli–Lagrange equation

$$p_{l,r} = H_{l,r}(t) - \rho g z - \rho \frac{\partial \Phi_{l,r}}{\partial t} - \frac{1}{2} \rho \left(\nabla \Phi_{l,r} \right)^2 \tag{8.65}$$

with Φ_l (resp. Φ_r) the velocity potential of the left-hand side (resp. right-hand side) and H_l (resp. H_r) the Bernoulli constant.

In most cases, the pressure is linearized in

$$p_{l,r} = -\rho g z - \rho \frac{\partial \Phi_{l,r}}{\partial t} \tag{8.66}$$

8.5.1   A Model Case: The Two-Dimensional Circular Cylinder

We consider a porous two-dimensional cylinder undergoing forced oscillatory motion in the x direction. The fluid domain is unbounded and at rest at infinity. See Figure 8.27. The motion amplitude is assumed to be small compared to the cylinder radius a and the equations are linearized.

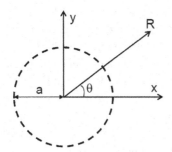

Figure 8.27 Two-dimensional porous cylinder.

The fluid domain can be decomposed into an inner domain (inside the cylinder) and an outer domain. In these domains, the velocity potential can be expressed as

$$\Phi_i(R,\theta,t) = \sum_{m=1}^{\infty} A_m(t)\, a \left(\frac{R}{a}\right)^m \cos m\theta \qquad\qquad R \le a \ (8.67)$$

$$\Phi_e(R,\theta,t) = \sum_{m=1}^{\infty} B_m(t)\, a \left(\frac{a}{R}\right)^m \cos m\theta \qquad\qquad R \ge a \ (8.68)$$

with $B_m = -A_m$ so that $\partial\Phi_i/\partial R = \partial\Phi_e/\partial R$ at $R = a$.

When the approximation $\cos\theta\,|\cos\theta| \simeq 8\cos\theta/(3\pi)$ is applied only the $m = 1$ terms of Φ_i and Φ_e need to be retained.

The discharge equation takes the form

$$\dot{A}_1(t) = \frac{dA_1}{dt} = -\frac{2(1-\tau)}{3\pi\mu\tau^2 a}\left(A_1(t) - U(t)\right)\left|A_1(t) - U(t)\right| \qquad (8.69)$$

with $U(t)$ the cylinder velocity. This equation can be integrated in time for given $U(t)$.

In the case of a sinusoidal motion, an analytical solution can be obtained. Writing

$$U(t) = A\omega\cos\omega t = \mathfrak{R}\left\{A\omega e^{-i\omega t}\right\} \qquad\qquad (8.70)$$

$$A_1(t) = \mathfrak{R}\left\{A\omega\,(1+b)\,e^{-i\omega t}\right\} \qquad\qquad (8.71)$$

with A the motion amplitude and ω the frequency, and applying again the linearization

$$\mathfrak{R}\left\{f\,e^{-i\omega t}\right\}\left|\mathfrak{R}\left\{f\,e^{-i\omega t}\right\}\right| \simeq \frac{8}{3\pi}\|f\|\,\mathfrak{R}\left\{f\,e^{-i\omega t}\right\} \qquad (8.72)$$

where $\|\ \|$ means the modulus of the complex number, one gets the equation

$$b\,\|b\| = i\,C\,(1+b) \qquad\qquad (8.73)$$

where

$$C = \left(\frac{3\pi}{4}\right)^2 \frac{\mu\tau^2}{1-\tau}\frac{a}{A}. \qquad\qquad (8.74)$$

Equation (8.73) has for solution

$$b = \frac{\sqrt{C^2+4}-C}{2}\left(-C + i\sqrt{C\,\frac{\sqrt{C^2+4}-C}{2}}\right) \qquad (8.75)$$

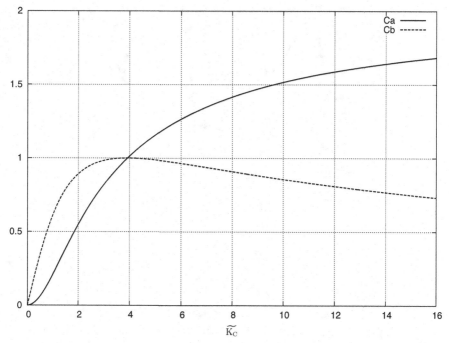

Figure 8.28 Two-dimensional porous cylinder. Added mass and damping coefficients vs. $\widetilde{K_C}$.

The hydrodynamic force is finally obtained as

$$F_x = \Re \left\{ i \rho \pi a^2 A \omega^2 (C_a + i C_b) e^{-i\omega t} \right\} \tag{8.76}$$

with the added mass (C_a) and damping (C_b) coefficients

$$C_a = 2 - C \left(\sqrt{C^2 + 4} - C \right) \tag{8.77}$$

$$C_b = \sqrt{\frac{C}{2}} \left(\sqrt{C^2 + 4} - C \right)^{3/2} \tag{8.78}$$

shown in Figure 8.28 versus the parameter $\widetilde{K_C} = (1 - \tau) A / (2 \mu \tau^2 a)$, which is a Keulegan-Carpenter number combined with the porosity parameter: This means that it is equivalent to increase the motion amplitude or to decrease the open-area ratio, the same hydrodynamic coefficients are obtained at equal $\widetilde{K_C}$ value.

At zero motion amplitude, both the added mass and damping coefficients are nil: it is as though the porous cylinder does not exist. As the amplitude increases from zero, the added mass coefficient steadily increases up to its asymptotic value of 2, which corresponds to the case of a solid cylinder full of water. In the low amplitude range the damping coefficient dominates the added mass coefficient, until the two curves cross each other. The crossing corresponds to the maximum value of the damping coefficient, which is equal to one, half the solid case added mass coefficient. This feature has been observed for all geometries, in two and three dimensions (in infinite fluid, i.e., without a free surface): the maximum value of the damping coefficient is

Figure 8.29 Experimental octagonal stabilizers (10% and 24% porosity ratios).

always half the solid added mass coefficient and the two curves always intersect at that point.

It might come as something of a surprise that an added mass arises when locally only a drag load is accounted for. This reflects the fact that the local flow and the global flow (here the imposed velocity) are shifted in time.

Experiments on perforated three-dimensional cylinders are reported in Molin & Legras (1990). The cylinders are open-ended, and not strictly circular but octagonal, with a height of 1 m and an equivalent radius of 0.5 m. They are located at mid-depth of the 3 m deep basin. As can be seen in the photographs of Figure 8.29, both slotted and perforated models were tested, with two different open-area ratios (10% and 24% in the perforated case, 10% and 20% in the slotted one). The wall thickness was 1 mm and the opening diameter 2 mm (in the perforated case).

Figure 8.30, taken from Molin & Legras (1990), shows measured and calculated values of the added mass (left) and damping (right) coefficients, in the 10% perforated case. The calculated values are obtained through a 3D extension of the analytical model presented above.

The motion amplitudes on the horizontal axis are full-scale values, to be referred to an equivalent diameter of 60 m (the experimental scale being 1:60). So 5 m means a Keulegan–Carpenter number $K_C = 2\pi A/D$ around 0.5. There is some scatter in the experimental results, but it is clear that the experimental added mass does go to zero or nearly zero as the motion amplitude decreases. The agreement between experimental and numerical added mass coefficients is rather good over the whole range of motion amplitude. As for the damping coefficient, the agreement deteriorates as the motion

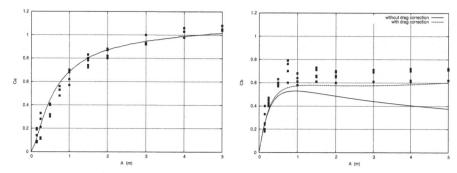

Figure 8.30 Perforated stabilizer at 10% porosity. Added mass (left) and damping (right) coefficients, vs. motion amplitude, from measurements (symbols) and from calculations.

amplitude increases. This has been identified to be due to flow separation at the lower and upper edges of the cylinder, an effect which is not accounted for by the theoretical model. The dotted curve is simply obtained by adding, to the calculated damping force, a viscous term in the form $F_D = -1/2 \rho C_D S V |V|$, with S the frontal area and a drag coefficient C_D equal to 5.

Very similar results were obtained in the slotted case, confirming that open-area ratios matter a lot more than the shape of the openings, as assumed in the theoretical model.

8.5.2 Hydrodynamic Coefficients for Plates and Disks

These cases have relevance for mud mats and hatch covers as shown in Figure 8.24, or for slotted bilge keels such as on the N'Kossa barge.

Two-dimensional perforated plates in unbounded fluid are considered first.

The plate motion being $A \sin \omega t$ and the hydrodynamic force being expressed as

$$F = \rho \pi b^2 A \omega^2 [C_a \sin \omega t - C_b \cos \omega t] \tag{8.79}$$

Figure 8.31 shows the added mass C_a and damping C_b coefficients vs the "porous Keulegan-Carpenter number" here defined as $\widetilde{K_C} = (1 - \tau) A/(2 \mu \tau^2 b)$ with A the amplitude of motion and b the half-width of the plate. These results have been obtained through the method of matched eigenfunction expansions, with the 2D plate sufficiently far away from the free surface and from the sea floor that the fluid domain may be considered as unbounded.

It can be seen that the damping coefficient peaks at 0.5, that is half the solid added mass value, where the two curves intersect. As written earlier this feature has always been observed for perforated bodies in unbounded fluid.

Figure 8.32, taken from Molin *et al.* (2008), shows the hydrodynamic coefficients (normalized with ρa^3, a being the disk radius) of a perforated disk, from experiments and from computations. The disk has a diameter of 60 cm, a thickness of 2 mm, and the open-area ratio is 20%. It is located at mid-distance between the free surface and the tank floor, the water depth being 50 cm. The oscillation period is 1.2 s. Negligible

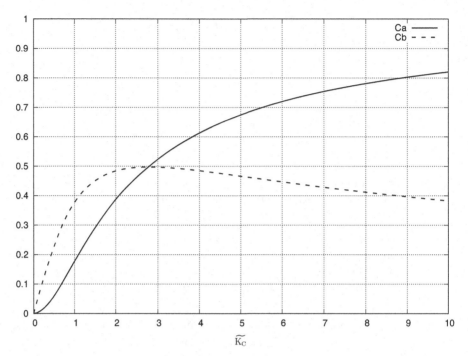

Figure 8.31 Two-dimensional porous plate in unbounded fluid. Added mass and damping coefficients vs. $\widetilde{K}_C$.

effects result from the proximities of the free surface and basin floor. Numerical added mass and damping coefficients are again obtained through the method of matched eigenfunction expansions (see Molin & Nielsen, 2004). From the figure, it can be seen that the agreement between experimental and numerical coefficients is satisfactory only at very low motion amplitudes. Again the discrepancy is due to flow separation at the disk edge and to the resulting drag force, not accounted for by the model. A supplementary drag term has been introduced, in the form

$$F_v = -\frac{1}{2}\, \rho\, C_D\, \pi\, a^2\, V_R\, |V_R| \tag{8.80}$$

with $C_D = \alpha\, K_C^{-1/3}$ and $K_C = \pi\, A/a$ (A the motion amplitude and a the radius), as proposed by Sandvik *et al.* (2006). The $K_C^{-1/3}$ dependence of the drag coefficient is based on the asymptotic analysis of Graham (1980). The relative velocity V_R in equation (8.80) is taken as the disk velocity minus the averaged relative fluid velocity through the disk; this results in a modification of the added mass coefficient as well. (It should be noted that when this viscous correction is applied, the hydrodynamic coefficients depend no longer on the $\widetilde{K}_C$ value alone, but both on $\widetilde{K}_C$ and on K_C.)

It can be seen that, with an α value of 6, a good match is obtained for the damping coefficient, and that the calculated added mass is somewhat reduced, but not sufficiently to agree with the experimental values. It must be mentioned that Sandvik *et al.* (2006) got a best fit with a much lower α value, equal to 2; this may be attributed to a

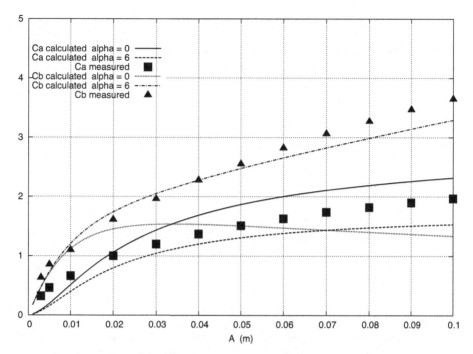

Figure 8.32 Perforated disk at porosity 20%. Immersion 25 cm. Period 1.2 s. Measured and calculated added mass and damping coefficients.

higher thickness over diameter ratio in their experiments (see He *et al.*, 2008). So the α value needs to be adjusted with care.

Extensive tests and calculations are reported by Mentzoni & Kristiansen (2019a, 2019b) on the 2D perforated plate case. They obtain a best fit with an α value of 6.5.

8.5.3   Water Entry of Perforated Bodies

The water entry loads are much reduced for perforated bodies, compared to their solid counterparts.

The water entry of a perforated wedge has been tackled by Molin & Korobkin (2001). Figure 8.33 shows the reduction factor of the vertical hydrodynamic load, as a function of the porosity τ, for three different values of the deadrise angle β. (Straight lines at the origin are asymptotic results for small open-area ratios). It can be observed that the reduction factor is appreciable: for instance, with a porosity of 20% the impact load is decreased seven-fold for the flattest wedge!

8.5.4   Introduction of Inertia Effects

When the number of openings is small, and there are wide solid areas, inertia effects may come into play. In fact, it is easy to add, to the discharge equation (8.62), an inertia component, related to the relative acceleration. This has been applied by Molin

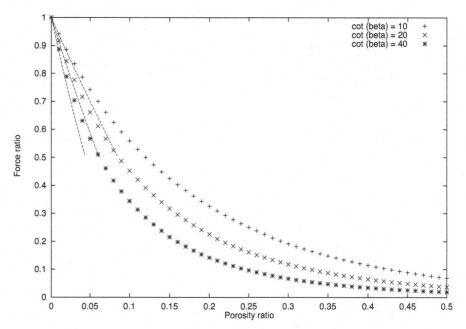

Figure 8.33 Water entry of a perforated wedge. Vertical force as a function of the porosity ratio τ, for different deadrise angles (taken from Molin & Korobkin, 2001).

& Remy (2015) in their study of the hydrodynamic performance of a Tuned Liquid Damper with a slotted screen. Forced motion experiments are performed on a rectangular tank with a vertical slotted wall at mid-length. The number of vertical slots over the tank width is varied, from 12 slots down to one single central slot. Experimental added mass and damping coefficients are compared with numerical values.

In the numerical model, the discharge law (8.62) at the porous wall is complemented with an inertia term and becomes

$$\mathrm{i}\,\omega\,(\varphi_l-\varphi_r) = \frac{4}{3\pi}\frac{1-\tau}{\mu\,\tau^2}\,\|\varphi_y-A\,\omega\|\,(\varphi_y-A\,\omega)-\mathrm{i}\,\omega\,(1-\tau)^2\,\pi\,C_a(\tau)\,\frac{d}{4N_S}\,(\varphi_y-A\,\omega)$$

(8.81)

with φ_l (resp. φ_r) the velocity potential on the left-hand (resp. right-hand) side of the screen, N_S the number of slots over the width d of the tank, and $C_a(\tau)$ the added mass coefficient given as (Morse & Ingard 1968, see also Crowley & Porter, 2012):

$$C_a(\tau) = \frac{8}{(1-\tau)^2\,\pi^2}\,\ln\left[\frac{1}{2}\tan\frac{\pi\,\tau}{4}+\frac{1}{2}\cot\frac{\pi\,\tau}{4}\right] = -\frac{8}{(1-\tau)^2\,\pi^2}\,\ln\left[\sin\frac{\pi\,\tau}{2}\right]$$

(8.82)

The added mass coefficient C_a is here referred to an area $\pi\,c^2$ with c half the slat width $c = (1-\tau)\,B/(2\,N_S)$.

The experimental tank has a length L of 80 cm, a width B of 50 cm, and the water-height h is 32 cm. The natural frequencies of the first three modes of the tank without screen (obtained from $\omega_n^2 = g\,\lambda_n\,\tanh\lambda_n h$, with $\lambda_n = n\pi/L$) are $\omega_1 = 5.72$ rad/s,

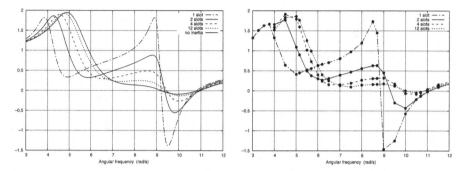

Figure 8.34 TLD with slatted screens. Added mass coefficient. Forced motion amplitude 4 mm. Calculated (left) and measured (right) values (from Molin & Remy, 2015).

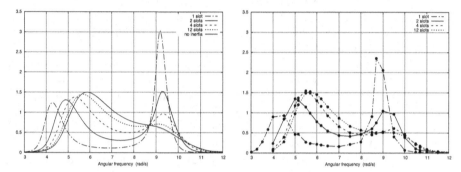

Figure 8.35 TLD with slatted screens. Damping coefficient. Forced motion amplitude 4 mm. Calculated (left) and measured (right) values (from Molin & Remy, 2015).

$\omega_2 = 8.72$ rad/s, and $\omega_3 = 10.74$ rad/s. The open-area ratio of the screen is kept constant at 18% as the number of slots is varied. The tests are run at angular frequencies from 3 rad/s to 12 rad/s, encompassing the frequencies of the first three modes, and at amplitudes of 1 mm, 4 mm, 8 mm, and 12 mm.

Figures 8.34 and 8.35 show the added mass and damping coefficients, from the experiments and from the numerical model, for the different screens, and for a forced motion amplitude of 4 mm. It can be seen that the overall agreement is rather good. In particular, the down-shift of the natural frequency of the first mode, as the number of openings is reduced, is correctly reproduced. More details can be found in Molin & Remy (2015) where the case of circular openings is also considered. See also Molin & Remy (2016) where the open-area ratio is varied from 10% to 90% with only one central opening.

8.6   References

BAARHOLM R., FALTINSEN O.M. 2004. Wave impact underneath horizontal decks, *J. Mar. Sci. Technol.*, **9**, 1–13.

BATTISTIN D., IAFRATI A. 2003. Hydrodynamic loads during water entry of two-dimensional and axisymmetric bodies, *J. Fluids and Structures*, **17**, 643–664.

BINGHAM H.B., MADSEN P.A., FUHRMAN D.R. 2009. Velocity potential formulations of highly accurate Boussinesq-type models, *Coast. Eng.*, **56**, 467–478.

BLOCKI W. 1980. Ship safety in connection with parametric resonance of the roll, *Int. Shipbuilding Progress*, **27**, 36–53.

CHANTREL J-M. 1984. Instabilités paramétriques dans le mouvement des corps flottants. Application au cas des bouées de chargement, PhD thesis, Ecole nationale supérieure de mécanique, Nantes (in French).

CHATJIGEORGIOU I.K., MOLIN B. 2013. Third-order interactions, wave run-up and hydrodynamic loading on a vertical plate in an infinite wave field, *Appl. Ocean Res.*, **41**, 57–64.

CHAU F.P., EATOCK TAYLOR R. 1992. Second-order wave diffraction by a vertical cylinder, *J. Fluid Mech.*, **240**, 571–599.

COINTE R., ARMAND J-L. 1987. Hydrodynamic impact analysis of a cylinder, *J. Offshore Mech. Arct. Eng*, **109**(3), 237–243.

CROWLEY S., PORTER R. 2012. The effect of slatted screens on waves, *J. Engineering Mathematics*, **76**, 33–57.

DE LAUZON J., GRGIC M., DERBANNE Q., MALENICA Š. 2015. Improved generalized Wagner model for slamming, Proc. 7th Int. Conf. on Hydroelasticity, Split, Croatia.

DERN J-C. 1972. Unstable motion of free SPAR buoys in waves, *Proc. 9th ONR Symposium on Naval Hydrodynamics*, Paris.

DINGEMANS M.W. 1997. *Water Wave Propagation Over Uneven Bottoms. Part 1 - Linear Wave Propagation*, World Scientific.

DNVGL 2017. Environmental conditions and loads. Recommended practice DNVGL-RP-C205.

FALTINSEN O.M. 2000. Hydroelastic slamming, *J. Mar. Sci. Technol.*, **5**, 49–65.

FALTINSEN O.M., ZHAO R. 1997. Water entry of ship sections and axisymmetric bodies. In *Proceedings of AGARD FDP Workshop on High speed Body Motion in Water*, Kiev, Ukraine.

FALTINSEN O.M., NEWMAN J.N., VINJE T. 1995. Nonlinear wave loads on a slender vertical cylinder, *J. Fluid Mech.*, **289**, 179–198.

FERRANT P. 1998. Fully non linear interactions of long-crested wave packets with a three dimensional body, *Proc. 22nd ONR Symposium on Naval Hydrodynamics*, Washington D.C.

FERRANT P., MALENICA Š., MOLIN B. 1999. Nonlinear wave loads and runup on a vertical cylinder, Chapter 3 in *Advances in Fluid Mechanics, Nonlinear Water Wave Interaction*, WIT Press.

GODA Y., HARANAKA S., KITAHATA M. 1966. Study on impulsive breaking wave forces on piles. Report of the Port and Harbour Technical Research Institute, **6** (5), 1–30 (in Japanese).

GRAHAM J.M.R. 1980. The forces on sharp-edged cylinders in oscillatory flow at low Keulegan-Carpenter numbers, *J. Fluid Mech.*, **97**, 331–346.

GREENHOW M., LI Y. 1987. Added masses for circular cylinders near or penetrating fluid boundaries – Review, extension and application to water-entry, -exit and slamming, *Ocean Eng.*, **14**, 325–348.

GRUE J. 2002. On four highly nonlinear phenomena in wave theory and marine hydrodynamics, *Applied Ocean Res.*, **24**, 261–274.

HASLUM H.A., FALTINSEN O.M. 1999. Alternative shape of SPAR platforms for use in hostile areas, *Proc. 31st Offshore Technology Conference*, Houston, paper 10 953.

HE H., TROESCH A.W., PERLIN M. 2008. Hydrodynamics of damping plates at small KC numbers, in *Proc. IUTAM Symposium on Fluid-Structure Interaction in Ocean-Engineering*, 93–104.

HUSEBY M., GRUE J. 2000. An experimental investigation of higher-harmonic wave forces on a vertical cylinder, *J. Fluid Mech.*, **414**, 75–103.

JAMOIS E. 2005. Interaction houle-structure en zone côtière, Ph.D. report, Aix-Marseille II University (in French).

JARLAN G.E. 1961. A perforated vertical wall breakwater, *The Dock and Harbour Authority*.

KOO B.J., KIM M.H., RANDALL R.E. 2004. Mathieu instability of a spar platform with mooring and risers, *Ocean Eng.*, **31**, 2175–2208.

KOROBKIN A.A. 2005. Analytical models of water impact, *Eur. J. Appl. Math.*, **16**, 1–18.

KOROBKIN A.A. 2013. A linearized model of water exit, *J. Fluid Mech.*, **737**, 368–386.

KOROBKIN A.A., KHABAKHPASHEVA T., MAKI K.J. 2014a. Water exit problem with prescribed motion of a symmetric body, *Proc. 29th Int. Workshop Water Waves & Floating Bodies*, Osaka (www.iwwwfb.org).

KOROBKIN A.A., KHABAKHPASHEVA T., MALENICA Š., KIM Y. 2014b. A comparison study of water impact and water exit models, *Int. J. Nav. Archit. Ocean Eng.*, **6**, 1182–1196.

KOROBKIN A.A., PUKHNACHOV V.V. 1988. Initial stage of water impact, *Annu. Rev. Fluid Mech.*, **20**, 159–185.

KRISTIANSEN T., FALTINSEN O.M. 2017. Higher harmonic wave loads on a vertical cylinder in finite water depth, *J. Fluid Mech.*, **833**, 773–805.

LIU S., JOSE J., ONG M.C., GUDMESTAD O.T. 2019. Characteristics of higher-harmonic breaking wave forces and secondary load cycles on a single vertical circular cylinder at different Froude numbers, *Mar. Struct.*, **64**, 54–77.

LONGUET-HIGGINS M.S., COKELET E.D. 1976. The deformation of steep surface waves on water - I. A numerical method of computation, *Proc. Royal Soc. Lond.*, **350**, 1–26.

LUGNI C., BROCCHINI M., FALTINSEN O.M. 2006. Wave impact loads: the role of the flip-through, *Phys. Fluids*, **18**, 122101.

MALENICA Š. 1994. Diffraction de troisième ordre et interaction houle-courant pour un cylindre vertical en profondeur finie, thèse de doctorat de l'Université Paris VI (in French).

MALENICA Š, CHEN X.B., OROZCO J.M., XIA J. 2006. Parametric roll - Validation of a numerical model, *Proc. 7th Int. Conf. Hydrodynamics*, ICHD 2006, Ischia.

MALENICA Š., MOLIN B. 1995. Third-harmonic wave diffraction by a vertical cylinder, *J. Fluid Mech.*, **302**, 203–229.

MENTZONI F., KRISTIANSEN T. 2019a. Numerical modeling of perforated plates in oscillating flow, *Applied Ocean Res.*, **84**, 1–11.

MENTZONI F., KRISTIANSEN T. 2019b. A semi-analytical method for calculating the hydrodynamic force on perforated plates in oscillatory flow, *Proc. of the ASME 2019 38th International Conference on Ocean, Offshore and Arctic Engineering*, OMAE2019, Glasgow.

MOLIN B. 2011. Hydrodynamic modeling of perforated structures, *Applied Ocean Res.*, **33**, 1–11.

MOLIN B., KIMMOUN O., LIU Y., REMY F., BINGHAM H.B. 2010. Experimental and numerical study of the wave run-up along a vertical plate, *J. Fluid Mech.*, **654**, 363–386.

MOLIN B., KIMMOUN O., REMY F., CHATJIGEORGIOU I.K. 2014. Third-order effects in wave-body interaction, *European J. Mech. B/Fluids*, **47**, 132–144.

MOLIN B., KOROBKIN A.A. 2001. Water entry of a perforated wedge, *Proc. 17th Int. Workshop Water Waves & Floating Bodies*, Hiroshima (www.iwwwfb.org).

MOLIN B., LEGRAS J.-L. 1990. Hydrodynamic modeling of the Roseau tower stabilizer, *Proc. 9th Int. Conf. on Offshore Mechanics and Arctic Engineering (OMAE)*, Houston, Vol. I, part B, 329–336.

MOLIN B., KIMMOUN O., REMY F., JAMOIS E. 2006. Non-linear wave interaction with a long vertical breakwater, *Proc. 7th Int. Conf. Hydrodynamics*, ICHD 2006, Ischia.

MOLIN B., LEGRAS J.-L. 1990. Hydrodynamic modeling of the Roseau tower stabilizer, *Proc. 9th OMAE Conference*, Houston, pp. 329–336.

MOLIN B., NIELSEN F.G. 2004. Heave added mass and damping of a perforated disk below the free surface, *Proc. 19th Int. Workshop Water Waves & Floating Bodies*, Cortona (www.iwwwfb.org).

MOLIN B., REMY F. 2015. Inertia effects in TLD sloshing with perforated screens, *J. Fluids and Structures*, **59**, 165–177.

MOLIN B., REMY F. 2016. Hydrodynamic coefficients of a rectangular tank with a baffle, in *Proc. ISOPE Conf.*, Rhodos.

MOLIN B., REMY F., KIMMOUN O., JAMOIS E. 2005. The role of tertiary wave interactions in wave-body problems, *J. Fluid Mech.*, **528**, 323–354.

MOLIN B., REMY F., RIPPOL T. 2008. Experimental study of the heave added mass and damping of solid and perforated disks close to the free surface, in *Maritime Industry, Ocean Engineering and Coastal Resources,* Taylor & Francis, 879–887.

MORSE P.M., INGARD K.U. 1968. *Theoretical Acoustics*. McGraw-Hill.

NAKOS D.E. 1990. Ship wave patterns and motions by a three-dimensional Rankine panel method, Ph.D. thesis, MIT.

OULED HOUSSEINE C., MOLIN B., ZHAO W. 2020. Third-order wave run-up in diffraction-radiation problems, in *Proc. 35th Int. Workshop on Water Waves and Floating Bodies*, Seoul (www.iwwwfb.org).

PARK D.-M., KIM Y., SONG K.-H. 2013. Sensitivity in numerical analysis of parametric roll, *Ocean Engineering*, **67**, 1–12.

RADDER A.C. 1979. On the parabolic equation method for water-wave propagation, *J. Fluid Mech.*, **95**, 159–176.

RAINEY R.C.T. 1989. A new equation for wave loads on offshore structures, *J. Fluid Mech.*, **204**, 295–324.

SANDVIK P.C., SOLAAS F., NIELSEN F.G. 2006. Hydrodynamic forces on ventilated structures, *Proc. Sixteenth Intern. Offshore and Polar Eng. Conf.*, ISOPE, San Francisco, Vol. 4, 54–58.

SCOLAN Y.-M., KOROBKIN A.A. 2001. Three-dimensional theory of water impact. Part 1. Inverse Wagner problem, *J. Fluid Mech.*, **440**, 293–326.

SCOLAN Y.-M., LE BOULLUEC M., CHEN X.B., DELEUIL G., FERRANT P., MALENICA Š., MOLIN B. 1997. Some results from numerical and experimental investigations on the high frequency responses of offshore structures, in *Proc. 8th Int. Conference on the Behaviour of Offshore Structures*, BOSS, Delft.

TENG B., KATO S. 2002. Third order wave force on axisymmetric bodies, *Ocean Engineering*, **29**, 815–843.

VAN DYKE M.D. 1964. *Perturbation methods in fluid mechanics*, Applied Mathematics and Mechanics: an International Series of Monographs, **8**, Academic Press.

WAGNER H. 1932. Über Stoss- und Gleitvorgänge an der Oberfläche von Flüssigkeite, *ZAMM*, **12**, 193–215.

WIENKE J., OUMERACI H. 2005. Breaking wave impact force on a vertical and inclined slender pile – theoretical and large-scale model investigations, *Coast. Eng.*, **52**, 435–462.

ZAKHAROV V.E. 1968. Stability of periodic waves of finite amplitude on the surface of a deep fluid, *J. Appl. Mech. Tech. Phys.*, **9**, 190—194.

ZHAO R., FALTINSEN O.M. 1993. Water entry of two-dimensional bodies, *J. Fluid Mech.*, **246**, 593–612.

ZHAO R., FALTINSEN O.M., AARNSES J. 1996. Water entry of arbitrary two-dimensional sections with and without flow separation, *Proc. 21st Symposium on Naval Hydrodynamics*, Trondheim, Norway.

ZHAO W., TAYLOR P.H., WOLGAMOT H.A., EATOCK TAYLOR R. 2019. Amplification of random wave run-up on the front face of a box driven by tertiary wave interactions, *J. Fluid Mech.*, **869**, 706–725.

9   Model Testing

9.1   Introduction

It is customary to perform model tests, at one stage or another of an offshore project. There might be different reasons for doing model tests:

- recourse to the experimental tool because theoretical or numerical models are lacking;
- determination/checking of certain physical parameters to be introduced into a numerical model (for instance wind or current load coefficients);
- ultimate check of a design;
- commercial promotion.

The last two are significant points: Many people within the industry are still unwilling to rely on numerical models and have more confidence (too much so in some cases) in physical modeling.

For researchers in hydrodynamics, wave tanks are valuable exploratory tools that allow their theories and numerical models to be validated and, sometimes, new or little known phenomena to be (re-)discovered. There is a strong interaction between progress in experimental techniques and progress in theoretical/numerical modeling.

This complementarity of physical and numerical models has become of more acute importance since the offshore industry has moved on to deep and ultra deep water: It is no longer possible, in existing facilities, to model, at a reasonable scale, the floating support together with its mooring lines and risers. Separate tests must be carried out on the different components, and their results assembled within a numerical model. Alternatively, a numerical model can be directly connected to the physical tests, for instance to input the reactive loads from the mooring lines to the support. This technique (sometimes known as "real-time hybrid model testing" or "software in the loop") is nowadays routinely used to input the wind loads upon floating wind turbines.

9.2   Modeling Principles and Scale Effects

9.2.1   Main Principles

The leading idea is to use a scaled model of the actual structure. With the scale being denoted as λ, all dimensions of the model relate to the full-scale ones through

$d = \lambda D$ (no geometric distortion), areas through $s = \lambda^2 S$, and volumes through $v = \lambda^3 V$. Most wave basins use fresh water (the Ifremer ocean basin at Brest being an exception). This means that the buoyancy force relates to that of the full-scale structure through

$$f_B = \frac{\rho_{fw}}{\rho_{sw}} \lambda^3 F_B \qquad (9.1)$$

with ρ_{fw} the density of fresh water ($\rho_{fw} = 1000$ kg/m^3) and ρ_s the density of sea water ($\rho_s \simeq 1025$ kg/m^3).

Hydrostatic balance imposes that masses obey the same relationship:

$$m = \frac{\rho_{fw}}{\rho_{sw}} \lambda^3 M = \mu \, \lambda^3 M \qquad (9.2)$$

Length and mass scales being thus defined, it remains to determine the time scale.

When dealing with tests in waves, it is intuitive that wave lengths must follow the same length scale as the model. In regular waves, the wave length L is related to the period T via the dispersion equation:

$$L = \frac{g \, T^2}{2 \, \pi} \tanh 2 \, \pi \, \frac{h}{L} \qquad (9.3)$$

where the period appears squared (h being the water depth).

It ensures that the wave period in the tank must be taken as the full-scale period multiplied by the square root of the scale factor:

$$t = \sqrt{\lambda} \, T. \qquad (9.4)$$

Velocities and accelerations are then related by:

$$v = \sqrt{\lambda} \, V \qquad (9.5)$$

$$\gamma = \Gamma. \qquad (9.6)$$

In the basin, the acceleration is the same as in full scale, a result that makes sense since gravity is the reference. The choice of this scaling law, known as **Froude** scaling, is closely associated with the presence of a free surface.

Froude scaling has been applied for more than a century, by naval architects, to the determination of forward speed resistance of hull forms. This resistance has two main components: the wave resistance associated with the wave pattern generated by the ship, and the frictional resistance associated with the viscous boundary layer. To obtain the same wave pattern, the Froude number, defined as $F_r = U/\sqrt{gL}$ (with U the forward speed and L the ship length), must be kept identical. The wave resistance coefficient is then the same in model scale and in full scale. The frictional coefficients differ, but they can be corrected via the equivalent flat plate principle.

Whenever the wave response of a structure is involved, and/or free surface effects cannot be neglected, Froude scaling law must be applied. This is the case of most model tests performed in offshore engineering or naval architecture. But there are cases when Froude scaling need not be applied, for instance the hydroelastic response of risers under current loading.

Length	Time	Velocity	Acceleration	Mass	Pressure	Force	Moment
λ	$\sqrt{\lambda}$	$\sqrt{\lambda}$	1	$\mu\,\lambda^3$	$\mu\,\lambda$	$\mu\,\lambda^3$	$\mu\,\lambda^4$

Scale factors in Froude scaling. λ: length scale. μ: density ratio (fresh water/sea water)

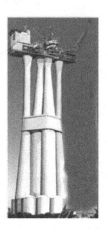  

Figure 9.1 Elastic model of the Troll GBS for ringing experiments (courtesy SINTEF Ocean and Equinor).

Rigid models are usually built from marine plywood, but other materials are also used such as wax (for ship models undergoing towing tests) or metal. Water tightness is obviously a main concern. When designing the model, some provision must be taken for adjustments of the weight, position of the center of gravity, and inertias through inserting additional weights. In many cases, several loading conditions are specified.

Elastic models are sometimes requested. Some ships, for instance containerships, have some flexibility, and their hull girders may undergo bending or torsion under nonlinear wave loads or bow slamming. Since the Young modulus E has the same dimension as a pressure, a geometric model should be made of a very soft material. The usual technique is to construct a segmented model, with an elastic backbone to which the successive hull sections are attached. Figure 9.1 shows a similar technique being applied to the Troll GBS.

9.2.2   Scale Effects

Application of Froude scaling creates a bias to all phenomena associated with viscosity. It has been seen that viscous effects are related to the Reynolds number:

$$\text{Re} = \frac{U\,D}{\nu} \tag{9.7}$$

with U the flow velocity and D a typical dimension.

The kinematic viscosity ν is about the same for fresh or sea water (in fact ν is more sensitive to temperature than to salinity). Froude scaling law applied to the characteristic dimension D ($d = \lambda\,D$) and to the velocity U ($u = \sqrt{\lambda}\,U$) leads to:

$$re \simeq \lambda^{3/2} Re \qquad (9.8)$$

Therefore, it is impossible to keep both the Froude and Reynolds numbers the same at model scale and at full scale.

At typical scale values around 1:50, the difference between model and full-scale Reynolds numbers is quite large.

As an example, take the case of a cylinder of diameter 0.50 m in a current of velocity 1 m/s: The Reynolds number is around $5 \ 10^5$: The flow regime is critical, the drag coefficient is very low, around 0.3, and the lift force is quite erratic with no dominant frequency. At a scale of 1:50, the Reynolds number drops down to 1 500: The flow is subcritical, the drag coefficient around 1.2, the lift force quasiperiodic at the Strouhal frequency. As a consequence, the drag and lift forces are biased. When vortex-induced vibration (VIVs) are not a concern and only steady drag forces matter, a remedy can be to reduce the diameter, here from 10 mm down to 2.5 mm.

Similarly, taking the case of an oscillatory flow with a full-scale period 10 s upon the same cylinder, the Stokes parameter $\beta = D^2/(vT)$ drops from 25,000 down to 70 at a 1:50 scale. The Keulegan–Carpenter number is conserved. Referring to Figures 4.17 through 4.22 in Chapter 4, it can be seen that both the drag and inertia coefficients are altered.

Another issue, related to the slow-drift surge motion of moored ship-shaped bodies, is the viscous damping upon the hull: At model scale, the oscillatory flow is laminar, implying that the wave flow and the slow-drift flow are decoupled. At full scale, the boundary layer is turbulent and the two flows are coupled, with an enhanced damping effect (see Section 7.7.5 in Chapter 7).

Other cases could be cited where it can be concluded that scale effects notably affect viscous dissipation processes. Whenever the considered system exhibits some resonant behavior, the question must be raised of a correct representation of the viscous damping effects in the basin.

Other problems may result from the fact that the atmospheric pressure is not scaled. Ideally, it should be decreased by the factor $\mu \lambda$ (see the table above). This matters in the case of floating structures with air cushion: The aeroelastic stiffness of the air cushion becomes too large and some ad hoc remedy must be devised such as adding an elastic membrane to the floater (see Figure 9.2). Another solution is to depressurize the whole facility: Some tanks, for instance MARIN's depressurized tank in Ede, offer this feature. The main application there relates to ship design, that is propeller cavitation.

In the case of wave impacts, a similar issue arises when some gas becomes entrapped: The air pocket is too stiff. This has been widely studied with regard to sloshing in LNG tanks, and some ad hoc remedies, based on the Bagnold model, have been proposed to exploit the test results (e.g., see Dias & Ghidaglia, 2018). On an other side, it has been established that it is important to keep the density ratio between fluid and gas to get the proper shape of the free surface at the impact time. For this reason, LNG sloshing, at model scale, is usually studied with water and a gas heavier than air (sulfur hexafluoride SF_6). Another experimental issue related to sloshing in LNG tanks is mass transfer between liquid and gas. Tests with near-boiling water have been done in some instances.

Figure 9.2 Tests on an air-cushioned model with an elastic membrane. The elastic membrane is there to compensate for the excessive aerostatic stiffness of the air cushion (courtesy Technip/Technip Energies).

9.2.3   Choosing the Scale

Model testing of offshore systems is most commonly done at scales between 1:80 and 1:40.

The choice of the scale must take into account several, often contradictory, considerations.

The first relates to the specified wave conditions. Maximum wave heights that can be generated by the wavemakers rarely exceed 80 cm, and they are usually reached at periods between 2 and 3 s (see Figure 9.8). In North Sea conditions with wave heights up to 30 m and peak periods around 16 s or 18 s, one can hardly operate at scales larger than 1:40. Some basins offer much smaller wave height capability, of the order of 40 or 60 cm, which leads to scales of less than 1:50.

A second consideration relates to the water depth. The deepest ocean basins are all about 10 m deep, with an occasional deeper pit to accommodate TLP tethers. When one wants to represent a complete production system from a FPSO with risers and mooring lines, the question arises on whether the whole system needs to be modeled or whether the mooring lines and risers may be truncated. For this issue, the width of the basin also matters since the mooring lines may expand far away on the sea floor.

Other considerations concern the size of the model, with respect to the dimensions of the basin: too large a model means spurious effects due to confinement, such as reflections from the wavemaker and from the sidewalls. A large model will also be expensive, possibly difficult to handle. With too small a model, it will be difficult to achieve and control the mass distribution ensuring the specified inertias and position of the center of gravity.

Too small a scale entails possible accuracy problems in the measurements and in the adjustment of the wave parameters.

Some particular scale values may have to be avoided because leading to the coincidence of a natural frequency of the system with a critical frequency of the basin, for instance the natural frequency of a transverse or longitudinal sloshing mode: If the natural period in surge of a moored FPSO is 3 minutes at full scale and the natural period of the first longitudinal seiching mode of the basin equal to 25 s, a 1:50 scale is to be avoided!

Finally, the larger the scale, the smaller the scale effects, even though the difference in Reynolds numbers will always be quite large and, most often, the flow in the basin will be in the laminar regime while turbulent at full scale. In towing tests of ship hulls, it is customary to add some roughness to the hull by the bow to speed up the transition to turbulence. When the bodies have sharp edges which impose the separation points, the mismatch in Reynolds numbers is not so much of a problem. With smooth bodies such as circular cylinders, there is no clear remedy: When the loading is dominated by drag, a possibility is to decrease the diameter, scaling the product $C_D D$.

9.3   Experimental Facilities

Experimental facilities were first constructed for the needs of the shipbuilding industry: Long and relatively narrow tanks equipped with carriages to tow the models, the main application being the experimental determination of the forward speed resistance of ships. Most existing towing tanks are fitted with a wavemaker at one end (and a beach at the other end) that can produce regular and irregular waves. Some towing tanks are several hundred meters long; for instance, the B600 towing tank of the French Navy, in Val de Reuil, is 545 m long and 15 m wide, with a depth of 7 m.

Because of their limited widths, towing tanks were quickly found to be inappropriate for the needs of the offshore industry, and dedicated offshore basins (or ocean basins) started appearing in the 1970s and 1980s.

The ocean basin of SINTEF Ocean, at Trondheim, was designed to perform tests on offshore systems at a scale of 1:50 in up to 500 m water depth: It is 80 m long, 50 m wide, and 10 m deep. By means of a false bottom, the water depth can be decreased to any value. The basin is fitted with wavemakers on two adjacent sides; the wavemaker on the long side is segmented in order to produce oblique and/or short-crested seas. Current can be generated in the longitudinal direction, the return flow taking place below the false bottom.

Many other offshore basins have been built, in the Netherlands (MARIN, see Figure 9.3), in Canada (St-John's), Texas (OTRC), Shanghai (Shanghai Jiao-Tong University), etc. The National University of Singapore now operates a Deepwater Ocean Basin, 60 m long, 48 m wide, 12 m deep, with a central pit adding another 50 m in depth. Figure 9.4 shows the ocean basin BGO-FIRST, at La Seyne-sur-Mer (France), of relatively modest size (40 m in length, 16 m in width), where many of the experimental results shown in this book were obtained.

Tanks of large size are also found in coastal engineering. They are occasionally used by the offshore industry for shallow depth or near shore applications. It is customary, in such tanks, to model the actual bathymetry and coastline profile.

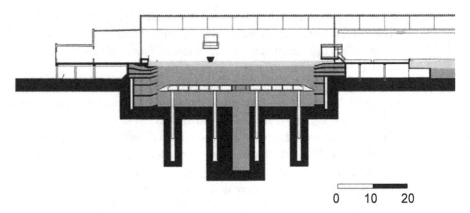

0   10   20

Figure 9.3 The Offshore Basin at MARIN.

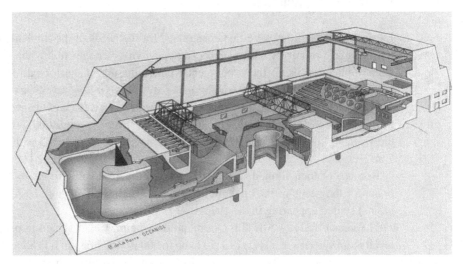

Figure 9.4 The BGO-FIRST wave tank at La Seyne-sur-Mer (courtesy Océanide).

9.4   Wave Generation and Absorption

9.4.1   Types of Wavemakers

Waves are generated by setting into motion a mechanical device, the **wavemaker**. The most common type of wavemaker is the "flap," articulated at its base and set in rotation, for example, by a hydraulic jack (the hinge point is not necessarily located at the tank floor; it may be higher or lower). In the case of a simple horizontal translation, the wavemaker is of "piston" type. Flaps and pistons may be "dry" or "wet" depending on whether the rear compartment is empty or full of water (see Figure 9.5).

The dry option has the main disadvantage that the actuating system must resist the hydrostatic load, that can amount to several tons per linear meter. The secondary

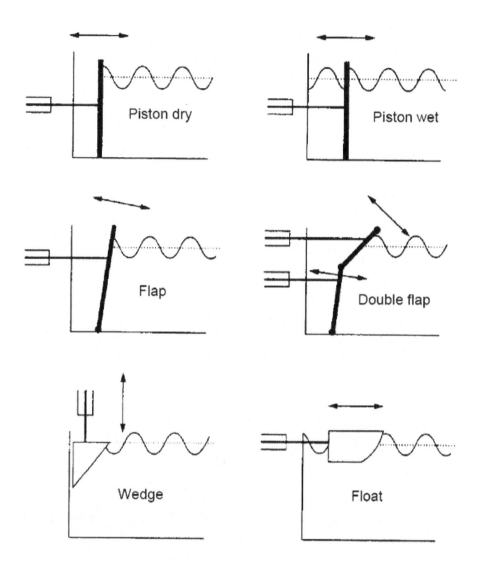

Figure 9.5 Different types of wavemakers.

problem of leaks taking place between the flap and the walls is easily solved by a pump returning the infiltrated water into the basin.

The wet mode has the drawback that the body of water in the rear compartment is also set in motion by the wavemaker: A strong sloshing response may take place, requiring some means of mitigation, for instance perforated walls or rubbles. The dynamic part of the force applied to the wavemaker is greater than in dry mode since both sides contribute to the added mass.

The efficiency of a wavemaker may be expressed as the transfer function relating the amplitude of the waves generated to the motion amplitude of the wavemaker. In the piston case, the transfer function is

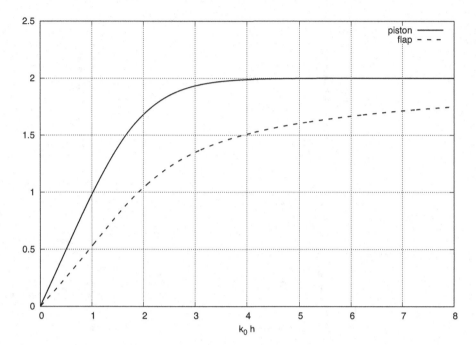

Figure 9.6 Transfer functions of the piston and flap wavemakers.

$$A/x_0 = \frac{4 \sinh^2 kh}{2\,kh + \sinh 2\,kh} \tag{9.9}$$

(see Appendix A).

In the flap case (with the articulation point at the tank floor):

$$A/x_0 = \frac{4 \sinh kh\,(kh\,\sinh kh - \cosh kh + 1)}{kh\,(2\,kh + \sinh 2\,kh)} \tag{9.10}$$

where the amplitude x_0 is taken at the free surface level.

In these equations, h is the water depth, k the wave number, linked to the frequency ω by the dispersion equation:

$$\omega^2 = gk\,\tanh kh. \tag{9.11}$$

Figure 9.6 shows the two transfer functions (9.9) and (9.10), plotted versus kh. At equal wavemaker amplitude x_0, the piston is more efficient than the flap, particularly in shallow conditions: As kh goes to zero, the piston is twice more efficient. As kh goes to infinity, both transfer functions have the same asymptotic 2 value, but the piston wavemaker is penalized since a larger mass of water is set into motion, requiring a larger driving force.

With regard to minimizing the driving force, the optimal wavemaker should move according to the $\cosh k(z+h)/\sinh kh$ profile of the horizontal motion of the wave particles. This argument leads to the double flap, where a secondary flap is embarked at the upper end of the first one. Depending on the wave length, the motions of the two flaps are determined to be as close as possible to the ideal profile. The most powerful wavemakers are usually double flaps, for instance at SINTEF Ocean or at the B600 in Val-de-Reuil (Figure 9.7).

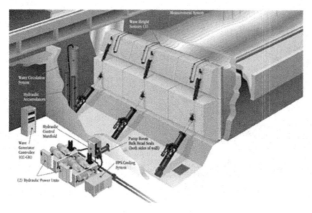

Figure 9.7 The biflap wavemaker at B600 towing tank at Val-de-Reuil (courtesy MTS).

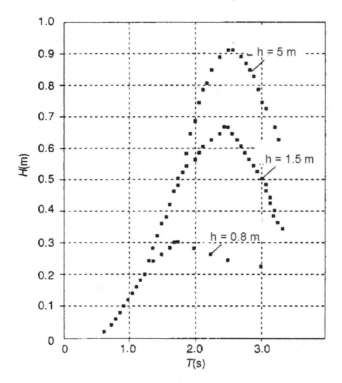

Figure 9.8 Maximum wave heights in the Ocean Basin of SINTEF Ocean, for three water depths.

There are other types of wavemakers, for instance the wedge in forced vertical (or close to vertical) motion, which has the advantage of minimizing the space occupied; or just a horizontal float in forced horizontal motion, such as at BGO-FIRST (see Figure 9.4), offering the advantage of combined current generation.

The heights of the waves generated depend on the stroke and power of the wavemaker and are also limited by wave breaking which occurs, at large depth, when the steepness H/L exceeds about 10%. Figure 9.8 shows the performances of the double flap in the Ocean Basin of SINTEF Ocean, in three different water depths.

Figure 9.9 Three-dimensional focused wave in Centrale Nantes Ocean Basin (courtesy Centrale Nantes).

Snake Wavemakers

Most offshore tanks are equipped with multiple flap wavemakers (so-called *snake wavemakers*), often along two adjacent sides, each flap being a few decimeters wide. When some phasing is introduced between the paddles motions, oblique waves are produced. To obtain waves propagating at an angle β with respect to the normal direction to the line of wavemakers, a phase shift equal to $kd \sin \beta$ must be introduced between the motions of adjacent units, d being their width that must be less than half the wave length. This also enables short-crested seas to be generated. Another advantage is that the mean wave direction can be varied without having to rotate the model and its moorings. Three-dimensional wave focusing can also be achieved, as Figure 9.9 shows.

9.4.2   Wave Absorbers

At the tank end opposite to the wavemaker, some means must be devised to destroy the waves and avoid their reflection: the wave absorber.

The most common wave absorber is just a beach: As the water depth decreases the waves overturn and break, most of their energy is then dissipated through turbulent mixing and viscous effects. When the bottom slope s is constant, a relevant parameter is the **Iribarren number** ξ, defined as:

$$\xi = s \sqrt{\frac{L_\infty}{H_b}} \tag{9.12}$$

where L_∞ is the wave length in infinite depth: $L_\infty = g T^2 / (2\pi)$, and H_b is the wave height at breaking point. Breaking occurs when ξ is less than about 2.3 and the reflection coefficient, defined as the ratio of the reflected wave height to the incoming wave height, is around $0.1 \, \xi^2$: To absorb low steepness waves, very mild slopes are necessary if one wants to keep the reflection coefficient C_R less than some target, 5% for instance. As breaking occurs at a depth comparable with the local wave height, it can be assumed that it is around that point that the slope matters the most. One strategy would then be to keep the Iribarren number less than, or equal to, some critical value

ξ_C for $H_b \sim h(x)$, leading to:

$$h'(x) \sqrt{\frac{L_\infty}{h(x)}} = \xi_C \tag{9.13}$$

giving, upon integration:

$$h(x) = \frac{\xi_C^2}{4} \frac{x^2}{L_\infty} = \frac{5 C_R}{2} \frac{x^2}{L_\infty} \tag{9.14}$$

where the abscissa x is counted from the beach toward the wavemaker: The beach profile should be parabolic, with a slope all the more gentle as the target reflection coefficient and target wave number are small. One can deduce the necessary beach length l_B, considering that it starts at a depth equal to half the wave length:

$$l_B = \sqrt{\frac{1}{5 C_R}} L_\infty \tag{9.15}$$

For a C_R value of 5%, this gives a beach length equal to two wave lengths. At a nominal wave period of 2.5 s, it is 20 m. In fact, the main beach at SINTEF Ocean is close to 20 m long (Figure 9.10).

When they are well-shaped, and sufficiently long, beaches are very efficient, ensuring reflection coefficients of less than 5% or 10% over a wide range of wave periods. Their efficiency may be improved by taking advantage of some other mechanisms of energy dissipation, for instance with some added roughness, or porosity (see Figure 9.11).

When the basin is not fitted with a false bottom, that can be raised or lowered, it becomes awkward to ensure a good performance of the beach for different water levels. An alternative to the beach, used in some tanks, is the **progressive wave absorber**, which consists in a succession of vertical perforated screens of decreasing porosity toward the end wall. These absorbers have been devised by the National Research Council of Canada (Jamieson & Mansard, 1987) through extensive experimental testing. They can easily be modeled numerically,[1] as Figure 9.12, taken from Molin & Fourest (1992), shows. It can be shown that 100% of the incoming wave energy can be absorbed with a single perforated plate located a quarter wave length from the

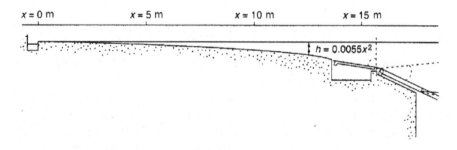

Figure 9.10 Profile of the beach at SINTEF Ocean.

---

[1] See the section on perforated structures in Chapter 8.

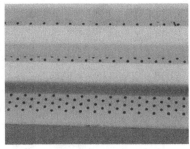

Figure 9.11 Details of the beach at Ifremer Brest, combining parabolic shape, roughness, and porosity (from Le Boulluec *et al.*, 2007. Copyright Ifremer - Bompais Xavier).

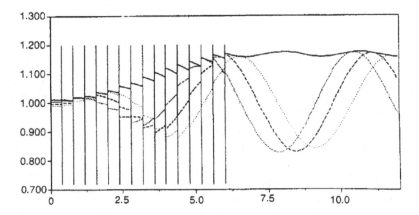

Figure 9.12 Wave attenuation through a progressive wave absorber (from Molin & Fourest, 1992).

end wall, the optimal open area ratio depending on the wave steepness. To absorb the wave energy in irregular waves of varying frequencies and steepnesses, a large number of screens, typically a dozen, and an overall length comparable to the longest wave lengths, are required. Such wave absorbers can be found at the Multidirectional Wave Basin of NRC at St-John's, Canada, at OTRC in Texas, and at CCOB in Santander. An advantage over the parabolic beach is that they are relatively insensitive to the water depth. A variant is a horizontal, or quasihorizontal, perforated plate, slightly immersed (Molin, 2001). This system may offer appreciable advantages when current is to be combined with waves.

Finally, there are dynamic absorbers: The absorber is just a wave paddle whose motion is driven to absorb the incoming waves. Some information is then required to identify the incoming waves: The free surface elevation needs to be measured some short distance from the paddle; another option is to incorporate pressure or force sensors. These techniques are well-mastered in coastal engineering facilities where the wavemakers need to be driven in order to absorb the reflected waves from the coastal structure studied. In Edinburgh, the FloWave Ocean Energy Research Facility is a circular basin, 25 m in diameter, lined all around with 168 paddles that can generate (and absorb) all kinds of waves propagating in any direction.

Model basins usually qualify their beaches through dedicated tests, in regular waves, that provide the reflection coefficient C_R, defined as the ratio A_R/A_I of the amplitude of the reflected wave to the amplitude of the incident wave. Sometimes the reflection coefficient is defined as an energy ratio, that is A_R^2/A_I^2. One should be aware that 1% of the energy reflected, which sounds like a good performance, means 10% of the amplitude reflected, which is a rather poor result!

The reflection coefficient can be obtained via various methods, the simplest one being using just two wave gauges. When the free surface elevation is taken to be:

$$\eta(x,t) = A_I \, \cos(k\,x - \omega\,t) + C_R\,A_I \, \cos(k\,x + \omega\,t + \phi) \qquad (9.16)$$

the local amplitude of the free surface motion is:

$$A(x) = A_I \, \sqrt{1 + C_R^2 + 2\,C_R \, \cos(2\,k\,x + \phi)} \simeq A_I \left(1 + C_R \, \cos(2\,k\,x + \phi)\right) \qquad (9.17)$$

When η is measured in two points of adequate separation (ideally a quarter wave length), the reflection coefficient is obtained.

There are other methods, for instance using more than two gauges and minimizing a difference, or just one gauge, by producing a wave packet, short enough that the incoming and reflected packets be well separated, but long enough that the packets be not too modified by dispersion. In towing tanks, one may also use a single gauge attached to the moving carriage: The incoming and reflected wave systems get separated by their different encounter frequencies. There are also techniques to separate incoming and reflected components in irregular waves (e.g., see Mansard & Funke, 1980).

Strictly the reflection coefficients depend both on the wave period and on the wave steepness. It is rare that model basins give complete information on their reflection coefficients (the dependence on steepness is often missing).

In towing tanks, the beach performance can be visually estimated (during calibration tests of regular waves) by looking at the separation contour between the wet and dry parts of the longitudinal walls: If it is a straight line, the reflection coefficient is nil.

9.5   Wave-Related Parasitic Phenomena

Doing model tests in regular waves sounds very simple: Just assign a sinusoidal motion to the wavemaker, wait for a steady state to be established, and start recording.

When looking at the records of the different sensors, it often looks as if a steady state is never achieved: One can see transients, long-lasting modulations, beating effects, superimposed low or high-frequency components, etc. A problem that arises then is the best choice of the time window to do the analysis. Sometimes this choice is quite subjective.

In this section, we describe some of the spurious phenomena that affect wave propagation in a confined tank.

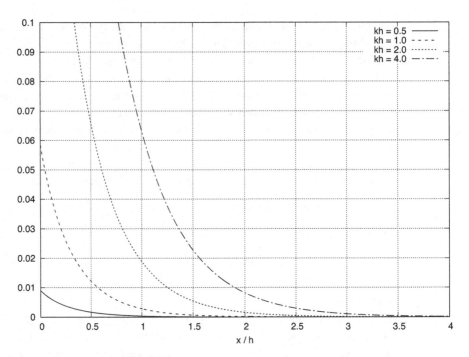

Figure 9.13 Piston wavemaker. Relative contribution of the evanescent modes to the free surface elevation.

9.5.1   Near Field of the Wavemakers

It is established in Appendix A that the wave system generated by the periodic motion of a wavemaker consists in a propagating mode together with a superposition of evanescent modes that decay exponentially with the distance from the wavemaker. It is generally considered that after twice the water depth, the contribution of the evanescent modes is negligible.

Figure 9.13 shows the relative contribution of the evanescent modes to the free surface elevation, in the case of a piston wavemaker, for different kh values. Twice the water depth ensures that the relative contribution is less than 1%, when kh is less than 4.

9.5.2   Modulation of the Wave Front

As the wavemaker gets started, the wave front progresses in the tank at the group velocity $C_G = \partial \omega / \partial k$. Some modulation appears in the first waves, and, far away from the wavemaker, it takes many waves for the amplitude to stabilize. This modulation is predicted by linear theory and is the result of dispersion.

Figure 9.14 shows the theoretical profile of the wave envelope, normalized by the steady state wave amplitude, in a coordinate system traveling at the group velocity. The parameter on the horizontal axis is $x/\sqrt{d\,L}$ with x the horizontal coordinate relative to the wave front, d the traveled distance from the wavemaker and L the wave length (the figure is for deep water).

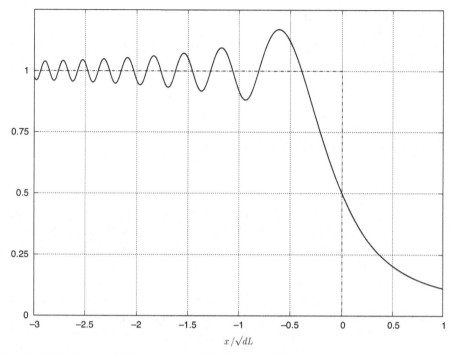

$x/\sqrt{dL}$

Figure 9.14 Modulation of the wave envelope at the wave front.

Free surface nonlinearities tend to increase the modulation, and it is frequent to see breaking (or crescent waves) at the wave front in steep wave conditions.

9.5.3   Benjamin–Feir Instability

It has been seen in Chapter 3 that regular waves are naturally unstable. When the steepness H/L is less than 10%, the main mechanism is the Benjamin–Feir insta- bility: Third-order free surface nonlinearities give rise to parasitic wave components at frequencies $\omega\,(1 \pm kA)$, which first grow exponentially with the distance from the wavemaker (Benjamin & Feir, 1967). The nonlinear Schrödinger equation then predicts a series of modulations and demodulations. When the wave steepness is mod- erate, usually the Benjamin–Feir instability does not appear unless the tank is very long or the wavemaker control quite poor. At high steepnesses, Benjamin–Feir insta- bility quickly leads to wave breaking, disorganizing the wave system (see Figures 3.25 and 3.26 in Chapter 3).

In irregular seas, third-order nonlinearities lead to a slow evolution of the spectral shape along the tank, with the energy moving to the lower frequencies.

The Benjamin–Feir instability usually is not a problem in ocean basins where the model is stationary at a relatively short distance from the wavemaker. It can be a matter of concern in towing tanks when the model is towed over the whole length of the tank and is likely to meet a slightly changing sea state.

9.5.4　Return Current

It has been seen in Chapter 3 that mass transport takes place in waves, the flow rate being given by:

$$Q = \frac{1}{2} A^2 \omega \coth kh. \tag{9.18}$$

In the absence of current generation, the mass transport is compensated by a return current, which has the depth-averaged velocity:

$$U_R = -\frac{1}{2} \frac{A^2 \omega}{h} \coth kh. \tag{9.19}$$

As pointed out in Chapter 3, this return current is far from negligible, particularly at shallow depth. Taking for instance a wave amplitude A equal to 30 cm, a wave period of 2 s, and a depth of 2 m, the return current velocity is 7 cm/s, that is 1 knot at a scale of 1:50!

The vertical profile of the return current has been a controversial matter. Within the scope of potential flow theory, there can be only one profile, that is constant velocity over the depth (see Figure 3.6). Viscous effects result in the appearance of boundary layers at the bottom and at the free surface: oscillatory boundary layers which have a thickness of the order of $\sqrt{\nu/\omega}$, and steady boundary layers with a thickness growing in time as $\sqrt{\nu t}$, t being the duration of the test. These effects are of concern only in very small (too small) experimental facilities, a few decimeters deep.

One way to get rid of the return current would be to cogenerate a current with the waves, which would oppose the return current. This is hardly ever done.

9.5.5　Free Harmonics

Whatever the wavemaker, when the boundary conditions at its wall and at the near free surface are expanded to second-order, forcing terms at the double frequency appear. As a result, a parasitic wave is generated with frequency 2ω and wave number k_2 linked to 2ω by the dispersion equation:

$$4\omega^2 = g k_2 \tanh k_2 h. \tag{9.20}$$

Away from the wavemaker (after evanescent components have died out), the free surface elevation, developed to second-order, has the form:

$$\eta(x,t) = A \cos(k x - \omega t) + C_L k A^2 \cos(2 k x - 2\omega t) + C_F k A^2 \cos(k_2 x - 2\omega t + \phi). \tag{9.21}$$

At large depth ($kh > 3$), the coefficient C_L of the locked wave component is 1/2. The coefficient C_F of the free wave depends on the type of wavemaker. Numerical results have been given by Sulisz & Hudspeth (1993) and Schäffer (1996) for piston and flaps (with different positions of the hinge point). It appears that the ratio C_F/C_L of the free to locked harmonics is usually between 0.5 and 1 (Figure 9.15).

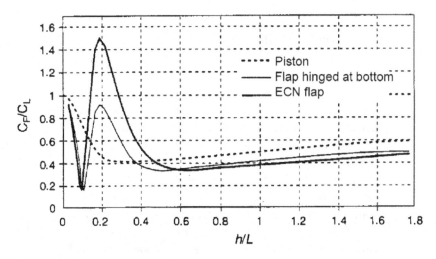

Figure 9.15 Double frequency wave components. Ratio C_F/C_L for different types of wavemakers (from Stassen, 1999).

It is to be noted that the second-order free waves propagate at a lower speed than the primary waves. When the model is some distance away from the wavemaker, the test may be completed prior to arrival of the second-order free waves.

Since the free wave content can be calculated, an additional double frequency signal can be given as an input into the wavemaker in order to cancel out the free wave emission.

When the wavemaker signal is multichromatic, second-order free waves are generated at the sum $\omega_i + \omega_j$ and difference $\omega_i - \omega_j$ frequencies of the wave components. For piston and flap wavemakers, their amplitudes and phases can be calculated (Schäffer, 1996) and the wavemaker can then be driven in order to cancel them out. This feature is offered by some model basins. It must be emphasized that these corrections assume a steady wave system to have been achieved in the tank, they ignore the transient period when the waves start being generated and progress toward the beach and the persistence of the natural modes excited in this time span.

9.5.6   Natural Modes

Due to confinement, some particular modes of deformation of the free surface, the natural modes of the tank, may preferably occur. In Chapter 6, it has been seen that, in a tank of length L, width B, and at a water height h, the natural sloshing modes have the following velocity potentials:

$$\Phi_{mn}(x,y,z,t) = \frac{A_{mn} \, g}{\omega_{mn}} \frac{\cosh \nu_{mn}(z+h)}{\cosh \nu_{mnn} \, h} \cos \lambda_m x \, \cos \mu_n y \, \sin(\omega_{mn} t + \theta) \quad (9.22)$$

with $\lambda_m = m\pi/L$, $\mu_n = n\pi/B$, m and n integers, $\nu_{mn}^2 = \lambda_m^2 + \mu_n^2$, and the resonant frequencies are given by:

$$\omega_{mn}^2 = g \, v_{mn} \, \tanh v_{mn} h \qquad (9.23)$$

the associated free surface elevation being:

$$\eta_{mn}(x,y,t) = A_{mn} \, \cos \lambda_m x \, \cos \mu_n y \, \cos(\omega_{mn} t + \theta) \qquad (9.24)$$

The particular cases $m = 0$, $n = 1,2,\ldots$ correspond to pure transverse modes; the cases $n = 0$, $m = 1,2,\ldots$ to longitudinal modes; m and n nonzero to combined modes. Typically, the natural periods of the longitudinal modes are very high, beyond the range of wave periods (but possibly coinciding with the natural periods of the slow-drift responses). For instance with a length L equal to 80 m and a depth h of 3 m, the periods of the first three modes are equal to 29.6 s, 14.9 s, and 10 s, the associated wave lengths being $2L$, L, and $3L/2$. At such wave lengths, the absorbing beaches are totally ineffective, more or less equivalent to vertical walls.

In the absence of side absorbers, the transverse modes are hardly damped, and their natural periods are more likely to fall within the wave period range.

As with any mechanical system, the natural modes of the tank are triggered during the transient phases, for instance when the wavemaker is started or stopped, and in the time span when the wave front travels from the wavemaker to the beach. Being weakly damped, these modes last throughout the tests and beyond. It is customary to wait for quite some time, typically several dozens of minutes, in between two tests, for the basin to go back to calm, as the longitudinal seiching modes take a very long time to die out.

It is important to check that no natural frequency of the tested system falls close to a natural mode of the tank. For instance, for moored structures, that the natural frequencies of their slow-drift motion do not coincide with any mode of the tank.

9.5.7 Seiching

The problem of seiching has just been mentioned: After a test in waves, the first longitudinal sloshing modes of the tank persist for a long time and delay the following test. In fact, as shown here, they are present from the very beginning of the test, which they spoil to some extent.

As written earlier, the tank natural modes are triggered when the wavemaker is started. It is often considered that increasing the wavemaker motion progressively decreases the effect. This is true but long mode generation actually is a result of the transient phase that lasts until the wave front reaches the opposite end of the tank (and, similarly, at the end of the test, when the rear front travels from the wavemaker to the beach).

As an illustration, Figure 9.16 shows experimental time traces from experiments carried out at the towing tank of Centrale Nantes. The tank is 65 m long,[2] and the water depth during the tests is 2.8 m. The upper plot shows the horizontal motion of the wavemaker, in mm: About 40 cycles are generated at a 2.4-s period, with short

---

[2] Its length was later on increased to 140 m.

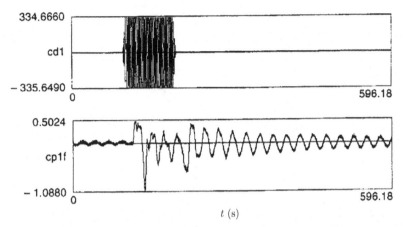

Figure 9.16 Regular wave tests in the towing tank of Centrale Nantes. Top: wavemaker motion (mm). Bottom: pressure (in cm of water) behind the beach at 1.3 m below the free surface.

linear ramps at the start and at the end. The motion amplitude of the wavemaker (a flap) is 32 cm, and the amplitude of the waves produced is 22.5 cm. The bottom plot shows the pressure measured behind the beach (in cm of water), at an immersion of 1.3 m. Some remaining pressure oscillations from the previous test can be seen before the wavemaker is turned on. The oscillation period is 25 s, which corresponds to the first sloshing mode of the tank.

Shortly after the wavemaker is started (before the wave front reaches the beach), a 4 mm rise in the pressure can be seen, followed by oscillations, at first somewhat irregular, then a slowly decaying sinusoidal signal, at a 25-s period, after the wavemaker has been turned off. Little contribution from the other sloshing modes can be seen.

In Molin (2001), it is established that the long mode generation is due to second-order effects. The proposed theoretical model is based on the concept of radiation stress, introduced by Longuet-Higgins & Stewart (1962) – see Mei *et al.*, 2005, Chapter 11. Second-order nonlinearities lead to a depression of the mean water level below the waves, and therefore, in a confined tank, to an increase elsewhere. As the wavemaker is turned on and the waves start progressing in the tank, the wave system is accompanied by a depression long wave (a negative Heaviside function) at the group velocity C_G. Owing to mass and momentum conservation, the wave front is preceded by a positive Heaviside function advancing at the critical velocity $\sqrt{gh}$, with an amplitude ratio equal to $C_G/\sqrt{gh}$. As it reaches the far end of the tank, this wave is reflected back to the wavemaker and back again to the beach and so on, until the wave front has arrived at the beach and only a combination of natural sloshing modes persists. When the wavemaker is stopped, the same process is repeated again.

Figure 9.17 shows the amplitudes of the first 3 long modes generated after the regular wave system has been established (therefore during the tests), in the form $\zeta_m h/A^2$, versus kh (there is no dependency on the length of the tank). Mode 1 appears to be dominant. It is remarkable that the amplitudes are nil at some particular kh values, which correspond to odd integer values of the ratio $\sqrt{gh}/C_G$. The figure shows that

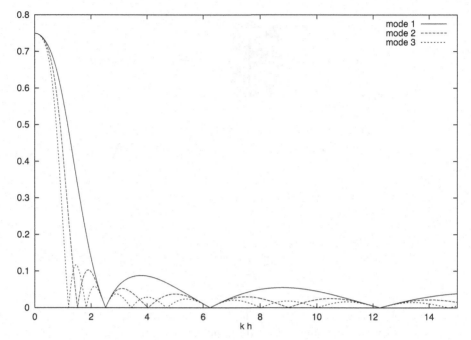

Figure 9.17 Amplitude $\zeta_m h/A^2$ of the first 3 seiching modes after establishment of the regular wave system in the tank.

seiching can become quite critical in shallow depth condition. The associated oscillatory current velocities are obtained as $U_m = \zeta_m \sqrt{g/h}$. They are small compared to the return current velocity except when $kh \leq 2$.

The theoretical model of Molin (2001) has been confirmed by detailed comparisons with the experiments in the towing tank of Centrale Nantes and also with a second-order numerical wavetank developed by Stassen (1999) (see also Molin *et al.*, 1999). In Molin (2001), it is proposed, and confirmed experimentally, that the seiching generation can be annihilated by an additional motion of the wavemaker at the seiching frequency, when the wavemaker is started and when it is stopped.

Dynamic absorbers efficiently reduce the persistence of seiching after the tests.

9.5.8   Reflections and Sidewall Effects

In the absence of the model, that is, during calibration tests, the only point of concern is the amount of reflection from the beach. As written above, it is not customary that model basins monitor reflection from the beach during calibration tests. Satisfactory performance of the beach can be verified visually from the time series of the wave elevation: If there are reflections, it will take them a time equal to twice the distance to the beach divided by the group velocity, to show up at the wave gauge, as a slight change in amplitude.

With the model in place, waves diffracted and radiated by the model will travel back to the wavemaker and be rereflected back to the model, unless the wavemaker has an absorption function. Again the time taken for the roundtrip to the wavemaker is twice the distance divided by the group velocity. When the model is too close to the wavemaker, the exploitable time window, before rereflections spoil the test, may be very short.

The diffracted and radiated waves traveling straight back toward the wavemaker are only part of the wave system. A more general form is:

$$\eta_R(x,y,t) = \sum_{n=0}^{N} C_{Rn} A_I \cos \frac{n\pi}{B} y \cos(\mu_n x + \omega t + \phi_n) \qquad (9.25)$$

where the series is truncated at the largest integer value of n such that $\mu_n = \sqrt{k^2 - n^2 \pi^2/B^2}$ be a real number. In a narrow tank, when the wave length $2\pi/k$ is larger than twice the width B, only $n = 0$ is admissible: Reflected waves travel straight back to the wavemaker. When the wave length decreases compared to the width other modes come into play, combining longitudinal propagation and sloshing in between the sidewalls.

When the wave number k is such that one of the μ_n is nil or quasinil (in other words when the frequency ω coincides or nearly coincides with one of the natural frequencies of the transverse sloshing modes), the diffracted + radiated wave system will transform into a local standing wave, in between the sidewalls, with an amplitude increasing in time. This will seriously bias the seakeeping behavior. In particular, the quantities related to the emitted wave field, for instance the radiation dampings and the drift forces are strongly affected. Figure 9.18 shows numerical and experimental results reported by Chen (1994). They present the heave response amplitude operator

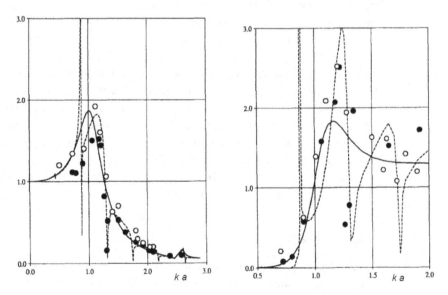

Figure 9.18 Floating hemisphere of radius 0.3 m in the middle of a 5-m-wide tank. Full line: theory (open ocean); dotted line: theory (with side walls); markers: experimental values. Right: heave RAO; left: horizontal drift force $F_d/(1/2\rho g A^2 a)$ (from Chen, 1994).

(RAO) and horizontal drift force for a floating hemisphere of radius 0.3 m in between two walls 5 m apart (the depth being 3 m). Spikes in the computations (using a Green function accounting for the sidewalls) can be seen at the sloshing frequencies. The discrepancies between the open sea and confined cases are very large. Note that the computations assume a steady state which may not be attained in the experiments.

One conclusion is that, when doing regular wave tests, one should avoid wave frequencies coinciding with natural frequencies of the transverse sloshing modes. Note that, for a symmetric model in the middle of the tank, only the even modes are of concern (in principle).

A remedy to the sidewall effects is to use side absorbers. Their design is not easy since, ideally, they should absorb the radiated and diffracted waves while letting the incident waves propagate without deterioration. Vertical perforated plates have been used occasionally.

9.6   Types of Waves and Generation Techniques

Prior to running the tests with the model, it is usual to carry out calibration tests, in order to achieve the specified sea states in the tank. Then the same calibrated signals are used with the model in place.

9.6.1   Regular Waves

In most facilities, regular waves are produced by a simple harmonic motion of the wavemaker, with linear ramps at the beginning and at the end of the input signal. The amplitude of motion of the wavemaker is the target wave amplitude divided by the transfer function of the wavemaker. This transfer function may be known from linear theory, or from past experience of the basin. It can also be updated during the calibration tests through an iterative procedure as usually done for the irregular waves. This would depend on how close to the specifications the wave heights should be: For instance to measure ultimate loads on a jacket structure, it is very important to achieve the design wave height; to measure drift forces in relatively small amplitude waves [to check the computed quadratic transfer function (QTFs)], what matters is precise knowledge of the generated wave amplitudes.

Some basins correct the wavemaker signal with a double frequency component in order to eliminate the second-order free waves.

9.6.2   Irregular Waves

Usually, the irregular sea state is defined as a wave spectrum $S(\omega)$, to be reproduced in the basin, at some distance from the wavemaker.

The most widely used technique is to construct the wavemaker signal as a sum of sinusoids:

$$x(t) = \sum_i x_i \; \cos(\omega_i t + \varphi_i) \tag{9.26}$$

the phases φ_i being picked up at random.

The target wave signal being written:

$$\eta_I(x,t) = \sum_i A_i \; \cos(k_i x - \omega_i t + \theta_i) \tag{9.27}$$

where the elementary amplitudes A_i are obtained from the spectrum S through

$$A_i^2 = 2 \, S(\omega_i) \, \Delta\omega_i \tag{9.28}$$

In some facilities, the amplitudes are further randomized.

When the wavemaker transfer function $A_i/x_i(\omega)$ is known, the amplitudes x_i are easily deduced. Since the wavemaker transfer function may be sensitive to diverse nonlinearities, and the wave spectrum may evolve somewhat on its way to the target location, the usual procedure is to build up the wavemaker signal through iterations, where the transfer functions are reconstructed at each iteration. The iterative procedure is stopped (usually after one or two iterations) when the target and achieved sea states are considered sufficiently close; the H_S and T_P values, together with the spectral shape, being the most critical parameters.

There may also be specifications such as the maximum achieved wave height being larger than some value (1.8 times H_S for instance) which would lead to changing the set of random numbers generating the phases.

Most often the carrier frequencies ω_i are multiples of a constant spacing $\Delta\omega$, equal to the frequency range $\omega_{max} - \omega_{min}$ divided by the number N of wave components. The wavemaker signal then repeats itself at a period $2\pi/\Delta\omega$. This period should be longer than the test duration (at least the part used for the analysis), meaning a very large number of frequencies if the test is to last 3 hours at full scale. On the other hand, one can see some interest in this periodicity of the wavemaker motion, which is to check whether the generated waves (and model response) identically repeat themselves as well: A poor repetition suggests spurious phenomena at hand in the basin, for instance reflections from the beach or natural basin modes coming into play.

The periodicity of the wavemaker motion can be avoided by picking up the carrier frequencies at random within equal (or not equal) intervals. The choice of the number of wave components can then be based on other criteria, for instance the mean frequency spacing should be much less than the smallest resonant frequency multiplied by the associated damping ratio (see Section 7.8 in Chapter 7).

As written in the previous section, spurious second-order free waves at the sum $\omega_i + \omega_j$ and difference $\omega_i - \omega_j$ of the carrier frequencies are generated by the wavemaker. Some basins apply corrections to the wavemaker motion in order to eliminate these spurious wave components.

There are other techniques to synthesize irregular waves, for instance white noise filtering.

9.6.3   Wave Packets

It is possible to run the wavemaker to produce, at the target location, a single wave, or a wave group, of specified shape. According to linear theory, if the transfer function between wavemaker motion and free surface elevation is known, a simple inversion and a convolution in time enable one to derive the wavemaker motion from the desired wave profile. In practice, particularly if the distance from wavemaker to target location is long, nonlinear effects will come into play and several adjustments may become necessary. Some shapes are also physically impossible to achieve, for instance if their steepnesses are too high, or the troughs deeper than the crests elevations.

The interests of this technique are manifold:

- With the phases φ_i in (9.26) well chosen, and nearly equal amplitudes x_i, it is possible, through a single test of short duration (one wave, or one short wave packet), to obtain the RAOs of the system. The drawback being that the generated wave being of large amplitude, some nonlinear effects may come into play and affect the response.
- The focusing technique permits the generation of waves of much higher amplitude and steepness than can be achieved when running the wavemaker to produce regular waves. Applications are studies of wave impacts or ship capsize in large breaking waves.
- A test of short duration means that spurious effects such as reflections from the beach and sidewalls, or arrival of free high-frequency harmonics, are avoided.
- It is possible to reproduce in the tank a given record of wave elevation over time, from in situ measurements, or tests in another facility.

NewWave Focused Wave Groups

It can been shown (Boccotti, 1983) that the average shape of the wave group surrounding the highest wave in a sea state is the autocorrelation function. This feature results from the wave elevation being assumed to be a linear Gaussian process.

Based on this, Jonathan & Taylor (1997) have proposed the following form of a focused wave group, baptized **NewWave**:

$$\eta(x,t) = \frac{\eta_{max}}{m_0} \sum_{i=1}^{N} S(\omega_i)\,\Delta\omega\,\Re\left\{ e^{i\,[k_i\,(x-x_0)-\omega_i\,(t-t_0)+\psi]} \right\} \tag{9.29}$$

where x_0 and t_0 are the location and instant of focus.

When the phase angle ψ is equal to 0, the group is "crest-focused," that is, $\eta(x_0,t_0) = \eta_{max}$. When $\psi = \pi$, the group is "trough-focused": $\eta(x_0,t_0) = -\eta_{max}$. Experimentally, the two time series are not strictly opposite to each other, because of nonlinear effects, mainly second-order. Likewise, the responses of the system are not strictly opposite, and some simple manipulations of the time series enable one to extract second-order quantities. When "up-crossing" ($\psi = \pi/2$) and "down-crossing" ($\psi = -\pi/2$) focused groups are run as well, third- and fourth-order quantities can be derived (Fitzgerald *et al.*, 2014). The underlying assumption is that the dynamics of the system are weakly nonlinear, that is to say, the Stokes expansion holds.

Applications of this technique to the ringing problem can be found in Fitzgerald *et al.* (2014) or Chen *et al.* (2018). See also Zhao *et al.* (2017) for application to gap resonance.

9.7 Current Generation

There are three different ways to simulate current in a wave tank:

- by towing the model
- by global circulation of the water in the tank
- by local generation.

9.7.1 Towing

Towing offers the advantage that the current velocity is known (opposite to the towing velocity), steady in time and in space, with zero turbulence intensity (provided sufficient time has elapsed since the previous test). No other current profile than constant over the depth can be achieved.

A major drawback is that, if there are catenary mooring lines, or other links to the sea floor (risers, tendons), the end points need to be towed as well. This means a complex and costly structure attached to the carriage with associated issues such as elastic response and vibrations, induced perturbations to the wave flow, etc. The test duration is limited by the usable length of the tank, over which the wave conditions are stationary.

9.7.2 Global Circulation

Several offshore tanks provide current generation, with a pumping system that sets the whole water mass into motion. This means a power-consuming pumping system, pipes returning the water to the other end of the tank, and a complex injection system to adjust the vertical profile and minimize turbulence intensity. In some facilities, the return flow takes place below the false bottom (for instance at SINTEF Ocean). Maximum attained velocities depend on the water height; they are typically a few decimeters per second (a few knots at full scale following Froude scaling).

Difficulties are to ensure a uniform flow over the width of the tank, and to control the vertical profile. Other issues are how to combine wave generation and current injection at one end, wave absorption and current suction at the other end. The current injection is necessarily located below the wavemakers, which makes it difficult to achieve a vertical profile with the maximum velocity at the free surface, as most often required. Some user-specified profiles may be hard or impossible to achieve, due to inherent instability. Finally, as the water circulates several times through the basin and through the piping system during the test, the turbulence intensity may increase over time.

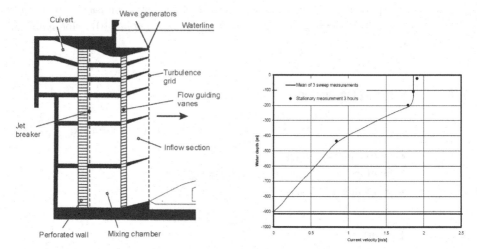

Figure 9.19 Current inflow system at MARIN Offshore Basin (left) and obtained vertical profile (right).

Figure 9.19 shows an example of sheared current profile, achieved in the Offshore Basin of MARIN.

9.7.3   Local Generation

Local generation, through an assembly of pipes and injection nozzles, can be found in some facilities. Issues are the time and space variabilities of the current generated, the perturbations caused to the wave flow, etc.

9.7.4   Current-Induced Modifications of the Wave Field

Quite often the same sea state (same spectrum $S(\omega)$) is specified, without and with current.

When current is simulated by towing, keeping the wavemaker signal unchanged produces the same waves, but the encounter frequencies vary with the towing velocity. If the encounter frequencies need to be kept the same, as many wave calibrations are necessary as specified current velocities.

When current is generated by global circulation, keeping the wavemaker signal unchanged conserves the frequencies, but the wave lengths are increased and the wave heights are decreased (with the current flowing in the same direction as the waves). New wave calibrations are also required to obtain the target wave heights.

A drawback of repeating the calibrations, besides the financial cost, is that the time histories of the free surface elevation are not directly comparable, due to the way the wavemaker signal is being synthesized. This means that the responses of the system, with and without current, cannot directly be compared in the time domain.

9.8 Wind

There are two ways to model the effects of wind: generate the wind directly or reproduce artificially the wind loads.

Some wave tanks are integrated into wind tunnels, for instance the BLWT2 (Boundary Layer Wind Tunnel Laboratory) at the University of Western Ontario or the LIOA (Laboratoire d'Interactions Océan Atmosphère) of Pytheas in Marseille. In such facilities, a wind of good quality is produced, and the turbulence level can be controlled. Due to their limited size, only very small-scale experiments can be performed (e.g., see Lacaze, 2015).

To generate wind over an ocean basin, a cluster of fans is often utilized. As for current, there are many difficulties associated with achieving a flow field that is steady and has the correct shear, with controlling the turbulence level, etc. In many instances, the fan signal is adjusted, not to produce the specified wind velocity but to produce the expected mean loads, previously determined by tests in a wind tunnel or by computational fluid dynamics (CFD).

An alternative method is to generate the wind loads artificially. For a long time, ship mooring tests have been performed with two fans on the deck, at the aft and bow, giving the proper wind loads in surge, sway and yaw, as determined from wind tunnel tests. This technique has gained renewed interest with offshore wind turbines. At a typical scale of 1:50, a 10-MW wind turbine would have a diameter of about 4 m: Generating an adequate wind field, over such a wide area, with fans, is virtually impossible. Moreover, using embarked rotors (or dynamic cables) has the advantage that the wind load signal can easily be varied in time, simulating gusts which fans fail to do. To correct for the support velocity (in the case of floating wind turbines), the *software in the loop* technique is now routinely used. The tests may even be coupled with real-time calculations of the wind loads, using a software such as Aerodyn (e.g., see Sauder *et al.*, 2016). Another advantage is that scale issues, due to the bias in Reynolds number, are eliminated.

9.9 Sensors

9.9.1 Free Surface Elevation

The most widely used sensors to measure the free surface elevation are wave gauges. They can be of resistive or capacitive type. They consist in thin tensioned wires, vertical, running through the free surface.

Resistive gauges consist in two parallel wires. What is actually measured is the electric conductivity of the water medium in between the wires, directly proportional to their wetted lengths.

In the case of capacitive gauges, there is only one wire which, with the surrounding fluid, acts as a capacitor, its capacity also proportional to its wetted length.

Most wave basins manufacture their own wave gauges.

Figure 9.20 Elastic barge model in waves with infrared motion measurement system (courtesy Océanide).

The sensitivities of the resistive and capacitive wave gauges are very good: Free surface variations less than one millimeter can be detected. The measurement accuracy depends on several parameters:

- The diameters of the wires: The smaller the diameters, the better the accuracy.
- The parallelism and separation of the wires, which may be affected by vibrations (VIVs).
- Their cleanliness, in particular, for capacitive gauges.
- Their dynamic responses which may not be as linear as their static ones.

Resistive and capacitive gauges require to be regularly calibrated, ideally on a daily basis, at the beginning and at the end of the tests. Calibrations are usually done in a quasi-static way, lifting up and down the gauge assembly to successive vertical positions.

Optical techniques have been developed to obtain the three-dimensional shape of the free surface; however, they have not passed the research stage. One of them is the schlieren technique, used in small-scale laboratories (Moisy *et al.*, 2009).

9.9.2   Model Response

All model basins nowadays use optical techniques to measure the motion of the models, Qualisys being the most widely used: A set of cameras track the positions in space of passive targets, with Qualisys small white spheres, placed on the models. Other systems use infrared light and pulsating active targets. Some can operate under water. Figure 9.20 shows an elastic barge model, consisting of 12 successive modules, every other module being instrumented with three infrared targets (Remy *et al.*, 2006).

Figure 9.21 Experimental model of a LNG tank equipped with pressure sensors (courtesy GTT).

The resolution of these systems is very good, less than one millimeter when the distance from target to cameras is not excessive. Their acquisition rate is up to several hundreds Hertz.

Accelerometers can be used in combination with optical systems, for the sake of redundancy, and/or in the case of elastic systems responding at very high frequencies, for instance under wave impacts.

9.9.3   Other Sensors

Load sensors are not specific to model testing, and there is not much to write about. Pressure gauges are used in connection with impacts loads, due to slamming for ships or sloshing in LNG tanks. Impact pressures are not easy to measure since they are very localized in time and space: The area of the sensor should be as small as possible, and the acquisition rate must be very high, larger than 20 kHz. Moreover, pressure gauges are sensitive to thermal effects, which can result in unrealistic negative values.

Figure 9.22 shows time traces of pressure measurements in a LNG tank model such as shown in Figure 9.21, instrumented with hundreds of pressure gauges. Note that the total time span in the horizontal axis is about 20 milliseconds (at model scale, where Froude similitude is applied). The oscillations following the first peak are due to gas entrapment.

It is not customary to measure the kinematics of the flow field in large-scale facilities, except for current calibration: The time-averaged velocity being usually obtained with rotor or acoustic current meters and the turbulence intensity with hot wire.

With lasers nonintrusive techniques have been developed, namely the Laser Doppler Velocimetry (LDV) and the Particle Imaging Velocimetry (PIV). They both require seeding of the fluid with small particles, 10 to 100 micrometers in size. With

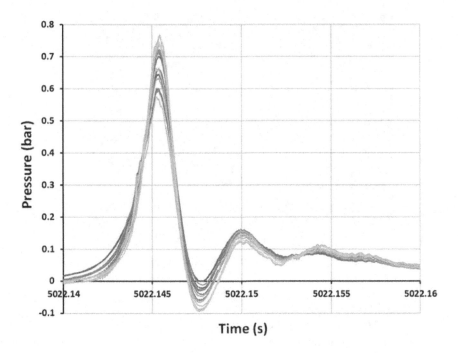

Figure 9.22 Time traces of pressure measurements in a LNG tank model (courtesy GTT).

LDV, two or three laser rays are set to cross at a given point in space. Each time a particle passes there, the rays are partially reflected and the particle velocity is obtained from the Doppler shifts. An application of LDV is measuring velocity profiles in boundary layers. With PIV, a part of the fluid domain is successively illuminated at two very close instants: From the displacements of the particles, the velocity field is obtained. PIV is widely used in university laboratories (see Figure 9.23).

9.10   Exploitation of the Measurements

The first step is to identify the exploitable time window, that is the time interval over which the measurements can be considered valid. In regular wave tests, for which a simple harmonic analysis is usually performed, a few cycles suffice. The matter is far different in irregular waves where the analysis requires a large number of cycles, to enable the extraction of statistical parameters; the study of slow drift motion, in particular, requires very long durations.

Typically, a certain portion of the beginning of the recording is to be eliminated, since it corresponds to the "transients": transients of the basin (the sea state takes some time to settle), and transients of the response of the model, the duration of which depends on its natural periods and on the associated damping ratios.

On the other hand, the wave system in the basin tends to deteriorate in time with the reflections from the beach, from the sidewalls, and from the wavemaker (if not absorbing), with the excitation of the tank natural modes, etc.

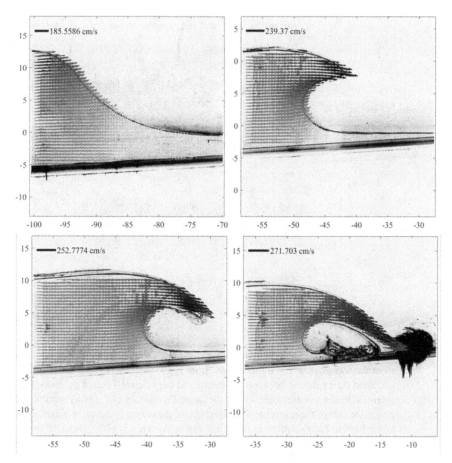

Figure 9.23 PIV measurements in an overturning wave at a sloping beach (courtesy O. Kimmoun).

The exploitable time window therefore extends from the time when the transients have died out, or are sufficiently attenuated, until the time when the wave system in the vicinity of the model has become too degraded. These two instants are not always easy to determine objectively, the analyst having for example a tendency to take as sole criterion the regularity of the signal, estimated visually or quantitatively. This criterion is not always reliable: The regularity of the signal does not guarantee the absence of reflections, for example.

9.10.1   Regular Wave Tests

In the most common case, the measured signal is more or less periodic, at the wave encounter frequency. There may be a transient portion, most often slowly decaying oscillations at a much lower frequency. To this signal is generally superimposed some noise, of various origins, requiring some filtering.

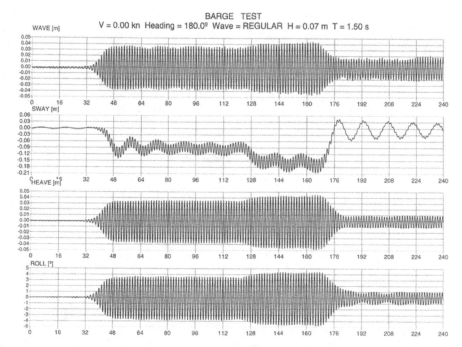

Figure 9.24 Tests on a barge model in beam waves. The model, 5 meters long, is moored 50 m from the wavemaker. The figure shows, from top to bottom, the time series of the free surface elevation (at a gauge in between the model and the sidewall) and of the sway, heave, and roll motions. Transients are noticeable at the natural frequency of the sway motion, at the beginning and at the end of the test (note the difference of the damping ratios, in waves and in the basin going back to calm). A change of regime can be seen at $t \simeq 130$ s, about 85 s after the arrival of the wave front. This duration corresponds to a roundtrip, at the group velocity (1.15 m/s at the period of 1.5 s) from the model to the wavemaker: Despite the large width of the basin (30 m), a significant part of the diffracted and radiated wave system is rereflected back to the model.

Figure 9.24 shows an example of recordings made during tests on a barge model, at Cehipar.

In rare cases, there may appear in the signals some components at subharmonics of the wave frequency, for example at the half frequency. Parametric instabilities may be the cause of such behavior (see Chapter 8).

A Fast Fourier Transform (FFT) analysis permits identification of the different frequency components. When the signal is periodic or very close to periodic, a Fourier analysis is better suited to extract the successive harmonics. This analysis can be performed via the following procedure (recommended by the International Towing Tank Conference (ITTC) association):

1. (Possibly) Splitting the time series into wave frequency component and low-frequency component by filtering.
2. Selection of a portion of the signal with a limited number of cycles.
3. Determination of the average value (if the initial filtering has not been done, adjustment of a locally linear law $A t + B$ to the low-frequency part).
4. Subtraction of the average value from the signal.

5. Choice of an integer number of oscillations and determination of the average period by an up-crossing or down-crossing criterion.
6. Fourier analysis on the basis of this period.

The determination of the average value of the signal generally requires a greater number of cycles than that required for the harmonic analysis, in particular, in the case of residual oscillations at low frequency.

9.10.2   Tests in Irregular Waves

Two types of analysis are usually performed: a statistical analysis and a spectral analysis.

The statistical analysis is similar to that applied to the wave elevation in Chapter 2. The average value, the standard deviation, and the extreme values of the signal are first determined over the time interval considered. Then, the signal is divided into successive cycles, on an up-crossing (or down-crossing) criterion. Are deduced the number of cycles, the average up-crossing or down-crossing period, the number of maxima and minima. These maxima and minima are then presented in the form of histograms, or distribution functions compared to the Rayleigh law or adjusted to a Weibull law.

Possibly, the signal has been previously decomposed, by filtering, into a low-frequency component and a wave frequency component (plus sometimes, in ringing problem for instance, a high-frequency component), and the statistical analysis is applied to each component separately. This procedure enables one to better identify the contribution of each component to the total signal, but makes the interdependencies disappear.

Spectral densities may be obtained by direct FFT of the time series, or via the determination of the autocorrelation and Fourier transform. Uncertainties on the resulting plots, related to the finite length of the record, to the chosen frequency step, and to the numerical treatments used, are rarely mentioned. The calculations of the moments of orders 0 and 2 give alternative estimates of the standard deviation and of the mean period, that can be checked to agree with the values obtained through the statistical analysis.

From the response spectrum, knowing the wave spectrum, one obtains, by dividing and taking the square root, the modulus of the RAO. This procedure assumes that the relationship between the input and the output is perfectly linear. To check the linearity, it is preferable to calculate the cross-spectrum, from the cross-correlation $E\{\eta_I(t) X(t+\tau)\}$. The modulus of the cross-spectrum divided by the square root of the product – wave spectrum × output spectrum provides the **coherence** and allows one to check the linearity: If the coherence is equal to 1, the linear relationship is established. In practice, it is considered that linearity holds when the coherence exceeds a certain threshold, for example 0.5 or 0.6. The cospectrum divided by the input spectrum then provides the complex transfer function (that is the modulus of the RAO and its phase).

An example is given in Figure 9.25, taken from the tests on coupling between sea-keeping and sloshing in tanks, reported in Section 6.11: A barge carrying two rectangular tanks is subjected to irregular beam seas. The input is the measured wave

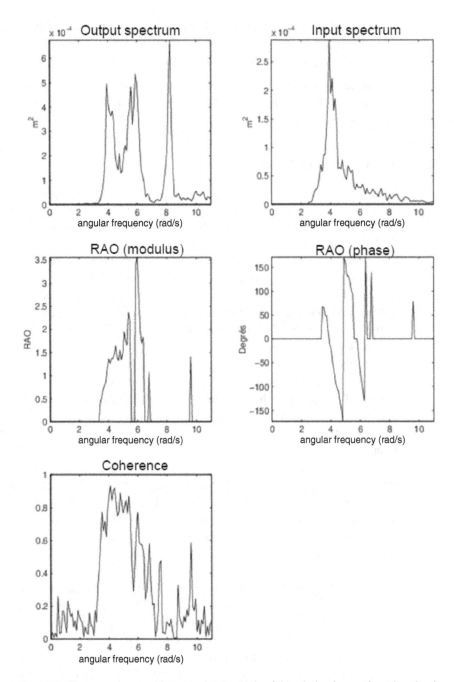

Figure 9.25 Tests on a barge with two tanks. Analysis of the relative free surface elevation in one of the tanks. Top left: spectrum of the relative free surface elevation. Top right: spectrum of the incident wave elevation. Bottom: coherence. Middle left: modulus of the RAO, plotted where the coherence exceeds 0.5. Middle right: phase of the RAO.

elevation measured away from the barge. The output is the relative free surface elevation inside the tank, at its wall. The output spectrum shows three peaks, around 4, 6, and 8 rd/s, while the wave spectrum has only one peak, at 4 rd/s. The coherence shows that it is only for frequencies between 3 and 7 rd/s that the two signals are linearly related. In particular, the peak at 8 rd/s is due to nonlinear effects: It corresponds to the second sloshing mode in the tank (the water height being 19 cm and the length of the tank 80 cm, the first mode is at 4.9 rd/s and the second at 8.3 rd/s), which is an even mode, therefore uncoupled to the movements of the barge in a linear theory.

9.11 Specification of Model Tests

It is difficult to define a standard test program. It depends very much on the case (fixed, or floating structure) and on the information sought through the tests (results directly usable for the design or for the calibration of a numerical model).

One may identify two constant concerns: minimize the cost, therefore the number of tests, and maximize confidence in the results obtained.

9.11.1 Wave Calibration Tests

Most basins conduct preliminary wave calibration tests to produce customer-specified sea conditions. Calibrations are done in the absence of the model. Once the model is in place, the same signals are applied to the wavemaker and the sea states are reproduced identically (in principle).

Adjusting regular waves rarely causes problems, but it is always good to check for example that they are not too degraded in the test zone and/or that reflections from the beach are small. In this respect, the visual examination of the time traces of the wave elevation is often useful. One can also verify whether the amplitudes of the harmonics (at double and triple frequencies) agree with the theoretical values of the Stokes model.

The calibration of the irregular waves usually requires some iterations, and often the final agreement between specified and achieved wave spectra is a bit disappointing. Furthermore, the information on the confidence interval to be given to the experimental spectrum is often lacking. As much as the shape of the spectrum, the distribution of the wave heights, with respect to the theoretical laws, is to be considered.

9.11.2 Decay Tests

For systems with dynamic response, it is always useful to carry out decay tests: The model is displaced from its equilibrium position, following one of its degrees of freedom, and released. In the absence of coupling with the other degrees of freedom, the ensuing motion is a simple damped sinusoid. Then, it can be checked that the natural period is as expected. When it is assumed that the damping consists in a linear plus quadratic components, that is, $B = B_L + B_Q |\dot{X}|$, the damping coefficients B_L and B_Q can be derived from the slowly decreasing amplitude. Through the equal energy dissipation principle, it can be established (see Appendix C) that, when the response is sinusoidal with an amplitude A, the quadratic damping B_Q is equivalent to a linear damping $B_{Leq} = 8B_Q A\omega/(3\pi)$. When the total damping is weak, the response appears as a slowly decreasing sinusoid with successive amplitudes X_n. From one cycle n to the following $n + 1$, the amplitude ratio is given by:

$$\frac{X_{n+1}}{X_n} = \exp\left\{-\left[B_L + \frac{8}{3\pi}B_Q X_n\omega_0\right]\frac{T}{2M}\right\} \simeq 1 - \left[B_L + \frac{8}{3\pi}B_Q X_n\omega_0\right]\frac{T}{2M} \quad (9.30)$$

This means that, when plotting $\Delta X_n/X_n$ versus X_n, the experimental points (should) align on a straight line. From its intercept with the vertical axis, the linear damping component B_L is deduced and, from its slope, the quadratic component B_Q (see Figure 6.23 in Section 6.8.2 of Chapter 6). This procedure is routinely applied to roll decay tests of ships and FPSOs, to extract the viscous damping.

In practice, the procedure may be complicated by the fact that several degrees of freedom may contribute to the response (for instance sway with roll, or surge with pitch); when the natural frequencies are far apart, this can be remedied by filtering.

Decay tests may also be performed in regular waves. Then such helpful information as the wave drift damping can be deduced (see Figure 9.24). It may also be that the natural periods be shifted from their still water values, due to viscous or wave drift added mass effects (see Section 7.8 in Chapter 7).

9.11.3 Tests in Regular Waves

For structures that can be considered as rigid, such as jackets or gravity base structure (GBS) in relatively shallow water, it is usual to base the design on tests (and calculations) in regular waves. "Rigid" here means that no ringing response should contribute to the stresses.

Whenever the considered system responds dynamically, it makes little sense to assess the design on the basis of tests in regular waves: If the wave frequency coincides with a natural frequency (or a subharmonic) of the system, an exaggerated dynamic response will appear; mean wave drift forces may take unrealistic values, while no slow-drift response will appear in regular waves.

Nevertheless regular wave tests may be quite useful:

- to check the RAOs and their sensitivity to the wave amplitude (at a given frequency);
- to obtain the wave drift forces and compare with the values delivered by a potential flow model; check their sensitivity to increasing wave amplitude;
- to derive the wave drift damping when combined with a decay test.

Some tests may also be duplicated with coflowing or adverse current, providing, among other information, an alternate access to wave drift damping.

Wind superposition is usually of little, or no interest.

9.11.4 Tests in Bichromatic Waves

Tests in bichromatic waves are occasionally specified, in order to extract QTFs. An unfortunate feature of bichromatic waves is that they are rather unstable, due to third-order free surface nonlinearities (Lo & Mei, 1985).

9.11.5 Tests in Irregular Waves

Irregular seas, as generated in offshore basins, are usually considered as sufficiently appropriate representations of actual sea states. Many facilities are now equipped with snake wavemakers and can generate short-crested seas, even closer to reality than long-crested seas. They can also generate two or more wave systems propagating at different angles, for instance a residual swell together with a local storm, such as occurs in the Gulf of Guinea.

Through the tests in irregular waves, the priority is often to reproduce the sea states considered for the design, in order to obtain results that are directly usable. This may lead to a very large number of tests since many parameters may have to be varied, for instance:

- the draft of the structure (ballast or fully loaded);
- the sea state, depending on whether one wants to simulate extreme or operational conditions;
- for a given sea state, the peak period of the spectrum which is often imprecise;
- the relative directions and velocities of the wind and current;
- the integrity of the mooring system: one or two lines broken for instance.

More and more frequently, the main goal of the tests is to validate and calibrate a numerical model, or a suite of numerical models, that have been used for the design. Nowadays, computer power is so vast and so cheap that numerical models such as Orcaflex or Simo or Deeplines can be run hundreds of time over, to cover all combinations of sea state, loading, mooring integrity, etc. The numerical model is tuned to the model tests (the "model of the model"), and then some parameters may be modified to account for scale effects (some drag coefficients, for instance, on the basis of experience or larger scale tests or CFD computations).

9.11.6 Other Tests

Forced Motion Tests

Most wave tanks are equipped with forced motion mechanisms. In naval engineering so-called PMMs (*Planar Motion Mechanisms*) have long been used to derive hydrodynamic coefficients (the "stability derivatives") needed for maneuvering software: While being towed the ship model undergoes forced oscillations in sway and yaw and the measured loads are converted into stability derivatives, equivalent to low-frequency added masses and dampings.

Many facilities are now equipped with "hexapods," similar to the one shown in Figure 9.26 that can oscillate the models through 6 degrees of freedom. The main application is forced motion tests in still water or in current to derive added mass and damping coefficients. Hexapods can also be used statically, to vary the heading while measuring wave or current loads for instance.

Hexapods are also used, in the upright position, for sloshing tests, for instance with models of LNG tanks as shown in Figure 9.21.

Figure 9.26 Forced motion tests on a square cylinder for galloping analysis (courtesy Océanide).

Mean Wind and Current Loads

Mean wind loads are usually obtained by tests in a wind tunnel where high Reynolds numbers can be attained. Current loads are also sometimes obtained by wind tunnel tests, but the results may be plagued by the mismatch in the boundary layer profiles: In the ocean, there is practically no boundary layer below the free surface, due to the difference in density between air and water. In the wind tunnel, a boundary layer exists at the floor and it is important that most of the model be out of the boundary layer. In confined wind tunnels, this is difficult to achieve.

Mean current loads are more frequently obtained in the basin, by towing (in towing tanks), or by generating the current. When Froude scaling is used, the loads to measure are small, and accuracy problems may occur in the measurements. As long as the wave resistance component is negligible compared to the viscous component, the towing or current velocity may be increased beyond the Froude scaled value. When the model is moored and the mooring stiffness is known, a better accuracy is usually attained from the mean offset than from summing the tensions in the mooring lines.

An alternative to model tests is CFD which can, to some extent, overcome the bias in Reynolds number.

9.12   References

BENJAMIN T.B., FEIR J.E. 1967. The disintegration of wave trains on deep water. Part 1. Theory, *J. Fluid Mech.*, **27**, 417–430.

BOCCOTTI P. 1983. Some new results on statistical properties of wind waves, *Applied Ocean Res.*, **5** (3), 134–140.

CHEN L.F., ZANG J., TAYLOR P.H., SUN L., MORGAN G.C.J., GRICE J., ORSZA-GHOVA J., TELLO RUIZ M. 2018. An experimental decomposition of nonlinear forces on a surface-piercing column: Stokes-type expansions of the force harmonics, *J. Fluid Mech.*, **848**, 42–77.

CHEN X.B. 1994. On the side wall effects upon bodies of arbitrary geometry in wave tanks, *Applied Ocean Res.*, **16**, 337–345.

DIAS F., GHIDAGLIA J.-M. 2018. Slamming: recent progress in the evaluation of impact pressures, *Annual Review Fluid Mechanics*, **50**, 243–273.

FITZGERALD C.J., TAYLOR P.H., EATOCK TAYLOR R., GRICE J., ZANG J. 2014. Phase manipulation and the harmonic components of ringing forces on a surface-piercing column, *Proc. R. Soc. Lond. A.*, **470** (2168), 20130847.

JAMIESON W.W., MANSARD E.P.D. 1987. An efficient upright wave absorber, *ASCE Specialty Conf. Coastal Hydrodynamics*, University of Delaware.

JONATHAN P., TAYLOR P.H. 1997. On irregular, nonlinear waves in a spread sea, *J. Offshore Mech. Arctic Engng.*, **119** (1), 37–41.

LACAZE J.B. 2015. Experimental and numerical study of hydrodynamic and aerodynamic coupled effects on a floating wind turbine, PhD thesis, Aix Marseille Université.

LE BOULLUEC M., KIMMOUN O., MOLIN B. 2007. Etude expérimentale pour l'optimisation des performances d'une plage d'amortissement parabolique, in *Actes des 11 èmes Journées de l'Hydrodynamique*, Brest (in French; http://website.ec-nantes.fr/actesjh/).

LO E., MEI C.C. 1985. A numerical study of water-wave modulation based on a higher-order nonlinear Schrödinger equation, *J. Fluid Mech.*, **150**, 395–416.

LONGUET-HIGGINS M.S. & STEWART R.W. 1962. Radiation stresses and mass transport in gravity waves with applications to surf-beats. *J. Fluid Mech.*, **13**, 481–504.

MANSARD E.P.D., FUNKE E.R. 1980. The measurement of incident and reflected spectra using a least squares method, *Proc. 17th Int. Conf. Coastal Engineering*.

MEI C.C., STIASSNIE M. & YUE D.K.P. 2005. Theory and applications of ocean surface waves, in, *Advanced Series on Ocean Engineering*, Vol 23, World Scientific, Singapore.

MOISY F., RABAUD M., SALSAC K. 2009. A synthetic Schlieren method for the measurement of the topography of a liquid interface, *Exp Fluids*, **46**, 1021–36.

MOLIN B. 2001. Numerical and physical wavetanks. Making them fit, *Ship Technol. Res.*, **48**, 2–22.

MOLIN B., FOUREST J.-M. 1992. Numerical modelling of progressive wave absorbers, *Proc. 7th Int. Workshop on Water Waves & Floating Bodies*, Val de Reuil (www.iwwwfb.org).

MOLIN B., STASSEN Y., MARIN S. 1999. Etude théorique et expérimentale des seiches parasites générées dans les bassins de houle, in *Actes des 7èmes Journées de l'Hydrodynamique*, 125–138 (in French; http://website.ecnantes.fr/actesjh/).

REMY F., MOLIN B., LEDOUX A. 2006. Experimental and numerical study of the wave response of a flexible barge, *Proc. 4th Int. Conf. Hydroelasticity in Marine Technology*, Wuxi, 255–264.

SAUDER T., CHABAUD V., THYS M., BACHYNSKI E., SAETHER L.O. 2016. Real-time hybrid model testing of a braceless semi-submersible wind turbine. Part 1: The hybrid approach, *Proc. ASME 2016 35th Int. Conf. Ocean, Offshore and Arctic Engineering*, OMAE2016, Busan.

SCHÄFFER H.A. 1996. Second-order wavemaker theory for irregular waves, *Ocean Engng.*, **23**, 47–88.

STASSEN Y. 1999. Simulation numérique d'un canal à houle bidimensionnel au troisième ordre d'approximation par une méthode intégrale, PhD thesis, Nantes University (in French).

SULISZ W., HUDSPETH R.T. 1993. Complete second-order solution for water waves generated in wave flumes, *J. Fluids & Structures*, **7**, 253–268.

ZHAO W., WOLGAMOT H.A., TAYLOR P.H., EATOCK TAYLOR R. 2017. Gap resonance and higher harmonics driven by focused transient wave groups, *J. Fluid Mech.*, **812**, 905–939.

Appendix A: Introduction to Potential Flow Theory

A.1   Introduction

In this chapter we recall general equations in incompressible fluid mechanics and in potential flow theory. The fluid that we deal with here is water. We first justify that it can be idealized as incompressible. It is also a fluid of low viscosity, and heavy.

By definition, incompressibility means that the density ρ of a fluid particle remains constant, hence that its Lagrangian derivative $d\rho/dt$ is nil.

The criterion for the assumption of incompressibility is that the flow velocities be much lower than the sound velocity. In water the sound velocity is around 1500 m/s. The flow velocities that we deal with, the wave induced velocities, are a few meters per second: Mach numbers are very low, and the incompressibility assumption is justified.

It must be noted, however, that sound velocity can drop drastically due to aeration, resulting from wave breaking, for instance. The minimum value is around 20 m/s for 50 % aeration (see Figure A.1). Such velocities can be encountered during impacts, waves breaking against dykes or against windmill foundations, for instance, or objects dropped in water. Then some acoustic effects can occur, one of these effects being to give an upper limit to the impact pressures.

All fluids are viscous. The kinematic viscosity of water (ν), at 20 degrees Celsius, is around 10^{-6} m^2/s. Viscosity implies boundary layers, which have thicknesses δ of order $\sqrt{\nu t}$, where t is a characteristic time. In waves t is the wave period: with a wave period of 10 s $\delta \sim 3$ mm. In ship forward speed resistance, the characteristic time is the distance from the bow divided by the ship velocity, so the same δ value is obtained for a ship model length of 5 m and 0.5 m/s forward speed. This is in laminar condition. In turbulent flow the boundary layer thicknesses are increased many fold.

Here we deal with bodies whose characteristic dimensions are a few meters or more, so overthicknesses due to boundary layers can be neglected. Unfortunately all boundary layers do not stay gently attached to the walls: they can separate. In fact, in steady uniform flow (in current), it is only in the case of streamlined bodies (ship hulls) that boundary layers do not detach. As shown in Figure 1.10 the flow separates massively in the case of circular cylinders.

Flow separation, and the resulting wakes, means that the vorticity does not remain confined within the boundary layers and that it contaminates the outer fluid domain. Then it is no longer possible to make such assumptions as perfect fluid and irrotational flow, which are the basic hypotheses of potential flow theory. Numerical modeling

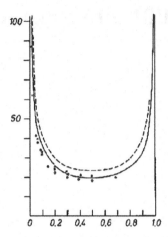

Figure A.1 Sound velocity (in m/s) in water vs aeration (from Müller *et al.*, 2003).

must resort to so-called CFD tools ("Computational Fluid Dynamics"), which solve the Navier–Stokes equations.

Under waves the situation differs: boundary layers can remain attached, even in the case of poorly streamlined shapes such as circular cylinders. In Chapter 1 it has been argued that this is the case when the Keulegan–Carpenter numbers are sufficiently low, leading to the classification of offshore structures into small bodies and large bodies.

Floating bodies are in hydrostatic equilibrium. This means that, following the Archimedes principle, their weights are counterbalanced by their buoyancies, equal to the weights of the displaced water. This is not the case of earth structures or aerial ones such as airplanes. For earth structures subjected to wind, the main aerodynamic loads are associated with viscosity (drag and lift) and they vary with the square of the wind velocity.

Similar loads take place in the case of marine structures. When the incoming flow is accelerated another load component appears, an **inertia** load, proportional to the flow acceleration. Likewise, a body accelerated in still water perceives hydrodynamic loads that oppose, not only its velocity but also its acceleration. For a translation this load component has the form $F = -M_a \ddot{X}$ where M_a is the **added mass**. In fact, for a given three-dimensional body in unbounded fluid, the added mass is not a scalar but a 6×6 matrix of added masses and added inertias: It makes sense that, in the case of an elliptical body, at a given acceleration, the hydrodynamic load will be smaller when the body is accelerated along its major axis than along the minor axis (in 2D). Likewise, for an asymmetrical shape, an acceleration along x can result in load components for the other degrees of freedom.

Naturally, these inertia loads also occur in aerodynamics, but they are usually negligible with regards to lift and drag, except of course in the case of Zeppelin or air balloons (aerial equivalent to submarines).

A.2   The Navier–Stokes Equations

Be $Oxyz$ a Galilean coordinate system, with the Oz axis vertically upwards, and $X(t), Y(t), Z(t)$ the trajectory of a fluid particle (Figure A.2). Its velocity is $\vec{V} = (u, v, w) = (dX/dt, dY/dt, dZ/dt)$. Be $f(X(t), Y(t), Z(t), t) = f(x, y, z, t)$ some quantity attached to the fluid particle (for instance, its density). The time derivative of f is

$$\frac{df}{dt} = \frac{d}{dt}\,[f(X(t), Y(t), Z(t), t)] = u\,\frac{\partial f}{\partial x} + v\,\frac{\partial f}{\partial y} + w\,\frac{\partial f}{\partial z} + \frac{\partial f}{\partial t} \tag{A.1}$$

or, in compact form:

$$\frac{df}{dt} = \vec{V} \cdot \nabla f + f_t \tag{A.2}$$

The time derivative thus defined df/dt is the **Lagrangian** derivative. The partial derivative $\partial f/\partial t$ (also noted f_t) is the **Eulerian** derivative.

With f the density ρ one gets

$$\frac{d\rho}{dt} = u\,\frac{\partial \rho}{\partial x} + v\,\frac{\partial \rho}{\partial y} + w\,\frac{\partial \rho}{\partial z} + \frac{\partial \rho}{\partial t} = \vec{V} \cdot \nabla \rho + \rho_t \tag{A.3}$$

By definition, incompressibility means $d\rho/dt \equiv 0$. This condition is less stringent than taking ρ as constant throughout the fluid domain. However potential flow theory requires ρ to be constant.

Taking f as, successively, each component u, v, w, of the velocity $\vec{V}$ one gets

$$\frac{du}{dt} = u\,\frac{\partial u}{\partial x} + v\,\frac{\partial u}{\partial y} + w\,\frac{\partial u}{\partial z} + \frac{\partial u}{\partial t} = \vec{V} \cdot \nabla u + u_t \tag{A.4}$$

$$\frac{dv}{dt} = u\,\frac{\partial v}{\partial x} + v\,\frac{\partial v}{\partial y} + w\,\frac{\partial v}{\partial z} + \frac{\partial v}{\partial t} = \vec{V} \cdot \nabla v + v_t \tag{A.5}$$

$$\frac{dw}{dt} = u\,\frac{\partial w}{\partial x} + v\,\frac{\partial w}{\partial y} + w\,\frac{\partial w}{\partial z} + \frac{\partial w}{\partial t} = \vec{V} \cdot \nabla w + w_t \tag{A.6}$$

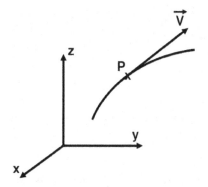

Figure A.2 Trajectory of a fluid particle.

Or, in compact form:

$$\frac{d\vec{V}}{dt} = (\vec{V} \cdot \nabla) \vec{V} + \vec{V}_t \tag{A.7}$$

Mass Conservation

Mass conservation writes

$$\frac{d\rho}{dt} + \rho \operatorname{div} \vec{V} = 0 \tag{A.8}$$

where "div" stands for "divergence" $\operatorname{div} \vec{V} = \partial u/\partial x + \partial v/\partial y + \partial w/\partial z$. When the fluid is incompressible, mass conservation reduces to

$$\operatorname{div} \vec{V} = u_x + v_y + w_z = 0 \tag{A.9}$$

Momentum Conservation

For a Newtonian and incompressible fluid, the stress tensor $\underline{\underline{\sigma}}$ writes

$$\underline{\underline{\sigma}} = -p \begin{pmatrix} 1 & 0 & 0 \\ 0 & 1 & 0 \\ 0 & 0 & 1 \end{pmatrix} + \mu \begin{pmatrix} 2u_x & u_y + v_x & u_z + w_x \\ v_x + u_y & 2v_y & v_z + w_y \\ w_x + u_z & w_y + v_z & 2w_z \end{pmatrix} \tag{A.10}$$

or, in compact notation:

$$\underline{\underline{\sigma}} = -p\,\underline{\underline{I}} + \mu \left(\nabla\vec{V} + {}^t\nabla\vec{V}\right) \tag{A.11}$$

In (A.10) p is the pressure, and $\mu = \rho v$ is the dynamic viscosity, v being the kinematic viscosity. From the stress tensor, the hydrodynamic force $d\vec{F}$ acting on a surface element dS with normal vector (into the fluid) $\vec{n}$ is given by

$$d\vec{F} = \underline{\underline{\sigma}}\,\vec{n}\,dS \tag{A.12}$$

The pressure p only acts in the normal direction whereas the viscous part of the stress tensor acts both normally and tangentially.

Conservation of momentum yields the equations

$$u_t + u\,u_x + v\,u_y + w\,u_z = -\frac{1}{\rho} p_x + v\,(u_{xx} + u_{yy} + u_{zz}) \tag{A.13}$$

$$v_t + u\,v_x + v\,v_y + w\,v_z = -\frac{1}{\rho} p_y + v\,(v_{xx} + v_{yy} + v_{zz}) \tag{A.14}$$

$$w_t + u\,w_x + v\,w_y + w\,w_z = -\frac{1}{\rho} p_z - g + v\,(w_{xx} + w_{yy} + w_{zz}) \tag{A.15}$$

or, in compact form

$$\frac{d\vec{V}}{dt} = \frac{\partial\vec{V}}{\partial t} + (\vec{V} \cdot \nabla) \vec{V} = -\frac{1}{\rho} \nabla p + \vec{g} + v\,\Delta\vec{V} \tag{A.16}$$

This vectorial equation, together with (A.9) that states mass conservation, are the Navier–Stokes equations for an incompressible (and Newtonian) fluid, in the laminar regime. There are four equations for four unknowns (the pressure p and the three components u, v, w of the velocity). No more equations are required (no need to resort to energy considerations).

These equations need to be complemented with boundary conditions at the boundaries of the fluid domain.

A.3 Potential Flow Theory

With the viscosity v set to zero, the Navier–Stokes equations become the Euler equations:

$$\frac{\partial \vec{V}}{\partial t} + (\vec{V} \cdot \nabla)\,\vec{V} = -\frac{1}{\rho}\,\nabla p + \vec{g} \tag{A.17}$$

$$\text{div}\,\vec{V} = 0 \tag{A.18}$$

The momentum equation can be transformed into

$$\frac{\partial \vec{V}}{\partial t} + (\text{Rot}\,\vec{V}) \wedge \vec{V} = -\nabla\left(\frac{V^2}{2} + \frac{p}{\rho} + g\,z\right) \tag{A.19}$$

where the density ρ is now assumed to be a constant. Kelvin's theorem states that, for a perfect fluid, if the flow vorticity is zero at some instant, it remains equal to zero: only viscosity can create (or dissipate) vorticity. Taking Rot $\vec{V}$ equal to zero, equation (A.19) reduces to

$$\frac{\partial \vec{V}}{\partial t} = -\nabla\left(\frac{V^2}{2} + \frac{p}{\rho} + g\,z\right) \tag{A.20}$$

That the vorticity be nil can be ensured a priori by stating that the velocity $\vec{V}$ is the gradient of a scalar quantity, the **velocity potential**, usually written Φ:

$$\vec{V} = \nabla\Phi \tag{A.21}$$

Mass conservation div $\vec{V} = 0$ gives

$$\text{div}\,(\nabla\Phi) = \Phi_{xx} + \Phi_{yy} + \Phi_{zz} = \Delta\Phi = 0 \tag{A.22}$$

The velocity potential verifies the Laplace equation.

As for equation (A.20) expressing momentum conservation, it becomes

$$\nabla\left(\frac{\partial \Phi}{\partial t} + \frac{V^2}{2} + \frac{p}{\rho} + g\,z\right) = 0 \tag{A.23}$$

or

$$\frac{\partial \Phi}{\partial t} + \frac{V^2}{2} + \frac{p}{\rho} + g\,z = C(t) \tag{A.24}$$

where the constant $C(t)$ can be incorporated into the velocity potential. Then the Bernoulli–Lagrange equation is obtained:

$$p = p_0 - \rho\,\frac{\partial \Phi}{\partial t} - \frac{1}{2}\rho\,(\nabla\Phi)^2 - \rho\,g\,z \tag{A.25}$$

Compared to the Euler equations, the number of unknowns has been divided by two: They are now the velocity potential Φ and the pressure p, in place of the three velocity components u, v, w and the pressure p. Moreover, the coupling between Φ and p is weak: in most practical cases, the velocity potential Φ is solved first, then the pressure p is used at a post-processing stage.

Boundary Conditions

With the Navier–Stokes equations, at the solid boundaries a **no-slip** condition is usually taken, stating that the flow velocity $\vec{V}$ and the body velocity $\vec{U}$ are locally equal. This is made possible thanks to the viscosity and to the boundary layers that can develop. With the viscosity set to zero, there can no longer be boundary layers, and the no-slip condition is replaced by a **no-flow** (or free slip) condition:

$$\vec{V} \cdot \vec{n} = \vec{U} \cdot \vec{n} \tag{A.26}$$

At interfaces, for instance the free surface, the same no-flow condition applies, with $\vec{U}$ the local velocity of the interface. Another condition that applies is the equality of the pressures at either side, under the condition that surface tension be negligible. For water waves, the role of surface tension is negligible as soon as the wave lengths are more than a few centimeters.

A.4   Some Problems Solved Within the Frame of Potential Flow Theory

A.4.1   Uniform Flow upon a Circular Cylinder

Consider a two-dimensional circular cylinder, of radius a, in a uniform current of velocity $U(t)$ along the Ox axis (see Figure A.3). In the absence of the cylinder the velocity potential is $\Phi_I = U x$. With the cylinder it is $\Phi = \Phi_I + \varphi$ where $\varphi(x, y, t)$, the perturbation potential, verifies the Boundary Value Problem:

$$\begin{aligned}
\Delta\varphi &= 0 & x^2 + y^2 \geq a^2 & \tag{A.27} \\
\frac{\partial\varphi}{\partial n} &= -\frac{\partial}{\partial n}(U x) & x^2 + y^2 = a^2 & \tag{A.28} \\
\nabla\varphi &\to 0 & x^2 + y^2 \to \infty & \tag{A.29}
\end{aligned}$$

Obviously, it is preferable to recast the BVP in polar coordinates (R, θ)

$$\begin{aligned}
\varphi_{RR} + \frac{1}{R}\varphi_R + \frac{1}{R^2}\varphi_{\theta\theta} &= 0 & R \geq a & \tag{A.30} \\
\frac{\partial\varphi}{\partial R} &= -U\cos\theta & R = a & \tag{A.31} \\
\nabla\varphi &\to 0 & R \to \infty & \tag{A.32}
\end{aligned}$$

It is easy to check that solutions of the Laplace equation in the form $f(R)\cos\theta$ are $R\cos\theta$ and $R^{-1}\cos\theta$. Only the second one decays at infinity and thus $\varphi = \alpha R^{-1}\cos\theta$, where the constant α is obtained from satisfying (A.31). Finally, the total velocity potential $\Phi = \Phi_I + \varphi$ is obtained as

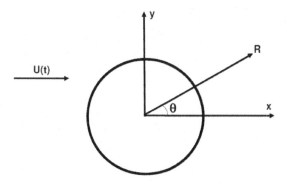

Figure A.3 Circular cylinder in uniform flow.

$$\Phi = U\left(R + \frac{a^2}{R}\right)\cos\theta \tag{A.33}$$

In fact, this is only one particular solution. A general solution is

$$\Phi(R,\theta,t) = U(t)\left(R + \frac{a^2}{R}\right)\cos\theta + \frac{\Gamma}{2\pi}\theta \tag{A.34}$$

where the last term is a circular flow around the cylinder, of circulation Γ. There is no contradiction with assuming the flow to be irrotational since the vorticity is confined inside the cylinder, out of the fluid domain. What must be retained is that Neumann problems such as the one considered here are ill-posed problems: Sometimes there is no solution, sometimes the solution is not unique.

There are mathematical techniques, known as conformal mapping, to transform circular cylinders into wing-like bodies, with sharp trailing edges, and then the circulation Γ is determined in order to satisfy a Kutta–Joukovsky condition, that is, the flow velocity must be finite at the trailing edge. Here we set the circulation equal to zero.

The streamlines can be visualized as isovalues of the stream function, which writes

$$\Psi = U\left(R - \frac{a^2}{R}\right)\sin\theta \tag{A.35}$$

They are shown in Figure A.4. They are identical whether the flow is from left to right or right to left.

The loads on the cylinder can be obtained from integrating the pressure, given by the Bernoulli equation as

$$p = p_0 - \rho\,\Phi_t - \frac{1}{2}\rho\,(\nabla\Phi)^2 \tag{A.36}$$

On the cylinder

$$p = p_0 - 2\rho\dot{U}\,a\,\cos\theta - 2\rho U^2\,\sin^2\theta \tag{A.37}$$

Only the unsteady term $-2\rho\dot{U}a\cos\theta$ contributes to the hydrodynamic force, which writes

$$\begin{pmatrix} F_x \\ F_y \end{pmatrix} = \begin{pmatrix} 2\rho\pi a^2\,\dot{U} \\ 0 \end{pmatrix} \tag{A.38}$$

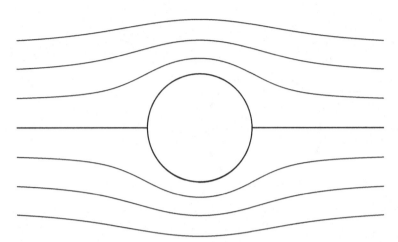

Figure A.4 Streamlines around a circular cylinder in uniform flow, from potential flow theory.

When the flow is not accelerated the hydrodynamic load is nil (d'Alembert paradox). When the flow is accelerated the load is 2 times the displaced mass times the acceleration, hence it is referred to as an **inertia** load.

A.4.2   Accelerated Cylinder in a Fluid at Rest

Consider now the parent problem of a moving cylinder through an unbounded fluid at rest at infinity. Be $x = X(t)$, $y = 0$ its trajectory. The boundary value problem to solve is now:

$$\Delta\Phi = 0 \qquad\qquad \text{in the fluid domain} \qquad (\text{A.39})$$

$$\frac{\partial\Phi}{\partial R} = \dot{X}\,\cos\theta \qquad\qquad R = a \qquad (\text{A.40})$$

$$\nabla\Phi \rightarrow 0 \qquad\qquad R \rightarrow \infty \qquad (\text{A.41})$$

This problem is identical with (A.30) through (A.32) with $-U$ replaced by $\dot{X}$. Hence the solution is

$$\Phi(R,\theta,t) = -\frac{a^2}{R}\,\dot{X}\,\cos\theta \qquad (\text{A.42})$$

The Bernoulli equation needs to be amended because the velocity potential is written in a moving coordinate system (but its gradient is still the fluid velocity with respect to a fixed coordinate system); hence $\partial/\partial t$ is no longer the true Eulerian derivative. The pressure now writes

$$p = p_0 - \rho\,\Phi_t - \frac{1}{2}\rho\,(\nabla\Phi)^2 + \rho\,\dot{X}\,\Phi_x \qquad (\text{A.43})$$

As in the fixed cylinder case, only the unsteady component of the pressure contributes to the load:

$$F_x = -\rho \pi a^2 \ddot{X} \tag{A.44}$$

The hydrodynamic force now opposes the cylinder acceleration. The multiplicative coefficient, $\rho \pi a^2$, has the dimension of a mass (per unit length), hence it is named **added mass**. In the case of a circular cylinder the added mass is identical with the displaced mass.

A.4.3   Moving Cylinder in an Accelerated Fluid

Since all equations for Φ are linear, we can superimpose the two previous problems and their solutions. For a circular cylinder with a trajectory $X(t)$, $Y(t)$ in a uniform flow of velocity $U(t)$, $V(t)$, the hydrodynamic loads write

$$
\begin{aligned}
F_x &= 2\rho \pi a^2 \dot{U} - \rho \pi a^2 \ddot{X} \tag{A.45} \\
F_y &= 2\rho \pi a^2 \dot{V} - \rho \pi a^2 \ddot{Y} \tag{A.46}
\end{aligned}
$$

A.4.4   Wave Generation by a Piston Type Wavemaker

Be a semi-infinite fluid domain, of constant depth h, bounded by a vertical wall (see Figure A.5), which undergoes small periodic oscillations with an amplitude x_0 and a frequency ω:

$$X(t) = x_0 \sin \omega t \tag{A.47}$$

We seek to obtain the characteristics of the waves produced (amplitude and wave length), and the hydrodynamic loads applied on the wavemaker. We assume that the wavemaker has been moving for a long time and that a steady state has been reached.

First we establish the boundary conditions at the free surface, which can be parametrized as $S(x,z,t) = 0$ or $z = \eta(x,t)$, the second form being more restrictive since it excludes overturning. However we will here restrict ourselves to waves of low steepness.

There are two boundary conditions at the free surface: We have to state that the pressure, on the fluid side, is equal to the atmospheric pressure, assumed to be constant; we also have to write that the fluid velocity and the velocity of the free

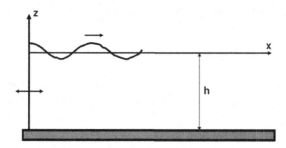

Figure A.5 Geometry.

surface, in the normal direction, are equal (the fluid cannot penetrate the free surface). The first condition is known as the **dynamic** condition, the second as the **kinematic** condition.

The dynamic condition is easily obtained from the Bernoulli equation:

$$g\,z + \Phi_t + \frac{1}{2}(\nabla\Phi)^2 = 0 \qquad \text{at } z = \eta(x,t) \tag{A.48}$$

The kinematic condition can be derived starting from $\vec{V}\cdot\vec{n} = \vec{U}\cdot\vec{n}$ with $\vec{V}$ the fluid velocity, $\vec{U}$ the velocity of the free surface, and $\vec{n}$ the normal vector. It is equivalent and faster to write that $S(x,z,t) = z - \eta(x,t) = 0$ is a property of the free surface and that its Lagrangian derivative is nil.

Then it comes

$$\frac{dS}{dt} = \nabla S \cdot \nabla\Phi + S_t = 0 \tag{A.49}$$

giving, since $S_x = -\eta_x$, $S_z = 1$ and $S_t = -\eta_t$:

$$\Phi_z - \Phi_x\,\eta_x - \eta_t = 0 \qquad \text{at } z = \eta(x,t) \tag{A.50}$$

Now we take advantage of the assumption that the wave steepness and free surface elevation η are low enough that the free surface conditions can be "linearized", meaning that the quadratic terms therein are dropped and that the free surface equations are no longer imposed in $z = \eta$ but in $z = 0$. They reduce to:

- kinematic condition:

$$\Phi_z - \eta_t = 0 \qquad \text{at } z = 0 \tag{A.51}$$

- dynamic condition:

$$g\,\eta + \Phi_t = 0 \qquad \text{at } z = 0 \tag{A.52}$$

Eliminating η we get:

$$g\,\Phi_z + \Phi_{tt} = 0 \qquad \text{at } z = 0 \tag{A.53}$$

Similarly and consistently the no-flow condition at the wavemaker is imposed no longer in $x = X(t)$, but in $x = 0$:

$$\Phi_x(0,z,t) = x_0\,\omega\,\cos\omega t \tag{A.54}$$

As already mentioned we assume that a steady state has been reached so that all quantities (velocity potential, free surface elevation, etc.) vary in time at frequency ω. Writing the velocity potential

$$\Phi(x,z,t) = \varphi_C(x,z)\,\cos\omega t + \varphi_S(x,z)\,\sin\omega t = \Re\left\{\varphi(x,z)\,e^{-i\omega t}\right\} \tag{A.55}$$

the reduced potential $\varphi = \varphi_C + i\,\varphi_S$ has to satisfy the Boundary Value Problem:

$$\Delta\varphi \ = \ 0 \qquad x \geq 0 \qquad -h \leq z \leq 0 \qquad (A.56)$$

$$\varphi_z \ = \ 0 \qquad x \geq 0 \qquad z = -h \qquad (A.57)$$

$$\varphi_x \ = \ x_0\,\omega \qquad x = 0 \qquad -h \leq z \leq 0 \qquad (A.58)$$

$$g\,\varphi_z - \omega^2\,\varphi \ = \ 0 \qquad x \geq 0 \qquad z = 0 \qquad (A.59)$$

Since the "linearized" fluid domain is rectangular, one may apply the eigen-function expansion method. The first step is to look for elementary solutions of the form $F(x)\,G(z)$, which verify as many as possible of the conditions above. Introducing $\varphi(x,z) = F(x)\,G(z)$ into the Laplace equation we get:

$$F''\,G + F\,G'' = 0 \qquad (A.60)$$

or, under the condition that F and G are non zero:

$$\frac{F''}{F} + \frac{G''}{G} = 0 \qquad (A.61)$$

F''/F is a function of x, whereas G''/G is a function of z, so they have to be two opposite constants: $F''/F = -G''/G = C$. The cases when the constant C is positive, negative, or zero, must be distinguished:

- $C = k^2$

Then we have:

$$F'' - k^2\,F = 0 \qquad\qquad G'' + k^2\,G = 0 \qquad (A.62)$$

Solutions for $F(x)$ are $\exp(k\,x)$ and $\exp(-k\,x)$, and for $G(z)$ they are $\cos k\,z$ and $\sin k\,z$. We suppose k to be positive; then $\exp(k\,x)$ must be discarded. To satisfy the no-flow condition at the bottom $\cos k\,z$ and $\sin k\,z$ are regrouped as $\cos k(z+h)$. Finally, the free surface condition $g\,\varphi_z - \omega^2\,\varphi = 0$ gives:

$$-g\,k\,\sin kh - \omega^2\,\cos kh = 0 \qquad (A.63)$$

usually written as

$$\omega^2 = -g\,k\,\tan kh \qquad (A.64)$$

We are looking for k values that satisfy this equation. A graphical resolution can be done by writing the equation in the form

$$\tan kh = -\frac{A}{kh} \qquad (A.65)$$

where $A = \omega^2\,h/g$. Solutions in kh are intersections of the hyperbola branch $y = -A/x$ with $y = \tan x$ (see Figure A.6). There is a discrete set of intersections $k_n\,h$ $(n = 1,\ldots,\infty)$, where, asymptotically, when n increases, $k_n h \simeq n\,\pi$.

- $C = 0$

F and G are linear functions of x and z but then the free surface condition cannot be satisfied.

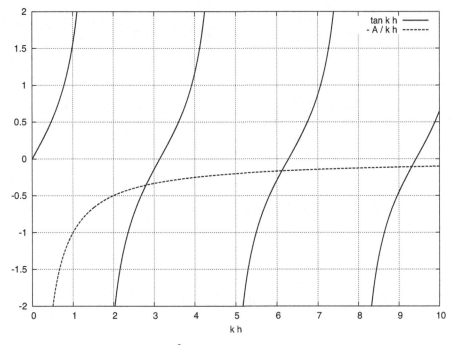

Figure A.6 Graphical resolution of $\omega^2 = -g\,k\,\tan kh$.

- $C = -k^2$

  We now have:

$$F'' + k^2\,F = 0 \qquad\qquad G'' - k^2\,G = 0 \qquad\qquad (A.66)$$

Solutions for $F(x)$ are $\cos kx$ and $\sin kx$, or $\exp(\mathrm{i}\,kx)$ and $\exp(-\mathrm{i}\,kx)$. To make the choice we must follow the visual observation that waves are progressing away from the wavemaker. Retaining $\cos kx$ or $\sin kx$ alone would produce standing waves, retaining $\exp(-\mathrm{i}\,kx)$ would produce a wave progressing from right to left, toward the wavemaker. So we choose $\exp(\mathrm{i}\,kx)$.

Associated solutions for $G(z)$ are $\exp(k\,z)$ and $\exp(-k\,z)$, combined in $\cosh k(z + h)$ to satisfy the bottom boundary condition.

The free surface condition now gives

$$\omega^2 = g\,k\,\tanh kh \qquad\qquad (A.67)$$

which can again be solved graphically (see Figure A.7), and which admits only one solution, be k_0.

To summarize, we have found a set of functions $F_n(x)\ G_n(z)$ that satisfy all equations of the Boundary Value Problem except for the no-flow condition at the wavemaker. To satisfy this condition we regroup the $F_n(x)\,G_n(z)$ as

$$\varphi(x,z) = -x_0\,\omega \left\{ \frac{\mathrm{i}\,A_0}{k_0}\,\frac{\cosh k_0(z+h)}{\cosh k_0 h}\,\mathrm{e}^{\mathrm{i}\,k_0 x} + \sum_{n=1}^{\infty} \frac{A_n}{k_n}\,\cos k_n(z+h)\,\mathrm{e}^{-k_n x} \right\}$$

$$(A.68)$$

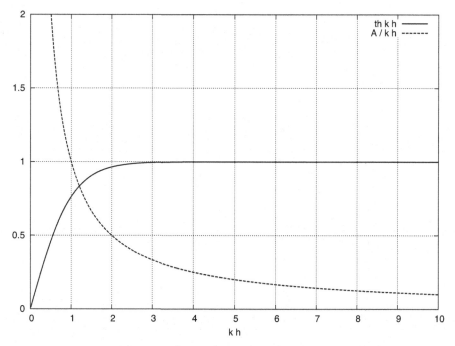

Figure A.7 Graphical resolution of $\omega^2 = g\,k\,\tanh kh$.

It can be shown that the set of z functions $[\cosh k_0(z+h),\ \cos k_n(z+h), (n = 1, \infty)]$ is complete over the interval $[-h\ 0]$. It can easily be checked that the set is orthogonal, that is, for instance

$$\int_{-h}^{0} \cosh k_0(z + h)\ \cos k_n(z + h)\ \mathrm{d}z \equiv 0 \qquad (A.69)$$

The no-flow condition at the wavemaker gives:

$$A_0\,\frac{\cosh k_0(z + h)}{\cosh k_0 h} + \sum_{n=1}^{\infty} A_n\,\cos k_n(z + h) = 1 \qquad (A.70)$$

which must be satisfied for $-h \le z \le 0$.

All we have to do to obtain A_0 is multiply both sides with $\cosh k_0(z + h)$ and integrate from $z = -h$ up to $z = 0$. We get:

$$A_0 = \frac{2\,\sinh 2k_0 h}{2k_0 h + \sinh 2k_0 h} \qquad (A.71)$$

Likewise, multiplying both sides with $\cos k_n(z + h)$ and integrating, we get:

$$A_n = \frac{4\sin k_n h}{2k_n h + \sin 2k_n h} \qquad (A.72)$$

When one is only interested in the wave system away from the wavemaker, the coefficients A_n for $n \ge 1$ are of no interest since the contribution of these terms to the velocity potential (and to the free surface elevation) exponentially decreases

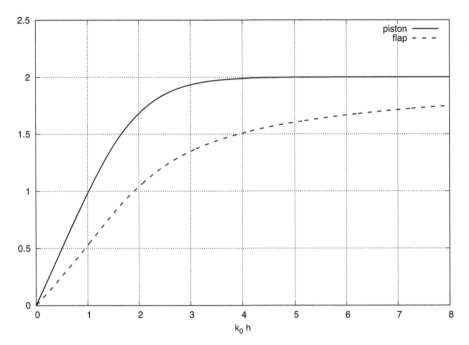

Figure A.8 RAOs of piston type and flap type wavemakers.

away from the wavemaker. For this reason they are usually named **local** modes or **evanescent** modes.

The free surface elevation is obtained through:

$$\eta(x,t) = -\frac{1}{g}\,\Phi_t(x,0,t) \qquad (A.73)$$

Some distance from the wavemaker it reduces to

$$\eta(x,t) = \Re\left\{x_0\,A_0\,\tanh k_0 h\,\mathrm{e}^{\mathrm{i}\,(k_0 x - \omega t)}\right\} = \frac{4\,\sinh^2 k_0 h}{2\,k_0 h + \sinh 2\,k_0 h}\,x_0\,\cos(k_0 x - \omega t)$$
$$(A.74)$$

The free surface profile is a sinusoid with a wave length $2\pi/k_0$, where k_0 is given by the dispersion equation (A.67), and with an amplitude A_I equal to the wavemaker motion amplitude x_0 multiplied by $4\,\sinh^2 k_0 h/(2\,k_0 h + \sinh 2\,k_0 h)$. This transfer function, or Response Amplitude Operator (RAO), is shown in Figure A.8.

When $k_0 h$ is much higher than 1 (in practice, $k_0 h > 3$, meaning a water depth greater than half a wave length), the RAO is equal to 2: the wave amplitude is twice the wavemaker motion amplitude. Conversely, when $k_0 h$ decreases to zero, the RAO becomes equal to $k_0 h$.

In the same way, the case of a flap type wavemaker can be tackled. When the axis of rotation is at the floor level, the right-hand side in equation (A.70) just has to be replaced with $(z + h)/h$, x_0 being now the motion amplitude at the free surface level.

The resulting performance is also shown in Figure A.8. In shallow depth, the piston type wavemaker is twice as efficient as the flap, but in deep water they become equivalent, the advantage of the flap being that the hydrodynamic inertia loads are much less. This is why piston type wavemakers are mostly encountered in coastal engineering facilities, whereas flaps are preferentially used in towing tanks and deep water offshore facilities.

Hydrodynamic loads upon the wavemaker

To obtain the hydrodynamic loads, one just needs to integrate the pressure which, within a linearized theory, reduces to

$$p = p_0 - \rho \, \Phi_t - \rho g \, z \tag{A.75}$$

the integration, still within a linearized approach, being done from $z = -h$ up to $z = 0$.

The loads then consist in the still water hydrostatic load, plus a dynamic contribution given by

$$F(t) = \mathcal{R} \left\{ f \, e^{-i \omega t} \right\} = \mathcal{R} \left\{ -i \, \rho \omega \int_{-h}^{0} \varphi(0, z) \, dz \, e^{-i \omega t} \right\} \tag{A.76}$$

The integration gives:

$$f = -\rho \omega^2 \, x_0 \left(\frac{A_0}{k_0^2} \tanh k_0 h - i \sum_{n=1}^{\infty} \frac{A_n}{k_n^2} \sin k_n h \right) \tag{A.77}$$

or

$$F(t) = -\rho \omega \frac{A_0}{k_0^2} \tanh k_0 h \, \dot{X}(t) - \rho \sum_{n=1}^{\infty} \frac{A_n}{k_n^2} \sin k_n h \, \ddot{X}(t) \tag{A.78}$$

The hydrodynamic load divides into a component that opposes the velocity $\dot{X}$ of the wavemaker, and a component that opposes its acceleration $\ddot{X}$. The term $\rho \omega \, A_0 \tanh k_0 h / k_0^2$ has the dimension of a mass (per unit width) divided by time, and is thus a damping term; it is associated with the wave generation and referred to as **radiation damping**. As for the term $\rho \sum A_n \sin k_n h / k_n^2$, it is a mass (per unit length), hence called **added mass**.

Figure A.9 shows the added mass coefficient C_a and the damping coefficient C_b, defined as

$$C_a = \sum_{n=1}^{\infty} \frac{A_n}{k_n^2 h^2} \sin k_n h \qquad C_b = \frac{A_0}{k_0^2 h^2} \tanh k_0 h \tag{A.79}$$

The damping coefficient is dominant in the low $k_0 h$ range, confirming the efficiency of the piston wavemaker.

Lastly, it can be checked that the mean power dissipated by the wavemaker is equal to the power carried away (or energy flux) by the waves. The power dissipated by the wavemaker motion is

$$P_b = \frac{1}{2} B \omega^2 \, x_0^2 = \frac{1}{2} \rho \omega \frac{A_0}{k_0^2} \tanh k_0 h \, \omega^2 \, x_0^2 = \frac{1}{2} \rho g \omega \frac{A_0}{k_0} \tanh^2 k_0 h \, x_0^2 \tag{A.80}$$

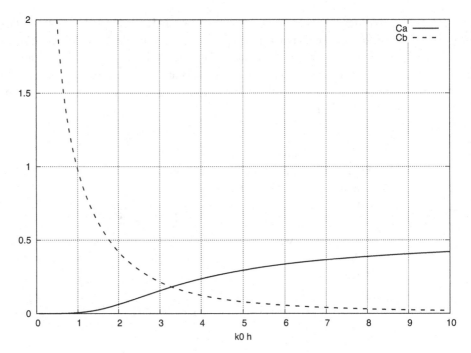

Figure A.9 Piston wavemaker. Added mass (C_a) and damping (C_b) coefficients.

The energy flux is (see Chapter 3):

$$\overline{F}_L = \frac{1}{2} \rho g \, C_G \, A_I^2 \tag{A.81}$$

where $A_I = x_0 A_0 \tanh k_0 h$ is the wave amplitude and C_G the group velocity equal to $\omega / (k_0 A_0)$. It is straight-forward to check that this agrees with equation (A.80).

A.5   Reference

MÜLLER G., WOLTERS G., COOKER M.J. 2003. Characteristics of pressure pulses propagating through water-filled cracks, *Coast. Eng.*, **49**, 83–98.

Appendix B: Hydrostatics

In this Appendix we restrict our interest to the loads resulting from the hydrostatic component of the pressure, that is,

$$p = p_0 - \rho g \, z \qquad (B.1)$$

with p_0 the atmospheric pressure (which does not play any role) and $z = 0$ the undisturbed free surface plane.

The considered bodies are assumed to be rigid (the hydrostatics of deformable bodies is a complicated and somewhat controversial matter).

The main goal here is to express the change in the hydrostatic loads when the body undergoes some displacement along one or several of its 6 degrees of freedom, in particular, to assess whether the structure is statically stable under some external load.

First we consider fully submerged bodies, then we move on to the case of surface-piercing bodies. In all cases, the reference point to express the torque is the center of gravity.

B.1 Fully Submerged Body

The considered geometry is shown in Figure B.1.

The hydrostatic torque is obtained by integrating the pressure over the wetted hull S_B. It consists in a force $\vec{F}$ and a moment $\vec{C}$:

$$\vec{F} = -\iint_{S_B} p \, \vec{n} \, \mathrm{d}S = -\iint_{S_B} (p_0 - \rho g \, z) \, \vec{n} \, \mathrm{d}S \qquad (B.2)$$

$$\vec{C} = -\iint_{S_B} p \, \overrightarrow{GP} \wedge \vec{n} \, \mathrm{d}S = -\iint_{S_B} (p_0 - \rho g \, z) \begin{pmatrix} x - x_G \\ y - y_G \\ z - z_G \end{pmatrix} \wedge \vec{n} \, \mathrm{d}S \qquad (B.3)$$

$\vec{C}$ being written at the center of gravity. It must be noted that, in this Appendix, the normal vector $\vec{n}$ is taken in the outer direction (into the fluid).

These equations can be transformed into volume integrals through applications of the gradient theorem and of the rotational formula:

$$\iint_{S_B} f \, \vec{n} \, \mathrm{d}S = \iiint_V \nabla f \, \mathrm{d}V \qquad (B.4)$$

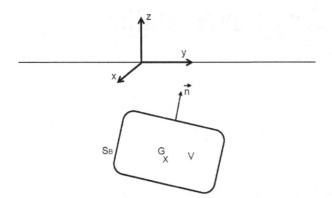

Figure B.1 Geometry.

$$\iint_{S_B} \vec{n} \wedge \vec{f} \, dS = \iiint_V \mathrm{Rot} \, \vec{f} \, dV \tag{B.5}$$

with V the interior domain.

Then it comes
- For the force $\vec{F}$:

$$\vec{F} = \rho \, g \iiint_V \begin{pmatrix} 0 \\ 0 \\ 1 \end{pmatrix} dV = \rho \, g \, \forall \, \vec{k} \tag{B.6}$$

with $\forall$ the volumic displacement and $\vec{k}$ the vertical unit vector. This is the **Archimedes theorem:** A body **completely** immersed receives an upward load equal to the weight of the displaced fluid.

- For the moment $\vec{C}$:

$$\vec{C} = \iiint_V \mathrm{Rot} \left\{ (p_0 - \rho g \, z) \begin{pmatrix} x - x_G \\ y - y_G \\ z - z_G \end{pmatrix} \right\} dV \tag{B.7}$$

which can be transformed, following

$$\mathrm{Rot} \, (A \, \vec{B}) = A \, \mathrm{Rot} \, \vec{B} + \nabla A \wedge \vec{B} \tag{B.8}$$

into

$$\vec{C} = -\rho g \left\{ \iiint_V \begin{pmatrix} 0 \\ 0 \\ 1 \end{pmatrix} \wedge \begin{pmatrix} x - x_G \\ y - y_G \\ z - z_G \end{pmatrix} dV \right\}$$

$$\vec{C} = \left\{ \frac{1}{\forall} \iiint_V \begin{pmatrix} x - x_G \\ y - y_G \\ z - z_G \end{pmatrix} dV \right\} \wedge \left(\rho g \, \forall \, \vec{k} \right) = \vec{GB} \wedge \vec{F} \tag{B.9}$$

where B, with coordinates (x_B, y_B, z_B), is the centroid of V, or **center of buoyancy**.

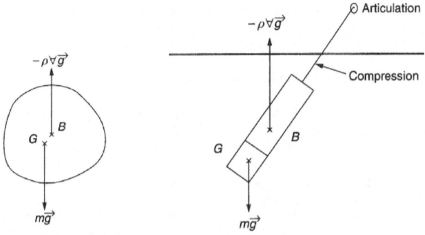

Figure B.2 Geometry.   **Figure B.3** Submerged body with external load.

This means that the hydrostatic load $\rho g \,\forall\, \overrightarrow{k}$ is applied at the center of buoyancy. For a rigid body, the position of the center of buoyancy is invariable in a body-fitted coordinate system.

B.1.1   Stability

Case without External Load

The loads coming into play are the body weight $-m\, g\, \overrightarrow{k}$ applied at the center of gravity G and the buoyancy force $\rho g \,\forall\, \overrightarrow{k}$ applied at the center of buoyancy B (see Figure B.2).

For the combined force to be nil, the body mass must be equal to the displaced mass of water:

$$m = \rho \,\forall \tag{B.10}$$

For the combined moment to be nil, the center of gravity and the center of buoyancy must be on the same vertical line. The equilibrium is stable if the center of buoyancy is above the center of gravity, unstable in the other case. When the body is inclined by an angle α to the vertical, the restoring moment is

$$C = -\rho g \,\forall \overline{GB} \,\sin\alpha \tag{B.11}$$

Case with External Load

This is, for instance, the case of Figure B.3.

The difference with the previous case is that the displaced mass is no longer equal to the body mass. As a result, the equilibrium may be unstable even though the center of buoyancy be above the center of gravity. This can be experienced by pushing down an upright floating bottle partially filled.

Conversely, bottom-hinged articulated towers are stable even though their centers of gravity be above their centers of buoyancy. This is due to their excess of buoyancy.

B.2   Bodies Piercing the Free Surface

A difference with the previous case is that the wetted surface is not closed anymore (see Figure B.4).

First we prove that the atmospheric term p_0 in the pressure equation does not play any role: it applies on the aerial parts of the structures as well as the immersed parts; the gradient theorem may be applied and the resultant load is zero.

We now state $p_0 \equiv 0$ and the pressure reduces to $-\rho g\, z$. The wetted hull S_B can then be closed by adjoining to S_B the **waterplane**, denoted S_{WP}, that is the continuation of the horizontal free surface inside the **waterline**. The gradient and rotational theorems may again be applied and the same results as in the fully submerged case are obtained: the buoyancy force is upward, equal to weight of displaced water, and is applied at the center of buoyancy.

The difference is that the geometry of the wetted part of the hull varies when the body undergoes some displacement in the vertical direction, or in roll and pitch. For each position, the immersed volume and the position of the center of buoyancy need to be determined.

B.2.1   Stability Curves

A practical problem is the roll (or pitch) stability of the floating structure, for instance, under wind loading. One usually considers that the external load consists in an overturning moment. The problem, then, is to derive the hydrostatic restoring moment, as a function of the inclination.

For each heeling angle α the waterplane must first be determined (keeping the immersed volume the same), then the new position $B(\alpha)$ of the center of buoyancy giving the restoring moment, usually expressed as

$$C(\alpha) = \rho g\, \forall\, GZ(\alpha) \tag{B.12}$$

where $GZ(\alpha)$ is the lever arm.

Figure B.5 illustrates the successive positions of the free surface and of the center of buoyancy.

Rules for ships and offshore structures demand that the restoring moments be calculated in intact and damaged conditions ("damaged" meaning, for instance, that some compartments are flooded), and to check that they are sufficient to ensure stability under external loads, such as a centenal wind squall.

Figure B.6 shows the hydrostatic restoring moment corresponding to the "hull" of Figure B.5: It is positive up to an inclination of about 70 degrees. The figure also

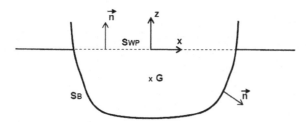

Figure B.4 Geometry.

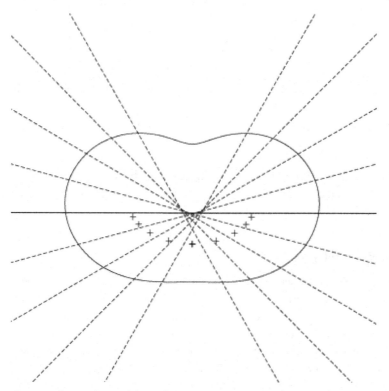

Figure B.5 Successive positions of the waterplane and of the center of buoyancy, for heeling angles from −60 to +60 degrees.

shows an inclining moment, due to the wind loading. The two curves first intersect in α_1 (about 15 degrees in the figure), which is the inclination of static equilibrium, and then in α_2 (about 55 degrees), beyond which the body will capsize. An intuitive stability criterion would be that α_1 be well inferior to α_2. Rules usually state "dynamic" stability criteria, in the form

$$(A + B) \geq k \, (B + C) \qquad \text{with } k > 1 \qquad (B.13)$$

where A, B, and C are the areas marked in the figure. The physical interpretation is as follows: imagine the wind velocity to jump suddenly from zero to the design value. The structure will start heeling and stop at an angle α_e such that the work $B + C$ produced by the inclining moment be equal to the energy stored $A + B$ by the

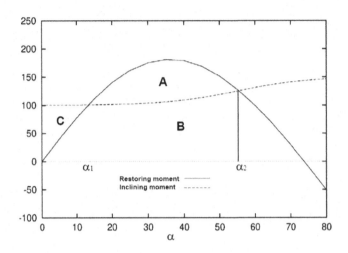

Figure B.6 Hydrostatic restoring moment, and wind inclining moment, vs the heeling angle α.

hydrostatic restoring moment (the kinetic energy being nil at α_e). This angle α_e must not exceed the second intercept α_2 of the two curves, hence the application of a safety coefficient k, typically equal to 1.3 or 1.4.

B.2.2   Linearized Hydrostatics

It is now assumed that the displacement of the structure, from its reference position, is "small" (of order ε), compared to its own dimensions. Then it is possible to linearize the hydrostatic restoring force and moment, and to express them from some geometric characteristics of the hull.

Here we are only concerned with the displacements in the vertical direction, in heel and in trim: there is no modification of the submerged volume under horizontal displacements nor rotation around the vertical axis. For convenience we assume these to be nil.

Be Δz_G the vertical displacement of the center of gravity, α and β the angles of rotation in roll and pitch. Since these angles are assumed to be small, we can take them around fixed axes and write the motion of a point attached to the hull as

$$\begin{pmatrix} \Delta x \\ \Delta y \\ \Delta z \end{pmatrix} = \begin{pmatrix} 0 \\ 0 \\ \Delta z_G \end{pmatrix} + \begin{pmatrix} \alpha \\ \beta \\ 0 \end{pmatrix} \wedge \begin{pmatrix} X \\ Y \\ Z \end{pmatrix} \qquad (B.14)$$

where X, Y, Z are its coordinates in a body-fitted coordinate system centered in G. Note that here, as in Chapter 6, we use two coordinate systems, a fixed one (xyz) with $z = 0$ the mean free surface, and a body-fitted one (XYZ). In the static reference position they correspond via a vertical translation.

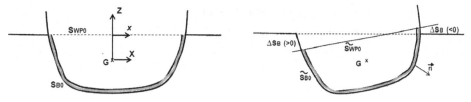

Figure B.7 Geometry

Be S_B the wetted part of the hull, consecutive to the displacement (B.14). It can be decomposed algebraically in the following way

$$S_B = \widetilde{S}_{B0} + \Delta S_B \tag{B.15}$$

where $\widetilde{S}_{B0}$ is the transform of S_{B0} and ΔS_B is the complementary part at the waterline (Figure B.7).

Let us first consider the hydrostatic force. It can be decomposed as

$$\overrightarrow{F} = \rho g \iint_{S_B} z \, \overrightarrow{n} \, dS = \rho g \iint_{\widetilde{S}_{B0}} z \, \overrightarrow{n} \, dS + \iint_{\Delta S_B} z \, \overrightarrow{n} \, dS \tag{B.16}$$

The second integral over ΔS_B can be discarded: it is of order ε^2. The first integral can be rewritten:

$$\rho g \iint_{\widetilde{S}_{B0}} z \, \overrightarrow{n} \, dS = \rho g \iint_{\widetilde{S}_{B0} \cup \widetilde{S}_{WP_0}} z \, \overrightarrow{n} \, dS - \rho g \iint_{\widetilde{S}_{WP_0}} z \, \overrightarrow{n} \, dS \tag{B.17}$$

$$= \rho g \, \forall_0 \, \overrightarrow{k} - \rho g \iint_{\widetilde{S}_{WP_0}} z \, \overrightarrow{n} \, dS \tag{B.18}$$

where $\widetilde{S}_{WP_0}$ is the transform of S_{WP_0} due to the roto-translation.

On this surface we have

$$z = \Delta z_G + \alpha \, Y - \beta \, X \qquad\qquad \overrightarrow{n} \simeq \overrightarrow{k} \tag{B.19}$$

The vertical force component is therefore obtained as

$$F_z = \rho g \, \forall_0 - \rho g \, \Delta z_G \iint_{S_{WP_0}} dS - \rho g \, \alpha \iint_{S_{WP_0}} Y \, dS + \rho g \, \beta \iint_{S_{WP_0}} X \, dS \tag{B.20}$$

The first term is the still water buoyancy force. The restoring force is therefore

$$\Delta F_z = -K_{33} \, \Delta z_G - K_{34} \, \alpha - K_{35} \, \beta \tag{B.21}$$

where the **hydrostatic stiffnesses** K_{33}, K_{34}, K_{35} are given by

$$K_{33} = \rho g \iint_{S_{WP}} dS = \rho g \, S_{WP} \tag{B.22}$$

$$K_{34} = \rho g \iint_{S_{WP}} Y \, dS = \rho g \, I_Y \tag{B.23}$$

$$K_{35} = -\rho g \iint_{S_{WP}} X \, dS = -\rho g \, I_X \tag{B.24}$$

I_X and I_Y are the **first moments of inertia** of the waterplane.

In a similar way the moment $\vec{C}$ can be rewritten

$$\vec{C} = -\rho g \int\!\!\int\!\!\int_{\widetilde{V_0}} \text{Rot}\left\{z\,\overrightarrow{GP}\right\} dV - \rho g \int\!\!\int_{\widetilde{S_W}_{P_0}} z\,\overrightarrow{GP}\wedge\vec{n}\ dS + \rho g \int\!\!\int_{\Delta S_B} z\,\overrightarrow{GP}\wedge\vec{n}\ dS$$
(B.25)

where the last term is again of order ϵ^2 and neglected.

The first term is transformed as

$$\vec{C_1} = -\rho g \int\!\!\int\!\!\int_{\widetilde{V_0}}\begin{pmatrix}-y\\x\\0\end{pmatrix} dV = -\rho g \int\!\!\int\!\!\int_{V_0}\begin{pmatrix}\alpha\,Z - Y\\\beta\,Z + X\\0\end{pmatrix} dV$$
(B.26)

In the reference position, the center of gravity and center of buoyancy are on the same vertical line. Therefore

$$\int\!\!\int\!\!\int_{V_0} X\ dV = \int\!\!\int\!\!\int_{V_0} Y\ dV \equiv 0$$
(B.27)

and

$$\vec{C_1} = -\rho g\ V_0\ \overline{GB}\begin{pmatrix}\alpha\\\beta\\0\end{pmatrix}$$
(B.28)

The second term in (B.25) is obtained as

$$\vec{C_2} = -\rho g \int\!\!\int_{S_{WP}}(\Delta z_G + \alpha\,Y - \beta\,X)\begin{pmatrix}Y\\-X\\0\end{pmatrix} dS$$
(B.29)

and can be written via the stiffnesses K_{34} and K_{35} and via the **second moments of inertia** I_{XX}, I_{YY}, I_{XY} of the waterplane:

$$I_{XX} = \int\!\!\int_{S_{WP}} X^2\ dS \qquad\qquad I_{YY} = \int\!\!\int_{S_{WP}} Y^2\ dS$$
(B.30)

$$I_{XY} = -\int\!\!\int_{S_{WP}} X\,Y\ dS$$
(B.31)

Then:

$$\vec{C_2} = -\begin{pmatrix}K_{34}\,\Delta z_G + \rho g\,I_{YY}\,\alpha + \rho g\,I_{XY}\,\beta\\K_{35}\,\Delta z_G + \rho g\,I_{XX}\,\beta + \rho g\,I_{XY}\,\alpha\\0\end{pmatrix}$$
(B.32)

Finally, the total restoring moment takes the form:

$$\vec{C} = \vec{C_1} + \vec{C_2} = -\begin{pmatrix}K_{34}\,\Delta z_G + K_{44}\,\alpha + K_{45}\,\beta\\K_{35}\,\Delta z_G + K_{55}\,\beta + K_{54}\,\alpha\\0\end{pmatrix}$$
(B.33)

where

$$K_{34} = K_{43} \quad = \quad \rho g \, I_Y \tag{B.34}$$

$$K_{35} = K_{53} \quad = \quad -\rho g \, I_X \tag{B.35}$$

$$K_{44} \quad = \quad \rho g \, (\forall_0 \, \overline{GB} + I_{YY}) \tag{B.36}$$

$$K_{55} \quad = \quad \rho g \, (\forall_0 \, \overline{GB} + I_{XX}) \tag{B.37}$$

$$K_{45} = K_{54} \quad = \quad \rho g \, I_{XY} \tag{B.38}$$

For hulls such that the waterplane has at least one axis of symmetry the cross-inertia I_{XY} is nil. All ships and most offshore structures have a port-starboard symmetry (implying also $I_y \equiv 0$).

As can be seen the difference with the fully immersed case comes from the water-plane inertia terms, which express the displacement of the center of buoyancy in the body-fitted coordinate system.

The following notations are frequently used:

$$\forall_0 \, \overline{GB} + I_{YY} = \forall_0 \, (\overline{GB} + \overline{BM}_\alpha) = \forall_0 \, \overline{GM}_\alpha \tag{B.39}$$

$$\forall_0 \, \overline{GB} + I_{XX} = \forall_0 \, (\overline{GB} + \overline{BM}_\beta) = \forall_0 \, \overline{GM}_\beta \tag{B.40}$$

M_α and M_β are the **metacenters** in roll and pitch. Geometrically M_α (respectively M_β) is the center of curvature of the curve described by the center of buoyancy when the hull is inclined in roll (resp. pitch) (see Figure B.8). This is the reason why BM_α and BM_β are named **metacentric radii** and are also written as ρ_α and ρ_β. $\overline{GM}_\alpha$ and $\overline{GM}_\beta$ are the **metacentric heights** (often written h_α and h_β). When a is the algebraic distance $\overline{BG} = -\overline{GB}$, then

$$h_\alpha = \rho_\alpha - a \qquad\qquad h_\beta = \rho_\beta - a \tag{B.41}$$

At variance with the fully submerged case, a floating structure can be stable even though its center of gravity be above its center of buoyancy, provided that its metacentric heights be positive (that the metacenters be above the center of gravity).

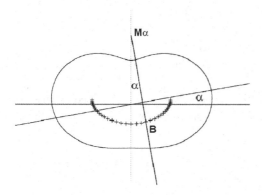

Figure B.8 Metacenter.

These results may be related to equation (B.12) giving the restoring moment as a function of an arbitrary inclining angle α: the metacentric height is just the derivative of the lever arm with respect to α:

$$GM_\alpha = \frac{\mathrm{d}}{\mathrm{d}\alpha} (GZ)\Big|_{\alpha=0} \tag{B.42}$$

For the academic hull considered in Figures B.5 and B.6, we see that the linearized moment is a reasonable approximation up to a heel angle of about 20 degrees.

Matrix of Hydrostatic Stiffnesses

To summarize, linearized hydrostatics lead to the restoring torque expressed via the stiffness matrix

$$\mathbf{K_H} = \begin{pmatrix} 0 & 0 & 0 & 0 & 0 & 0 \\ 0 & 0 & 0 & 0 & 0 & 0 \\ 0 & 0 & K_{33} & K_{34} & K_{35} & 0 \\ 0 & 0 & K_{34} & K_{44} & K_{45} & 0 \\ 0 & 0 & K_{35} & K_{45} & K_{55} & 0 \\ 0 & 0 & 0 & 0 & 0 & 0 \end{pmatrix} \tag{B.43}$$

such that:

$$\begin{pmatrix} F_x \\ F_y \\ F_z \\ C_x \\ C_y \\ C_z \end{pmatrix} = -\mathbf{K_H} \begin{pmatrix} \Delta x_G \\ \Delta y_G \\ \Delta z_G \\ \alpha \\ \beta \\ \gamma \end{pmatrix} \tag{B.44}$$

where $(\Delta x_G, \Delta y_G, \Delta z_G)$ is the displacement of the center of gravity and (α, β, γ) the rotations in roll, pitch, and yaw.

Expressions for the different terms of this matrix have been given in equations (B.22), (B.23), (B.24) and (B.34) through (B.38).

Trim Induced by an External Moment

The results obtained above give the restoring torque due to a displacement $(\Delta z_G, \alpha, \beta)$ of the structure.

Conversely, the question arises of the displacement induced by an external load.

Let us consider the case of a pure moment C_y about the y axis. The result displacement $(\Delta z_G, \alpha, \beta)$ verifies

$$K_{33} \Delta z_G + K_{34} \alpha + K_{35} \beta = 0 \tag{B.45}$$

$$K_{34} \Delta z_G + K_{44} \alpha + K_{45} \beta = 0 \tag{B.46}$$

$$K_{35} \Delta z_G + K_{45} \alpha + K_{55} \beta = -C_y \tag{B.47}$$

Taking K_{34} and K_{45} to be nil (port-starboard symmetry), this system reduces to

$$K_{33}\,\Delta z_G + K_{35}\,\beta \;=\; 0 \tag{B.48}$$

$$K_{35}\,\Delta z_G + K_{55}\,\beta \;=\; -C_y \tag{B.49}$$

from which we obtain the trim angle

$$\beta = -\frac{K_{33}}{K_{33}\,K_{55} - K_{35}^2}\,C_y \tag{B.50}$$

and the vertical displacement of the center of gravity

$$\Delta z_G = \frac{K_{35}}{K_{33}\,K_{55} - K_{35}^2}\,C_y \tag{B.51}$$

Center of Flotation

In the case just considered, we have obtained a vertical displacement of the center of gravity:

$$\Delta z_G = -\frac{K_{35}}{K_{33}}\,\beta \tag{B.52}$$

For a point at distance X from G, the vertical displacement is:

$$\Delta z = \Delta z_G - \beta\,X = -\left(\frac{K_{35}}{K_{33}} + X\right)\beta \tag{B.53}$$

It is zero in X_F given by

$$X_F = -\frac{K_{35}}{K_{33}} = \frac{\iint_{S_{WP_0}} X\,dS}{\iint_{S_{WP_0}} dS} \tag{B.54}$$

Likewise, for a moment in roll, there is no vertical displacement at a distance Y_F given by

$$Y_F = \frac{K_{34}}{K_{33}} = \frac{\iint_{S_{WP_0}} Y\,dS}{\iint_{S_{WP_0}} dS} \tag{B.55}$$

The point F of the waterplane with coordinates (X_F, Y_F) is the **center of flotation**. It is the point that does not move under a pure external moment.

Effects of Liquid Cargo

Many ships, such as tankers or LNG carriers, hold liquid cargos. When the ship heels and the free surfaces in the tanks remain horizontal, the center of gravity of the ship plus cargo shifts to the lower side, reducing the hydrostatic restoring force.

The shift of the center of gravity is easily obtained from the inertias of the free surfaces in the tanks as

$$\begin{pmatrix} \Delta x_G \\ \Delta y_G \end{pmatrix} = -\sum_n \frac{\rho_n}{\rho\,V_0}\begin{pmatrix} I_{YYn} & I_{XYn} \\ I_{XYn} & I_{XXn} \end{pmatrix}\begin{pmatrix} \alpha \\ \beta \end{pmatrix} \tag{B.56}$$

where ρ_n is the density of the fluid contained in tank n and where I_{XXn}, I_{XYn}, and I_{YYn} are the second moment of inertias of the tank waterplane.

The corrected values of the hydrostatic stiffnesses are then

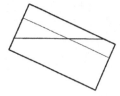 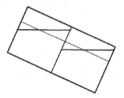

Figure B.9 Tank partitioning improves hydrostatic stability.

$$K'_{44} = K_{44} - \sum_n \rho_n \, g \, I_{YYn} \tag{B.57}$$

$$K'_{55} = K_{55} - \sum_n \rho_n \, g \, I_{XXn} \tag{B.58}$$

$$K'_{45} = K_{45} - \sum_n \rho_n \, g \, I_{XYn} \tag{B.59}$$

An efficient way to reduce the effects of liquid cargo is to partition the tanks (Figure B.9).

It must be borne in mind that liquid cargos also have "dynamic" effects on the wave response. This can be accounted for in different ways, for instance, by introducing, in the equations of motion, additional added mass and damping terms (see Chapter 6). Advantage may be taken of these effects to shift the roll natural frequency away from energetic wave frequencies and/or to introduce additional roll damping: many ships and offshore structures are equipped with **anti-roll tanks**.

B.2.3   Second-Order Hydrostatics

It is relatively easy, following the method used in the previous section (dividing the wetted surface S_B into $\widetilde{S}_{B0} + \Delta S_B$ and then using integral transforms), to extend the restoring torque to second-order. It is important to be very clear on the way the angular motion is described to second-order, and on the reference point used to express the hydrostatic torque.

Here we continue to take the center of gravity as the reference point. For the angular motion we use a formulation, correct to second-order, that does not assume any sequence of the rotations around the different axes.

We use the same coordinate systems, $Oxyz$ fixed with $z = 0$ the free surface, and $GXYZ$ body-fixed, such that at equilibrium G be on the Oz axis and the axes in $Oxyz$ and $GXYZ$ be parallel. To second-order, the correspondence between the two systems is taken as

$$\begin{pmatrix} x \\ y \\ z \end{pmatrix} = \begin{pmatrix} X \\ Y \\ Z + z_G^{(0)} \end{pmatrix} + \begin{pmatrix} \Delta x_G \\ \Delta y_G \\ \Delta z_G \end{pmatrix} + \begin{pmatrix} \alpha \\ \beta \\ \gamma \end{pmatrix} \wedge \begin{pmatrix} X \\ Y \\ Z \end{pmatrix}$$

$$+ \frac{1}{2} \begin{pmatrix} -\beta^2 - \gamma^2 & \alpha\,\beta & \alpha\,\gamma \\ \alpha\,\beta & -\alpha^2 - \gamma^2 & \beta\,\gamma \\ \alpha\,\gamma & \beta\,\gamma & -\alpha^2 - \beta^2 \end{pmatrix} \begin{pmatrix} X \\ Y \\ Z \end{pmatrix} \tag{B.60}$$

where $z_G^{(0)}$ is the vertical coordinate of the center of gravity in $Oxyz$.

All mathematical developments done, the hydrostatic restoring force is obtained as:

$$F_z = - \left[K_{33} \, \Delta z_G + K_{34} \, \alpha + K_{35} \, \beta \right]$$
$$- \frac{1}{2} \left[K_{33} \, (\alpha^2 + \beta^2) \, z_G^{(0)} + K_{34} \, \beta \, \gamma - K_{35} \, \alpha \, \gamma \right]$$
$$- \frac{1}{2} \rho g \int_{\Gamma_0} (\Delta z_G + \alpha \, Y - \beta \, X)^2 \, \tan \theta \, d\Gamma \qquad (B.61)$$

where Γ_0 is the waterline and θ is the angle between the outward normal vector and the horizontal plane, positive when the normal vector points upwards, negative in the opposite case (so, for most ship hulls, θ is negative or zero).

The moments in roll and pitch are obtained as

$$\begin{pmatrix} C_x \\ C_y \end{pmatrix} = - \begin{pmatrix} K_{34} \, \Delta z_G + K_{44} \, \alpha + K_{45} \, \beta \\ K_{35} \, \Delta z_G + K_{45} \, \alpha + K_{55} \, \beta \end{pmatrix} + \gamma \begin{pmatrix} K_{35} \, \Delta z_G + K_{45} \, \alpha + K_{55} \, \beta \\ -(K_{34} \, \Delta z_G + K_{44} \, \alpha + K_{45} \, \beta) \end{pmatrix}$$

$$- \frac{1}{2} \begin{pmatrix} K_{34} \, (\alpha^2 + \beta^2) \, z_G^{(0)} + K_{44} \, \beta \, \gamma - K_{45} \, \alpha \, \gamma \\ K_{35} \, (\alpha^2 + \beta^2) \, z_G^{(0)} - K_{55} \, \alpha \, \gamma + K_{45} \, \beta \, \gamma \end{pmatrix}$$

$$- (K_{33} \, \Delta z_G + K_{34} \, \alpha + K_{35} \, \beta) \, z_G^{(0)} \begin{pmatrix} \alpha \\ \beta \end{pmatrix}$$

$$- \frac{1}{2} \rho g \int_{\Gamma_0} (\Delta z_G + \alpha \, Y - \beta \, X)^2 \begin{pmatrix} Y \\ -X \end{pmatrix} \tan \theta \, d\Gamma \qquad (B.62)$$

When the off-diagonal stiffnesses K_{34}, K_{35}, K_{45} are zero (waterplane with double symmetry fore-aft and port-starboard), they simplify into

$$F_z = -K_{33} \, \Delta z_G - \frac{1}{2} K_{33} \, (\alpha^2 + \beta^2) \, z_G^{(0)} - \frac{1}{2} \rho g \int_{\Gamma_0} (\Delta z_G + \alpha \, Y - \beta \, X)^2 \, \tan \theta \, d\Gamma$$
$$\qquad (B.63)$$

$$\begin{pmatrix} C_x \\ C_y \end{pmatrix} = \begin{pmatrix} -K_{44} \, \alpha + K_{55} \, \beta \, \gamma - \frac{1}{2} K_{44} \, \beta \, \gamma - K_{33} \, z_G^{(0)} \, \Delta z_G \, \alpha \\ -K_{55} \, \beta - K_{44} \, \alpha \, \gamma + \frac{1}{2} K_{55} \, \alpha \, \gamma - K_{33} \, z_G^{(0)} \, \Delta z_G \, \beta \end{pmatrix}$$
$$- \frac{1}{2} \rho g \int_{\Gamma_0} (\Delta z_G + \alpha \, Y - \beta \, X)^2 \begin{pmatrix} Y \\ -X \end{pmatrix} \tan \theta \, d\Gamma \qquad (B.64)$$

where couplings between the four degrees of freedom, heave, roll, pitch, yaw, clearly appear. The coupling between heave and roll, for instance, may be responsible for parametric instabilities (see Chapter 8).

Appendix C: Damped Mass Spring System

Here we study the response of a linear mechanical system driven by the following equation

$$M \frac{\mathrm{d}^2 X}{\mathrm{d}t^2} + B \frac{\mathrm{d}X}{\mathrm{d}t} + K X = F(t) \tag{C.1}$$

X being the considered degree of freedom, M the mass (plus added mass), K the stiffness, B the damping, and $F(t)$ the external load.

C.1 Case without Excitation

Consider the equation,

$$M \ddot{X} + B \dot{X} + K X = 0 \tag{C.2}$$

with the initial conditions $X(0) = X_0$, $\dot{X}(0) = 0$.

The solution is looked for as $X(t) = A \exp(\lambda t)$, leading to:

$$M \lambda^2 + B \lambda + K = 0 \tag{C.3}$$

which has the solutions

$$\lambda = \frac{-B \pm \sqrt{B^2 - 4 K M}}{2 M} \tag{C.4}$$

The discriminant is equal to zero for $B = B_C = 2 \sqrt{K M}$. B_C is defined as the **critical damping**.

When B is greater than B_C, the two roots are real and negative. The system returns to its equilibrium position $X = \dot{X} = 0$ without oscillation.

When B is lower than B_C the roots are complex:

$$\lambda = -\frac{B}{2 M} \pm i \frac{\sqrt{4 K M - B^2}}{2 M} \tag{C.5}$$

and the response following an initial offset X_0 and a zero initial velocity is:

$$X(t) = X_0 \, e^{-\frac{B t}{2 M}} \left[\cos \omega_0 t + \frac{B}{2 M \omega_0} \sin \omega_0 t \right] \tag{C.6}$$

where $\omega_0 = \sqrt{4 K M - B^2}/(2 M) = \sqrt{K/M} \sqrt{1 - B^2/B_C^2}$ is the **natural frequency**.

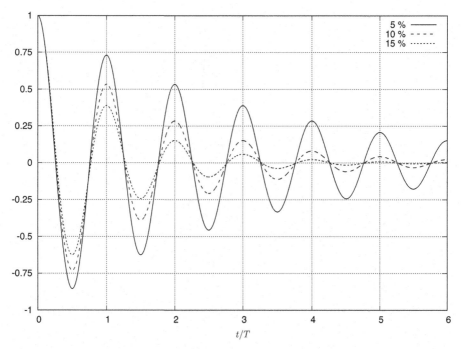

Figure C.1 Decaying time traces for damping ratios of 5 %, 10 %, and 15 %.

The response is a sinusoidal function with an amplitude decaying exponentially (see Figure C.1).

When the **damping ratio** $\zeta = B/B_C$ is small, the natural frequency ω_0 is well approximated by $\sqrt{K/M}$. The response following an initial displacement X_0 can be written as:

$$X(t) \simeq X_0 \, e^{-\zeta \, \omega_0 \, t} \, [\cos \omega_0 t + \zeta \, \sin \omega_0 t] \tag{C.7}$$

For a damping ratio ζ equal to 1 %, the decaying factor $\exp(-2\pi\zeta)$, from cycle to cycle, is equal to 0.94. With $\zeta = 5$ %, the decaying factor is 0.73. With $\zeta = 10$ % it is 0.53.

Most damping ratios encountered in the wave response of marine and offshore systems are lower than 10 %.

C.2   Response under Harmonic Excitation

Consider now

$$M \, \ddot{X} + B \, \dot{X} + K \, X = F_0 \, \cos \omega t. \tag{C.8}$$

We are only interested in the steady state response. Looking for $X(t)$ under the form:

$$X(t) = \Re \left\{ A \, e^{-i \omega t} \right\}, \tag{C.9}$$

we get

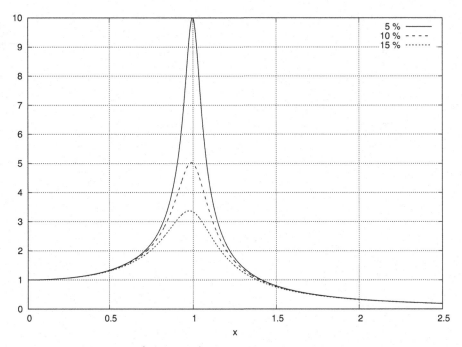

Figure C.2 Modulus of $[1 - x^2 - 2 \, \mathrm{i} \, \zeta \, x]^{-1}$ for ζ = 5 %, 10 % and 15 %.

$$A \; = \; \frac{F_0}{-M\,\omega^2 + K - \mathrm{i}\,B\,\omega} \qquad\qquad (C.10)$$

$$= \; \frac{F_0}{K \left[1 - \dfrac{\omega^2}{\omega_0^2} - 2\,\mathrm{i}\,\zeta\,\dfrac{\omega}{\omega_0}\right]} \qquad\qquad (C.11)$$

with ω_0 here defined as $\sqrt{K/M}$.

The modulus of the complex function $[1 - x^2 - 2\,\mathrm{i}\,\zeta\,x]^{-1}$ is shown in Figure C.2, for 3 different values of the damping ratio: ζ = 5 %, 10 % and 15 %. The maximum value is attained at $\omega = \omega_0 \, \sqrt{1 - 2\zeta^2}$ which is the **resonant frequency**. Strictly speaking the resonant frequency $\omega_0 \, \sqrt{1 - 2\zeta^2}$ is not equal to the natural frequency $\omega_0 \, \sqrt{1 - \zeta^2}$.

Three cases can be distinguished:

1. $\omega \ll \omega_0$ (stiff system):

Then we get

$$A \simeq \frac{F_0}{K} \qquad\qquad (C.12)$$

The response amplitude only depends on the stiffness.

2. $\omega \gg \omega_0$ (soft system):

Then

$$A \simeq -\frac{F_0}{M \omega^2} \tag{C.13}$$

The response amplitude only depends on the inertia.

3. $\omega \simeq \omega_0$ (resonance):

Then

$$A \simeq i \frac{F_0}{B \omega_0} \simeq \frac{i}{2 \zeta} \frac{F_0}{K} \tag{C.14}$$

The response amplitude is inversely proportional to the damping B.

The ratio $1/(2 \zeta)$ is known as the **dynamic amplification factor** (d.a.f.) with reference to the static response F_0/K.

C.3   Response under Random Excitation

Let $S_F(\omega)$ be the spectral density of the excitation.

The RAO being given by (C.10), the response spectrum follows from the excitation spectrum as

$$S_X(\omega) = \frac{S_F(\omega)}{(-M \omega^2 + K)^2 + B^2 \omega^2} \tag{C.15}$$

The variance (the square of the standard deviation) of the response is obtained as

$$\sigma_X^2 = \int_0^\infty \frac{S_F(\omega) \, d\omega}{(-M \omega^2 + K)^2 + B^2 \omega^2} \tag{C.16}$$

For a weakly damped system, most of the response takes place in a small neighborhood of the natural frequency (provided, of course, that there be a significant excitation there). Then an approximation of the standard deviation is obtained as

$$\sigma_X^2 \simeq S_F(\omega_0) \int_0^\infty \frac{d\omega}{(-M \omega^2 + K)^2 + B^2 \omega^2} = \frac{\pi \, S_F(\omega_0)}{2 \, B \, K} \tag{C.17}$$

The standard deviation σ_X is now conversely proportional to the **square root** of the damping ratio.

C.4   Mass Spring System with Quadratic Damping

Let the oscillator be

$$M \ddot{X} + B_Q \dot{X} |\dot{X}| + K X = F(t) \tag{C.18}$$

The technique is to replace the quadratic damping term $B_Q \dot{X} |\dot{X}|$ with an equivalent linear damping $B_L \dot{X}$ that ensures the same energy dissipation.

C.4.1   Harmonic Excitation

$$M \ddot{X} + B_Q \dot{X} |\dot{X}| + K X = F_0 \cos \omega t \tag{C.19}$$

Again we are only interested in the steady state

$$X(t) = \mathfrak{R} \left\{ A e^{-i \omega t} \right\} + \text{neglected higher-order harmonics.} \tag{C.20}$$

For a linear damping $B_L \dot{X}$ the energy dissipated over one cycle is

$$\Delta E_L = \int_0^T B_L \dot{X}^2 \, dt = \pi B_L |A|^2 \omega \tag{C.21}$$

With a quadratic damping, the dissipated energy is

$$\Delta E_Q = \int_0^T B_Q |\dot{X}|^3 \, dt = \frac{8}{3} B_Q |A|^3 \omega^2 \tag{C.22}$$

The equivalent linear damping is therefore

$$B_L = \frac{8}{3\pi} B_Q |A| \omega \tag{C.23}$$

The response is then found by solving a second-degree equation in $|A|^2$.
The technique employed here is known as **Lorentz linearization**.

C.4.2   Random Excitation

The technique is to assume a known form of the pdf (probability density function) of the response $X(t)$, usually a Gaussian law:

$$p(\dot{X}) = \frac{1}{\sqrt{2\pi} \, \sigma_{\dot{X}}} e^{-\dot{X}^2 / 2\sigma_{\dot{X}}^2} \tag{C.24}$$

The equivalent linear damping is then obtained through equating the mean dissipated energies:

$$\int_{-\infty}^{+\infty} B_Q |\dot{X}|^3 \, p(\dot{X}) \, d\dot{X} = \int_{-\infty}^{+\infty} B_L \dot{X}^2 \, p(\dot{X}) \, d\dot{X} = B_L \sigma_{\dot{X}}^2 \tag{C.25}$$

When a Gaussian distribution is assumed, this gives

$$B_L = \sqrt{\frac{8}{\pi}} \, \sigma_{\dot{X}} \, B_Q \tag{C.26}$$

This is known as **stochastic linearization**.

There is some contradiction in assuming a Gaussian response since, due to the nonlinearity implied by the quadratic damping, even though the excitation force be a normal process, the response will not be Gaussian. In most practical cases, the response is close enough to Gaussian for this stochastic linearization to be applicable.

The standard deviation $\sigma_{\dot{X}}$ is obtained from the response spectrum as

$$\sigma_{\dot{X}}^2 = \int_0^\infty \omega^2 \, \text{RAO}_X^2(\omega) \, S(\omega) \, d\omega \tag{C.27}$$

with $S(\omega)$ the wave spectrum.

It is then just a matter of iterating over the value of $\sigma_{\dot{X}}$ for a given wave spectrum, leading to sea state dependent RAOs.

C.5   Mass Spring System with Coulomb Damping

Coulomb damping means a constant force that directly opposes the velocity. This can be the case of friction due to pulleys or bearings in model tests. In dynamic positioning, it is the case of ideal thrusters that constantly oppose the slow-drift velocity.

Let the oscillator be

$$M \ddot{X} + F_C \operatorname{sign}(\dot{X}) + K X = F(t) \tag{C.28}$$

C.5.1   Harmonic Excitation

Using the same technique of equal energy dissipation, the equivalent linear damping is obtained as

$$B_L = \frac{4}{\pi} \frac{F_C}{|A| \omega} \tag{C.29}$$

where the response amplitude now appears at the denominator. This means that parasitic Coulomb damping (in model tests) will mostly show up at very small amplitudes (under very mild sea states).

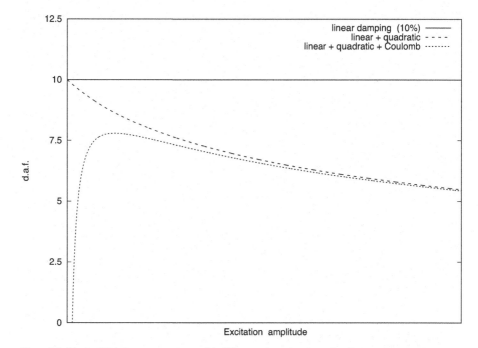

Figure C.3 Typical RAOs at resonance with different damping contributions.

There is no difficulty in combining all three kinds of damping (linear, quadratic, and Coulomb) together. A typical RAO at the resonant frequency is shown in Figure C.3.

C.5.2   Random Excitation

Using the technique of stochastic linearization, we get the equivalent linear damping as

$$B_L = \sqrt{\frac{2}{\pi}} \frac{F_C}{\sigma_{\dot{X}}} \tag{C.30}$$

Appendix D: The Boundary Integral Equation Method

This Appendix was written by Xiaobo Chen and Šime Malenica.

D.1 Generic Boundary Value Problem

Reference is made to Figure D.1 where the different boundaries of the fluid domain, together with the associated boundary conditions, are presented.

The generic boundary value problem (BVP) for the velocity potential $\varphi(\vec{x})$ is defined by the following set of equations:

$$\left.\begin{array}{ll} \Delta\varphi = 0 & -h < z \leq 0 \\ -\alpha\varphi + \frac{\partial\varphi}{\partial z} = Q_F & z = 0 \\ \frac{\partial\varphi}{\partial n} = Q_B & S_B \\ \frac{\partial\varphi}{\partial z} = 0 & z = -h \\ \varphi \to 0 & r \to \infty \end{array}\right\} \qquad (D.1)$$

where Q_F and Q_B are the nonhomogeneous forcing terms at the free surface and at the body surface, respectively, and α is the parameter associated with the oscillation frequency ($\alpha = \omega^2/g$ in the linear case, $\alpha = 4\omega^2/g$ in the second-order case ...).

The dispersion relation associated with the homogeneous free surface condition can be written in the form:

$$k_0 \tanh k_0 h = \alpha \qquad (D.2)$$

The solution of this equation gives the wave number k_0, which defines the characteristic wave length.

D.2 Boundary Integral Equation Method

The most widely used numerical methods for the resolution of the BVP (D.1) are based on the Boundary Integral Equation technique (BIE), where the velocity potential is generated from Green identities, using a distribution of singularities over the boundaries. The method requires an appropriate choice of Green function. For the time-harmonic problem, the Green function is classically defined as the real part of

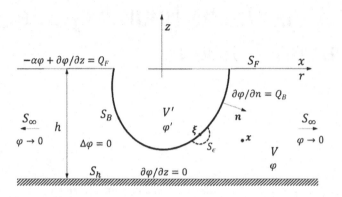

Figure D.1 Generic BVP definition.

$$G(\vec{x}; \vec{\xi}, t) = \mathrm{Re}\left\{ G(\vec{x}; \vec{\xi})e^{-i\omega t} \right\} \tag{D.3}$$

in which the spatial Green function $G(\vec{x}; \vec{\xi})$ is usually classified into two main types commonly denoted as Rankine and Kelvin Green functions.

D.2.1   Green Function of Rankine Type

The Rankine Green function is a fundamental solution of the Poisson equation in the whole physical space R^3:

$$\Delta G = \delta(\vec{x} - \vec{\xi}) \tag{D.4}$$

where $\delta(\vec{x} - \vec{\xi})$ denotes the Dirac delta function.

The simplest solution of (D.4) is easily found in the form:

$$G(\vec{x}; \vec{\xi}) = -\frac{1}{4\pi r} \tag{D.5}$$

where r is the distance between the field point $\vec{x}$ and the singularity point $\vec{\xi}$ i.e. $r = |\vec{x} - \vec{\xi}|$.

Physically the Green function represents the potential at any point $\vec{x} = (x, y, z)$ in the fluid due to a source placed at $\vec{\xi} = (\xi, \eta, \zeta)$. The direct application of Green's third identity leads to the following representation for the potential at any point in the fluid domain:

$$\varphi(\vec{x}) = \iint_{S_B + S_F + S_h + S_\infty} \left[G(\vec{x}; \vec{\xi}) \frac{\partial \varphi(\vec{\xi})}{\partial n_\xi} - \varphi(\vec{\xi}) \frac{\partial G(\vec{x}; \vec{\xi})}{\partial n_\xi} \right] dS_\xi, \qquad \vec{x} \in V \tag{D.6}$$

The main drawback of the BIE method based on the Rankine Green function is that the surface integrals apply throughout the boundaries of the fluid domain. In most cases, the integral over the control surface at infinity disappears and the integral over the seabed can be removed by adding an additional (image) source point at $(\xi, \eta, \zeta - 2h)$, but it is not possible to avoid the integral over the free surface! This is rather inconvenient, not only because the number of unknowns increases but also because the free surface extends to infinity and it is necessary to introduce additional assumptions in order to satisfy the radiation condition. For these reasons, when applicable the Kelvin type Green function is usually preferred.

D.2.2 Green Function of Kelvin Type

The Kelvin Green function $G\left(\vec{x}; \vec{\xi}\right)$ is defined as the solution of the following BVP:

$$\left.\begin{array}{rl} \Delta G = \delta(\vec{x} - \vec{\xi}) & -h < z \leq 0 \\ -\alpha G + \frac{\partial G}{\partial z} = 0 & z = 0 \\ \frac{\partial G}{\partial z} = 0 & z = -h \\ \sqrt{k_0 R}\left(\frac{\partial G}{\partial R} - i\, k_0 G\right) \to 0 & r \to \infty \end{array}\right\} \quad (D.7)$$

where k_0 is the wave number (D.2), and R denotes the horizontal distance $R = \sqrt{x^2 + y^2}$.

The Kelvin Green function satisfies, in addition to the Laplace equation, the homogeneous boundary condition at the free surface, the no-flow condition at the seabed, and the radiation condition at infinity. The Kelvin Green function can be derived via the Fourier transform technique. It is generally written in the following form:

$$G\left(\vec{x}; \vec{\xi}\right) = -\frac{1}{4\pi r} + H\left(\vec{x}; \vec{\xi}\right) \quad (D.8)$$

As can be seen an additional term $H\left(\vec{x}; \vec{\xi}\right)$ is added to the fundamental solution (D.5) in order to satisfy the boundary conditions at the free surface, at the seabed, and at infinity. The consequence is that the integrals over the free surface, over the sea floor, and at infinity in (D.6) disappear, and only the integral on the body surface remains.

Two types of BIE methods based on the Kelvin Green function are commonly used, namely the mixed distribution method and the source only distribution method. The mixed distribution method relies on the direct application of Green's second identity, and the final expression for the velocity potential φ at any point in the fluid is obtained in the form:

$$\varphi\left(\vec{x}\right) = - \iint_{S_B} \varphi(\vec{\xi}) \frac{\partial G(\vec{x}; \vec{\xi})}{\partial n_\xi} dS_\xi + \iint_{S_B} G(\vec{x}; \vec{\xi}) Q_B(\vec{\xi}) dS_\xi$$
$$- \iint_{S_F} G(\vec{x}; \vec{\xi}) Q_F(\vec{\xi}) dS_\xi \qquad \vec{x} \in V \quad (D.9)$$

where it can be noted that the symmetry property of the Green function $G(\vec{x}; \vec{\xi}) = G(\vec{\xi}; \vec{x})$ has been used in the derivation. The unknowns on the body (the dipole strengths) are evaluated from the following BIE, which is obtained after carrying out the limiting process for the points at the body surface $\vec{x} \in S_B$:

$$\frac{1}{2}\varphi(\vec{x}) + \iint_{S_B} \varphi(\vec{\xi}) \frac{\partial G(\vec{x}; \vec{\xi})}{\partial n_\xi} \mathrm{d}S_\xi = \iint_{S_B} G(\vec{x}; \vec{\xi}) Q_B(\vec{\xi}) \mathrm{d}S_\xi$$

$$- \iint_{S_F} G(\vec{x}; \vec{\xi}) Q_F(\vec{\xi}) \mathrm{d}S_\xi \quad \vec{x} \in S_B \qquad (\text{D.10})$$

The source only formulation follows from combining the BIE for the physical (exterior) fluid domain with the BIE for the fictitious interior problem. This is based on the fact that the Green function satisfies the BVP in the whole domain $z \leq 0$. The final expression for the velocity potential φ is obtained in the form:

$$\varphi(\vec{x}) = \iint_{S_B} \sigma(\vec{\xi}) G(\vec{x}; \vec{\xi}) \mathrm{d}S_\xi - \iint_{S_F} G(\vec{x}; \vec{\xi}) Q_F(\vec{\xi}) \mathrm{d}S_\xi \quad \vec{x} \in V + V'$$

$$(\text{D.11})$$

and the corresponding BIE for the source strength $\sigma(\vec{\xi})$ takes the form:

$$\frac{1}{2}\sigma(\vec{x}) + \iint_{S_B} \sigma(\vec{\xi}) \frac{\partial G(\vec{x}; \vec{\xi})}{\partial n_x} \mathrm{d}S_\xi = Q_B(\vec{x}) + \alpha \iint_{S_F} G(\vec{x}; \vec{\xi}) Q_F(\vec{\xi}) \mathrm{d}S_\xi \quad \vec{x} \in S_B$$

$$(\text{D.12})$$

In both cases the BIE is solved by discretization of the body surface into a certain number of panels. The most common approach assumes constant singularities over flat panels and the BIE is enforced at the centroids of the panels. Being theoretically equivalent, the two BIE formulations (mixed and source) may nevertheless yield slightly different numerical solutions. To compute the fluid velocity $\nabla\varphi$, which is required for solving the second-order problems, it is preferable to use the source method since only first-order derivatives of the Green function are needed.

As already mentioned, the generic BVP (D.1) covers all usual cases, that is, both the linear and the second-order problems, diffraction as well as radiation. It can be noted that the linear problem is significantly simpler due to the absence of the free surface integral. The complexity of the free surface integration depends on the behavior of the forcing term $Q_F(x, y)$, which extends to infinity. In particular, for the second-order problem, this integration is extremely complex because the integrand decays very slowly and is highly oscillatory. It is also important to note that, even though discretization of the free surface is necessary, it is for integration purposes only: no additional unknowns are introduced. This means that the unknowns (source and dipole strengths) are distributed on the body surface only regardless of whether Q_F be zero or not. Typical hydrodynamic meshes are shown in Figure D.2.

D.2.3   Artificial Dissipation

A drawback of potential flow theory is its assumptions of ideal fluid and irrotational flow, and the neglect of fluid viscosity and of its damping effect. This may lead to unrealistic results especially in the case of resonant conditions where large free surface deformations may occur. It is possible to model some non-potential flow effects

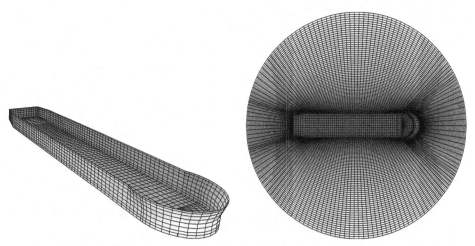

Figure D.2 Typical hydrodynamic mesh (left $Q_F = 0$, right $Q_F \neq 0$).

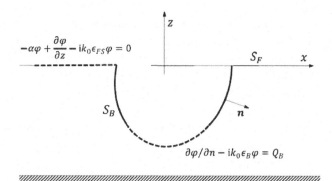

Figure D.3 Modified boundary conditions by introducing dissipation terms

by artificially modifying the boundary conditions. The generic case is shown in Figure D.3 where, for simplicity, the homogeneous free surface condition ($Q_F = 0$) is considered. Two different dissipation coefficients ($\epsilon_{FS}, \epsilon_B$) are introduced in the left-hand sides of the free surface and body boundary conditions.

Using the same Green function as before, in the case of the source formulation, the following representation of the velocity potential is obtained:

$$\varphi(\vec{x}) = \iint_{S_B+S_F} \sigma(\vec{\xi}) G(\vec{x};\vec{\xi}) dS_\xi \qquad \vec{x} \in V + V' \tag{D.13}$$

and the corresponding BIE for the source strength $\sigma(\vec{\xi})$ is:

$$\frac{1}{2}\sigma(\vec{x}) + \iint_{S_B+S_F} \sigma(\vec{\xi}) \left[\frac{\partial G(\vec{x};\vec{\xi})}{\partial n_x} - i k_0 \epsilon_B G(\vec{x};\vec{\xi}) \right] dS_\xi = Q_B(\vec{x}) \qquad \vec{x} \in S_B$$
$$\tag{D.14}$$

$$\sigma(\vec{x}) + i k_0 \epsilon_{FS} \alpha \iint_{S_B+S_F} \sigma(\vec{\xi}) G(\vec{x};\vec{\xi}) dS_\xi = 0 \qquad \vec{x} \in S_{FS} \tag{D.15}$$

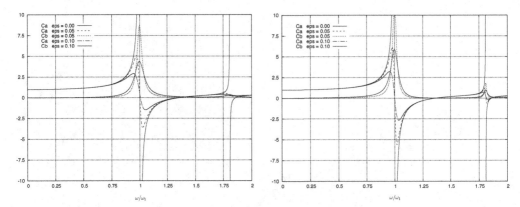

Figure D.4 Rectangular tank. Sway added mass and damping coefficients. With dissipation at the free surface (left) or at the wall (right).

It should be noted that the part of the free surface with the nonzero dissipation coefficient ϵ_{FS} needs to be discretized and additional unknowns (the source strengths) must be located there. This means that the size of the linear system increases.

Figure D.4 shows, in the case of a rectangular tank of length equal to twice the water height, the added mass (C_a) and damping (C_b) coefficients obtained following one or the other method. They are plotted vs ω/ω_1 where ω_1 is the first sloshing frequency (see Figure 6.34 in Chapter 6).

The determination of the appropriate value of the dissipation coefficient remains the main problem with this approach. It can be done through comparison with experimental results or with results from CFD simulations.

D.2.4   Irregular Frequencies

A drawback of the BIE methods based on the Kelvin Green function is the irregular frequencies, which may occur when the body pierces the free surface. This is a purely numerical problem related to the fact that, when the exterior problem is being solved, the interior problem is solved as well. With a source formulation, continuity of the potential across the body surface is imposed. This means that any eigensolution to the interior problem, verifying a homogeneous Dirichlet condition at the body surface, can be added up: unicity is lost. Practically, the result is that the matrix of the linear system becomes ill-conditioned in the vicinity of the irregular frequencies, which are the eigenfrequencies of the interior problem (with $\varphi = 0$ at the wall).

There are different methods to eliminate the problem, and the most intuitive one is based on a modification of the interior problem, in such a way that the eigenfrequencies do not exist any more, or are removed from the wave frequency range. This is usually done by modifying the boundary condition at the free surface, the most common choice being to impose $\varphi_z = 0$, that is the inner free surface is covered by a solid lid. Then it is necessary to discretize the interior free surface (see Figure D.5) and to set additional unknowns there. The corresponding extended BIE (EBIE) method can

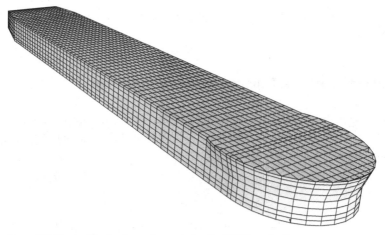

Figure D.5 Typical hydrodynamic mesh for the EBIE method.

be obtained straightforwardly using the Green identities as before. In the case of a source formulation, the potential is defined by:

$$\varphi(\vec{x}) = \iint_{S_B + S'_F} \sigma(\vec{\xi}) G(\vec{x}; \vec{\xi}) dS_\xi \qquad z \le 0 \tag{D.16}$$

and the corresponding EBIE for the source strength is:

$$\frac{1}{2}\sigma(\vec{x}) + \iint_{S_B + S'_F} \sigma(\vec{\xi}) \frac{\partial G(\vec{x}; \vec{\xi})}{\partial n_x} dS_\xi = Q_B(\vec{x}) \qquad \vec{x} \in S_B \tag{D.17}$$

$$\sigma(\vec{x}) + \iint_{S_B + S'_F} \sigma(\vec{\xi}) \frac{\partial G(\vec{x}; \vec{\xi})}{\partial n_x} dS_\xi = 0 \qquad \vec{x} \in S'_F \tag{D.18}$$

A drawback of the EBIE method is an increase in the number of unknowns.

In practice, it is useful to know the value of the lowest irregular frequency: If it is higher than the wave frequencies, no remedy is needed. A lower bound can be obtained based on the fact that, for any closed geometry surrounding the body, the irregular frequencies are lower. For shiplike bodies, the surrounding rectangular box can be taken as a reference, for which the lowest irregular frequency is given by

$$\omega_{11} = \sqrt{\frac{\pi g}{L}} \left[\frac{\sqrt{1 + (L/B)^2}}{\tanh\left(\frac{\pi D}{L}\sqrt{1 + (L/B)^2}\right)} \right]^{1/2} \tag{D.19}$$

with L the length, B the width, and d the draft. For instance, for a tanker 300 m long, 45 m wide, at a draft of 20 m, the first irregular frequency is larger than 0.88 rad/s

For a truncated circular cylinder of radius a and draft d:

$$\omega_1 = 1.55 \sqrt{(g/a)\coth(2.405\,d/a)} \tag{D.20}$$

where $2.405 = 1.55^2$ is the first root of $J_0(z) = 0$, with J_0 the Bessel function of the first kind.

D.3   Boundary Integral Equation Method with Current

Even though the basic principles remain the same, the presence of a current introduces some modifications. The generic BVP is unchanged except for the free surface condition, which is now:

$$-\nu\varphi + \frac{\partial \varphi}{\partial z} + 2\mathrm{i}\,\tau\frac{\partial \varphi}{\partial x} = Q_F \ , -\nu G + \frac{\partial G}{\partial z} + 2\mathrm{i}\,\tau\frac{\partial G}{\partial x} = 0 \qquad (\text{D}.21)$$

where $\nu = \omega_e^2/g$, ω_e the encounter frequency, and $\tau = U\omega_e/g$ the Brard number.

A procedure similar to the no current case leads to the following BIE for the mixed singularity distribution:

$$\frac{1}{2}\varphi(\vec{x}) + \iint_{S_B} \varphi(\vec{\xi})\frac{\partial G(\vec{x};\vec{\xi})}{\partial n_\xi}\mathrm{d}S_\xi + 2\mathrm{i}\,\tau\int_\Gamma \varphi(\vec{\xi})G(\vec{x};\vec{\xi})\mathrm{d}\eta$$

$$= \iint_{S_B} G(\vec{x};\vec{\xi})Q_B(\vec{\xi})\mathrm{d}S_\xi - \iint_{S_F} G(\vec{x};\vec{\xi})Q_F(\vec{\xi})\mathrm{d}S_\xi \quad \vec{x}\in S_B$$

$$(\text{D}.22)$$

where Γ denotes the body waterline oriented in the clockwise direction when viewed from above the free surface. As in the no current case the symmetry of the Green function $G(\vec{x};\vec{\xi};\tau) = G(\vec{\xi};\vec{x};-\tau)$ has been used in the derivation.

In the source formulation, the final expression for the velocity potential φ is:

$$\varphi(\vec{x}) = \iint_{S_B} \sigma(\vec{\xi})G(\vec{x};\vec{\xi})\mathrm{d}S_\xi - \iint_{S_F} G(\vec{x};\vec{\xi})Q_F(\vec{\xi})\mathrm{d}S_\xi \qquad \vec{x}\in V+V'$$

$$(\text{D}.23)$$

and the BIE for the source strength $\sigma(\vec{\xi})$ is:

$$\frac{1}{2}\sigma(\vec{x}) + \iint_{S_B} \sigma(\vec{\xi})\frac{\partial G(\vec{x};\vec{\xi})}{\partial n_x}\mathrm{d}S_\xi = Q_B(\vec{x})$$

$$+ \iint_{S_F} \frac{\partial G(\vec{x};\vec{\xi})}{\partial n_x}Q_F(\vec{\xi})\mathrm{d}S_\xi \qquad \vec{x}\in S_B \qquad (\text{D}.24)$$

Compared to the mixed singularity distribution, it can be observed that in the source formulation the integral along the waterline has disappeared. This makes the source formulation more attractive. It can also be noted that the free surface integral that occurs in the BIEs can be evaluated at a relatively low CPU cost because the nonhomogeneous term Q_F decays rapidly away from the body (at variance with the second-order diffraction problem), so that only a limited portion of the free surface needs to be discretized.

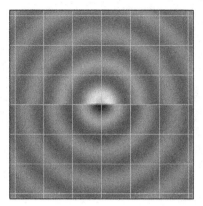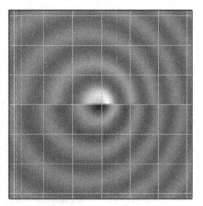

Figure D.6 Far-field wave systems without (left, $\tau = 0$) and with current (right, $\tau = 0.1$).

The Green function for the problem with current can be obtained by manipulating the zero current Green function in the following way:

$$G = e^{2i\tau\frac{\partial k}{\partial v}k(x-\xi)}\left\{G_0 - 2i\,\tau\left[\frac{\partial^2 G_0}{\partial v \partial x} + \frac{\partial k}{\partial v}k(x-\xi)G_0\right]\right\} \qquad \text{(D.25)}$$

where G_0 denotes the zero current Green function.

From the above representation of the Green function (D.25), a very important feature of the far-field waves can be inferred. Knowing that the zero current Green function produces only circular (ring) waves at infinity, it can be concluded that the presence of current will also produce ring waves, but now shifted in space due to Doppler effect. This fact limits the domain of validity of the present approach. For instance, the more complex wave system that occurs at high forward speed, with transverse and divergent waves, cannot be represented with this approach, the reason being that the quadratic velocity term in the full free surface boundary condition ($U^2 \partial^2 \varphi / \partial x^2$) is neglected. This means that care should be taken in the use of the approach above because, strictly speaking, not only must the parameter τ be small, but also the Froude number $F_n = U/\sqrt{gL}$ must be even smaller, that is, $F_n \ll \tau$.

Figure D.6 shows the wave patterns generated by a submerged source without and with small forward speed.

D.4   Numerical Evaluation of the Kelvin Green Function

The Kelvin Green function (D.8) defined by the set (D.7) of differential equations contains the unbounded space Rankine source and an additional term $H\left(\vec{x};\vec{\xi},v,h\right)$ ensuring that the boundary conditions on the free surface, at the seabed and the radiation condition at infinity are satisfied. In deep water an expression of the Kelvin Green function is

$$G(\vec{x};\vec{\xi},v,h=\infty) = -\frac{1}{4\pi\,r} - \frac{1}{4\pi r'} - 2vG_\infty(Z',R') - 2i\,\pi ve^{v(z+\zeta)}J_0(vR) \quad \text{(D.26)}$$

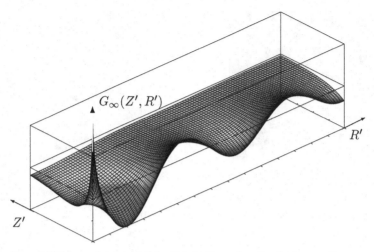

Figure D.7 Free surface term of the Kelvin Green function in deep water.

with

$$G_\infty(Z', R') = \int_0^\infty \frac{e^{-kZ'}}{k-1} J_0(kR') dk \qquad (D.27)$$

where r' is the distance between the field point $\vec{x}(x, y, z)$ and its symmetric point $\vec{\xi}'(\xi, \eta, -\zeta)$ above the free surface, that is, $r' = \sqrt{R^2 + (z + \zeta)^2}$ with $R = \sqrt{(x - \xi)^2 + (y - \eta)^2}$. The second term on the right-hand side of (D.26), given by the Fourier integral (D.27), is a real function of two variables $Z' = -\nu(z + \zeta)$ and $R' = \nu R$ (Figure D.7). It can be replaced by various analytical expressions, highly accurate, in different zones of the (Z', R') plane. For the purpose of numerical computations, the deep water Green function is further written as

$$G_\infty(Z', R') = \begin{cases} -e^{-Z'} J_0(R') \left[\ln(Z' + \sqrt{Z'^2 + R'^2}) - \ln 2 + \gamma \right] \\ +S(Z', R') \qquad\qquad 0 \le R' \le 0, 0 \le Z' \le 4 \\ \\ -\pi e^{-Z'} \widetilde{H}_0(R') + S(Z', R') \qquad\qquad\qquad \text{elsewhere} \end{cases}$$

$$(D.28)$$

where $\gamma = 0.577215\ldots$ is the Euler constant and $\widetilde{H}_0(\cdot)$ the zeroth-order Struve function. The function $S(Z', R')$ in (D.28) is smoothly varying and can be approximated by Chebychev polynomials over a series of partitions of the (Z', R') plane.

In finite depth, the so-called John's series formulation can be applied:

$$G(\vec{x}; \vec{\xi}, \nu, h) = -\frac{4i\pi k_0}{2k_0 h + \sinh 2k_0 h} \cosh k_0(z + h) \cosh k_0(\zeta + h) H_0(k_0 R)$$

$$- \sum_{m=1}^\infty \frac{8k_m}{2k_m h + \sin 2k_m h} \cos k_m(z + h) \cos k_m(\zeta + h) K_0(k_m R)$$

$$(D.29)$$

in which k_0 is determined by (D.2) and k_m by $k_m \tan k_m h = -\omega^2/g$, while $H_0(\cdot)$ and $K_0(\cdot)$ are the zeroth-order Hankel function of the first kind and the zeroth-order modified Bessel function of the second kind, respectively.

The John's series formulation (D.29) is well suited for numerical computations when $R/h > 1$. However, its convergence is poor for $R/h < 1$, and it diverges when $R/h \to 0$. Complementary formulations for $R/h < 1$ are needed. For this purpose, the Green function is written as

$$G(\vec{x}; \vec{\xi}, v, h) = R_n - 2k_0 G_h - \frac{4i\pi k_0}{2k_0 h + \sinh 2k_0 h} \cosh k_0(z+h) \cosh k_0(\zeta+h) \mathrm{J}_0(k_0 R) \tag{D.30}$$

in which R_n is defined by

$$R_n = -\frac{1}{4\pi r} - \frac{1}{4\pi r_1} - \frac{1}{4\pi r_2} - \frac{1}{4\pi r_3} - \frac{1}{4\pi r_4} - \frac{1}{4\pi r_5} \tag{D.31}$$

where the r_j are the distances between the field point $\vec{x}(x, y, z)$ and the mirror source points $\vec{\xi}_j(\xi, \eta, \zeta_j)$ of vertical coordinates

$$\begin{aligned} \zeta_1 &= -\zeta = \zeta' \\ \zeta_2 &= -\zeta - 2h \\ \zeta_3 &= \zeta - 2h \\ \zeta_4 &= \zeta + 2h \\ \zeta_5 &= -\zeta - 4h \end{aligned} \tag{D.32}$$

The real part of the free surface term in (D.30) is given by

$$G_h(Z_P, Z_M, \bar{R}) = \begin{cases} G_P(Z_P, \bar{R}, H) + G_M(Z_M, \bar{R}, H) \\ \quad + \tanh H G_\infty(Z', R') & 0 \le Z_P < 1 \\ G_M(2 - Z_P, \bar{R}, H) + G_M(Z_M, \bar{R}, H) & 1 \le Z_P \le 2 \end{cases} \tag{D.33}$$

where the following notations are used

$$\begin{aligned} \bar{R} &= R/h \\ Z_P &= |z + \zeta|/h \\ Z_M &= |z - \zeta|/h \\ H &= k_0 h \end{aligned} \tag{D.34}$$

The depth functions $G_P(Z_P, \bar{R}, H)$ and $G_M(Z_M, \bar{R}, H)$, given as Fourier integrals in Chen (1993), are smoothly varying functions as shown in Figure D.8. They are approximated by Chebychev polynomials of three variables for efficient computations. The depth functions $G_P(Z, \bar{R}, H)$ and $G_M(Z, \bar{R}, H)$ tend toward zero for $H = k_0 h \to \infty$. However, they vary rapidly for $H \to 0$ so that an improved formulation

$$\begin{aligned} G_P(Z, \bar{R}, H) &= -(\ln H)/H + \widetilde{G}_P(Z, \bar{R}, H) \\ G_M(Z, \bar{R}, H) &= -(\ln H)/H + \widetilde{G}_M(Z, \bar{R}, H) \end{aligned} \tag{D.35}$$

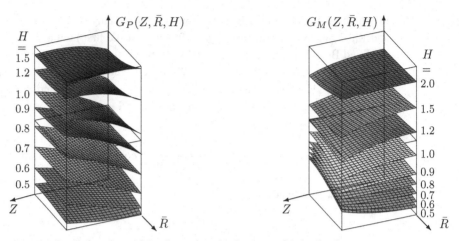

Figure D.8 Depth functions of the Green function in water of finite depth.

is adopted for $0 \leq H < 1/2$ only, and $\widetilde{G}_P(Z, \bar{R}, H)$ and $\widetilde{G}_M(Z, \bar{R}, H)$ are represented by Chebychev polynomials.

D.5   References

CHEN X.B. 1993. Evaluation de la fonction de Green du problème de diffraction/radiation en profondeur d'eau finie. Une nouvelle méthode rapide et précise, in *Actes des 4èmes Journées de l'Hydrodynamique*, Nantes (in French; http://website .ec-nantes.fr/actesjh/).

CHEN X.B. 2011. Offshore hydrodynamics and applications, *The IES Journal Part A: Civil & Structural Engineering*, **4**(3), 124–142.

NEWMAN J.N. 1985. Algorithms for the free surface Green function, *Journal of Engineering Mathematics*, **19**(1):57–67.

NOBLESSE F. 1982. The Green function in the theory of radiation and diffraction of regular water waves by a body, *Journal of Engineering Mathematics*, **16**(2), 137–169, 1982.

NOBLESSE F., CHEN X.B. 1995. Decomposition of free surface effects into wave and near-field components, *Ship Technology Research*, **42**, 167–185.

WEHAUSEN J.V., LAITONE E.V. 1960. Surface waves. In Handbuch der Physik, 446–778. Springer

Author Index

Subject Index